W9-CAO-478

CAUSALITY AND MODERN SCIENCE

CAUSALITY AND MODERN SCIENCE

by
Mario Bunge

Third Revised Edition

Dover Publications, Inc.
New York

Published in Canada by General Publishing Company, Ltd., 30 Lesmill Road, Don Mills, Toronto, Ontario.
Published in the United Kingdom by Constable and Company, Ltd., 10 Orange Street, London WC2H 7EG.

This Dover edition, first published in 1979, is an unabridged and corrected republication of the second (1963) edition as published by The World Publishing Company, Cleveland, in 1963 under the title *Causality*. The author has contributed a new preface for this edition.

International Standard Book Number: 0-486-23728-1
Library of Congress Catalog Card Number: 78-74117

Manufactured in the United States of America
Dover Publications, Inc.
180 Varick Street
New York, N.Y. 10014

Contents

PART II
WHAT CAUSAL DETERMINISM DOES NOT ASSERT

PART III
WHAT CAUSAL DETERMINISM DOES ASSERT

PART IV

THE FUNCTION OF THE CAUSAL PRINCIPLE IN SCIENCE

Preface to the Dover Edition

1. *The Demise of Causality*

Two decades ago, when I wrote this book, the principle of causality seemed dead. It had succumbed under the joint attack of the two great intellectual revolutions of the 1920s: the quantum theory and logical positivism. Since the former was at the time interpreted in the light of the latter, and positivism had always denounced causality as a myth, it is no wonder that the quantum theory was thought to have buried the principle of causality. From then on until recently indeterminism became fashionable.

Causalism, the view that all relations among events except the purely spatio-temporal events are causal, is surely untenable. Think not only of quantum jumps but also of genetic mutations and even of the events in our ordinary lives. Each of us is a result of both the random shuffling of parental genes as well as of a number of social circumstances that appear and disappear largely at random. Chance is not just our ignorance of causes: chance is a mode of being. So, any theory of events that ignores chance—and causalism does just that—must be false.

Causalism seemed then defunct when the first edition of this book appeared in 1959. And yet a thesis of this book was that the causal principle does hold a necessary if somewhat modest place in the ontology of contemporary science.

2. *The Resilience of Causal Thinking*

The verdict of philosophers concerning the demise of causality seems to have been premature. As the founder of wave mechanics has stated, "The search for causation is an instinctive tendency of the human mind" (de Broglie, 1977). Developmental psychology agrees (Piaget, 1930, Inhelder and Piaget, 1958). Even ethologists find reasons for suspecting that all vertebrates are capable of forming representations of causal links. After all, an animal that followed David Hume's view that there is no causation but only accidental constant conjunction, would be ill-equipped to face the challenges of its environment.

So, there must be something in causalism: surely not the whole truth but a good portion of it. Hence indeterminism cannot be an adequate substitute. We may be persuaded to give up the causal principle, but in the end we come back to some form of it, even if a considerably diluted one. As Anne in Jane Austen's *Persuasion* discovers, reason may prevail for a while against our instincts, but love conquers in the end—particularly when reason is found to be quite unreasonable. Indeed it is perverse to claim that causality is altogether mythical: that kicking the ball does not cause it to move, or that listening to propaganda day in and day out does not cause us to change our beliefs and act accordingly. To claim that these are just constant but otherwise accidental conjunctions is not just artificial: it is a plain mistake.

3. *Causality and Contemporary Science*

Consider a sequence of events often regarded as a model of acausality: light absorption. When a quantum physicist calculates the probability that an atom will absorb a photon, he treats this event as the cause of the atom's jumping to an excited energy level. That is, he calculates the probability that such a cause-effect relation obtains. He admits then that there are causes and effects even if he does not accept the assumptions that every event must have a cause, and that every effect has a single cause. Moreover he could argue that, although the quantum theory is basically stochastic, it also has a causal aspect summarized in the hamiltonian operator. (The commutator of the hamiltonian with a coordinate equals the corresponding component of the force acting on the micro-thing.)

Likewise not everything is random in biology. True, mutation is a chance event and gene recombination is random as well.

But this is only half of the story. The other half is that such randomly constituted genotypes are selected by the environment, and this selection is anything but random. In fact those organisms that are unfit are eliminated and so leave no offspring. For this reason it has been rightly said that the environment is the anti-chance or causal aspect of evolution (Dobzhansky, 1974).

What holds for physics and biology holds also, *mutatis mutandis*, for the social sciences. Surely chance is rampant in human affairs; if it were not we might be able to predict accurately the future state of our economies. But causation is no less a factor. Every person counts on certain causal links when intent on causing someone else to do this or that, or on preventing somebody from causing him to behave in a given manner. Surely, as in the case of physics, we can often modify only the probabilities of certain causal relations. But this may suffice for practical purposes and, in any case, it shows that causality is anything but dead in social science.

4. *The Comeback of Causality*

The philosophical climate I encountered when I started to write this book a quarter of a century ago has altered considerably with regard to the causal problem. For one thing we have come to form a more balanced view of science: we no longer believe that all of the natural sciences are ultimately reducible to physics, and that the latter is practically compressed into the quantum theory with its attendant indeterminacy. (Even the so-called "indeterminacy relations" have received a number of non-indeterministic interpretations: cp. Bunge, 1977a.)

For another thing the Hume-Comte-Mach estate is bankrupt. Most positivists are dead and very few philosophers follow in their wake—a deplorable fact considering that the antipositivistic reaction has been pushed so far that it has often turned into a reaction against science and even rationality in general. In any case, the positivistic prejudice against causality and, in general, against ontology, is no longer in force. Ontology is back and doing rather well considering that it had been pronounced dead so many times over the past two centuries. In particular, the investigation of the causal problem has become respectable once again.

A perusal of the philosophical literature will leave no doubt as to the vigorous comeback of the causal problem during the 1970s. There are not only hundreds of journal articles but also a number of distinguished monographs on the subject. First of all we have Wallace's scholarly and massive work (1972, 1974), undoubtedly the most exhaustive and authoritative historical-philosophical treatment of the subject. Then we have Mackie's profound criticism of the positivist doctrine of causation, and examination of the role of causal thinking (1975). An alternative, though also non-positivist view, is found in the provocative book by Harré and Madden (1975). Svechnikov's monograph on physical causality (1971) tells about the interest of the problem among dialectical materialists. Von Wright's work (1971) emphasizes the role of causality in action theory. Finally there is Suppes' quixotic attempt (1970, 1974) to formalize and rehabilitate Hume's noncausal doctrine on causation. In short, the student of the causal problem has now a good starting point—something he did not have a quarter of a century ago.

5. *The Two Aspects of the Causal Problem*

Like many other important philosophical problems, the causal problem is actually a whole cluster of problems. This set can be partitioned into two subsets:

1. The *ontological* problem of causality, i.e. what *is* causation: what are the characteristics of the causal link; to what extent are such links real; are there causal laws; how do causation and chance intertwine (and so on)?
2. The *methodological* problem of causality, i.e. what are the causation *criteria*; how do we recognize a causal link and how do we test for a causal hypothesis?

Every one of these subsets has a variety of problems. Just think of the difficult methodological problem of disclosing causal relations beneath statistical correlations (cf. Blalock and Blalock, 1968). Moreover the membership of every one of the above subclusters of problems keeps changing in the course of time. But somehow none of them ever gets depleted even though once in a while a hasty philosopher will proclaim the final solution to "the" causal problem—perhaps without having bothered to propose a careful formulation of the causal principle. This is not surprising in philosophy; at some time or other, matter, motion, space, time, law, chance, and many other metaphysical categories were decreed dead until it was eventually discovered that they refused to lie down. Likewise the problem of induction has been pronounced solved or even dissolved, although it is still very much alive among scientists. And at other times the problem of interpreting a theory in factual terms was proclaimed either finally solved or altogether insoluble or even meaningless. Philosophers seem to be poor forecasters and accident-prone. We seem to abide only by one law, namely Murphy's. ("If anything can go wrong, it will.")

This book focuses on the ontological problem of causality. It analyzes causal links and processes as well as noncausal relations among events, with a view to finding out the range of validity of the causal principle. And in the process it analyzes a number of doctrines on causation. It thus offers a philosophical analysis of causation and of causal doctrines but does not pay much attention to the methodological problem of causality. There is still much to be done. In particular, there is room for a full-fledged hypothetical-deductive theory of causation framed with the help of exact tools. Let me explain.

6. *Causation as Event Generation*

There are four main modes of conceiving of causation: to deny it, to equate it with uniform succession, to identify it with constant conjunction, and to regard it as a (or even the) way events are generated. This book takes the latter view. More precisely, it formulates and defends the following theses.

1. *The causal relation is a relation among events*—not among properties, or states, let alone among ideas. Strictly speaking

causation is not even a relation among things. When we say that thing *A* caused thing *B* to do *C*, we mean that a certain event (or set of events) in *A* generated a change *C* in the state of *B*.

2. Unlike other relations among events, the causal relation is not external to them, as are the relations of conjunction, or coincidence, and succession. *Every effect is somehow produced (generated) by its cause(s)*. In other words, causation is a mode of energy transfer.

3. *The causal generation of events is lawful* rather than capricious. That is, there are causal laws or, at least, laws with a causal range.

4. *Causes can modify propensities* (in particular probabilities) but *they are not propensities*. In the expression "Event *X* causes event *Y* with probability *p*" (or "The probability that *X* causes *Y* equals *p*"), the terms "causes" and "probability" are not interdefinable. Moreover strict causality is non-stochastic.

5. *The world is not strictly causal:* not all interconnected events are causally related. Causation is just one mode of determination. Hence determinism need not be conceived narrowly as causal determinism. Science is deterministic in a loose sense: it requires only lawfulness (of any kind) and non-magic.

7. *The State-space Approach to the Causal Relation*

To say that the causal relation is a relation among events is insufficient. Nor does it suffice to add that the causal relation is irreflexive, asymmetric, and transitive. We must specify further, but not so much that the desired generality may be lost. The way to deal with things and their changes in general, without presupposing any particular laws, is to adopt the state-space approach. (For a system of ontology conceived this way, and details on what follows, see Bunge, 1977b, 1979.)

Consider any concrete thing—whether particle or field, atom or chemical system, cell or organism, ecosystem or society. We may assume that every thing is, at each instant, in some state or other (relative to a given reference frame), and that no thing remains forever in the same state. One way of describing states and changes of state is as follows.

Draw the list of all (known) properties of a thing, and represent each property by some (mathematical) function. Such a list (or n-tuple) is the state function of the thing: call it F. The changeability of the thing is reflected in the time-dependence of F. As time goes by, the value of F at t, i.e. $F(t)$, moves in an abstract space. This space is called the *state-space* of the thing x, or $S(x)$ for short. Any event or change occurring in x can be represented by an ordered pair of points in $S(x)$ and visualized by an arrow joining those points. The collection of all such pairs of states is the *event space* of x, or $E(x)$. The set of all really possible (i.e. lawful) events in (or changes of) thing x, i.e. $E(x)$, is a subset of the cartesian product of $S(x)$ by itself. The statement that event e happens in or to thing x is abbreviated: $e \in E(x)$. A *process* in thing x is representable as a sequence (or list) of states of x or else as a list of events in x. A convenient representation of the set of all changes occurring in a thing x is obtained by forming all the ordered pairs $\langle t, F(t) \rangle$ of instants of time and the corresponding states of the thing concerned. Such a set, or $h(x) = \{\langle t, F(t) \rangle \mid t \in T\}$, can be called the *history* of x during the time span T.

Consider now two different things, or parts of a thing. Call them x and y, and call $h(x)$ and $h(y)$ their respective histories. Moreover call $h(y|x)$ the history of y when x acts on y. Then we can say that *x acts on y* if and only if $h(y) \neq h(y|x)$, i.e. if x induces changes in the states of y. The *total action* (or *effect*) of x on y is defined as the difference between the forced trajectory of y, i.e. $h(y|x)$, and its free trajectory $h(y)$, i.e.

$$A(x, y) = h(y|x) \cap \overline{h(y)}.$$

Likewise for the reaction of y upon x. And the *interaction* between x and y is of course the union of $A(x, y)$ and $A(y, x)$.

Finally consider an event e in a thing x at time t, and another event e' in a thing $y \neq x$ at time t'. (The events and the times are taken relative to one and the same reference frame.) Then we can say that e is a *cause* of e' just in case (a) t precedes t' and (b) e' is in (belongs to) the total action $A(x, y)$ of x upon y. In this case e' is called an *effect* of e.

Having defined the notions of cause and effect we may now

state the strict ***principle of causality***. A possible formulation is this: "Every event is caused by some other event." More precisely: "Let x be a thing with event space $E(x)$. Then for every $e \in E(x)$ there is another thing $y \neq x$, with event space $E(y)$ relative to the same reference frame, such that $e' \in E(y)$ causes e." It is a thesis of the present book that the principle of causality is never strictly true because it neglects two ever-present features: spontaneity (or self-determination) and chance.

The matters of space-time contiguity, continuity, etc., can also be handled with the help of the above notions. Moreover, it would be interesting to formulate all possible doctrines of causality within the state-space framework. This would facilitate their mutual comparison as well as their confrontation with the most general principles of science. This is just one among many open problems in the field of causality.

8. *Conclusion*

The causal problem, which was generally regarded as buried once and for all when this book first appeared, has become topical once again. True, we are no longer causalists or believers in the universal truth of the causal principle; we have curtailed it. However, we do continue to think of causes and effects, as well as of causal and noncausal relations among them. (In particular, we can often compute or measure the probability that a given event will have a certain effect.) Instead of becoming indeterminists we have enlarged determinism to include noncausal categories. And we are still in the process of characterizing our basic concepts and principles concerning causes and effects with the help of exact (i.e. mathematical) tools. This is because we want to explain, not just describe, the ways of things. The causal principle is not the only means of understanding the world but it is one of them. Therefore beware next time you hear that causality has now finally been eliminated.

I thank Robert Blohm and Thomas R. Shultz for some useful remarks.

MARIO BUNGE

Foundations and Philosophy of Science Unit
McGill University, Montreal H3A 1W7, Canada

BIBLIOGRAPHY

Blalock, Hubert M., Jr., and Ann B. Blalock, eds., *Methodology in Social Research* (New York: McGraw-Hill Book Co., 1968).

Broglie, Louis de, "Réflexions sur la causalité." *Annales de la Fondation Louis de Broglie*, 2 (1977): 69–71.

Bunge, Mario, "Conjunction, Succession, Determination, and Causation." *International Journal of Theoretical Physics* 1 (1968): 299–315.

Bunge, Mario, Review of Suppes' *Probabilistic Theory of Causality*. *British Journal of the Philosophy of Science* 24 (1973): 409–410.

Bunge, Mario, "The Interpretation of Heisenberg's Inequalities." In *Denken und Umdenken: Zu Werk und Wirkung von Werner Heisenberg*, ed. by H. Pfeiffer (München-Zürich: R. Piper & Co, 1977a).

Bunge, Mario, *The Furniture of the World, Treatise on Basic Philosophy* (Dordrecht-Boston: D. Reidel, 1977b), Vol. III.

Bunge, Mario, *A World of Systems, Treatise on Basic Philosophy* (Dordrecht-Boston: D. Reidel, 1979), Vol. IV.

Dobzhansky, Theodosius, "Chance and creativity in evolution." In *Studies in the Philosophy of Biology*, ed. by F. J. Ayala and T. Dobzhansky (Los Angeles and Berkeley: University of California Press, 1974).

Harré, Rom, and E. H. Madden, *Causal Powers: A Theory of Natural Necessity* (Oxford: Blackwell, 1975).

Inhelder, B. and Jean Piaget, *The Growth of Logical Thinking from Childhood to Adolescence* (New York: Basic Books, 1958).

Mackie, J. L., *The Cement of the Universe: A Study of Causation* (Oxford: Clarendon Press, 1975).

Piaget, Jean, *The Child's Conception of Physical Causality* (London: Kegan Paul, 1930).

Suppes, Patrick, *A Probabilistic Theory of Causality*. Acta Philosophica Fennica XXIV (Amsterdam: North-Holland, 1970).

Suppes, Patrick, *Probabilistic Metaphysics* (Uppsala: Filosofiska Studier, 1974), 2 vols.

Svechnikov, G. A., *Causality and the Relation of States in Physics* (Moscow: Progress Publishers, 1971).

Wallace, William, *Causality and Scientific Explanation*, vol. I, *Medieval and Early Classical Science* (Ann Arbor, Mich.: University of Michigan Press, 1972).

Wallace, William, *Causality and Scientific Explanation*, vol. II, *Classical and Contemporary Science* (Ann Arbor, Mich.: University of Michigan Press, 1974).

Wright, Georg Henrik von, *Explanation and Understanding* (Ithaca, N.Y.: Cornell University Press, 1971).

Preface to the Second Edition

The problem of causality has again become prominent in relation with the quantum theory. In fact, all recent quantum field theories include the principle of antecedence ("There is no output before there is an input"), and most of them also include the principle of contiguity, or local interaction ("Every action is exerted through some medium, e.g., a field"). The former hypothesis is often called "causality condition," although it can be argued (as is done in Chapter 3 of this book) that it is logically independent from the principle of causation, or regular production of the effect by the cause. Likewise, the principle of local interaction is often linked with the causal principle and is even called "condition of microscopic causality," although it is not a necessary ingredient of the latter, as I have tried to show in Chapter 3. In this way many physicists have come to adopt a curious dual attitude towards causality: when confronted with the elementary theory (first quantization), they tend to say that causality is finished, but as soon as they pass over to the advanced theory (field theory), they begin by stating what they misname causality conditions.

The above questions are not merely terminological: any discussion of determinism and indeterminism in modern science will be misguided if it involves confusions such as those of antecedence and contiguity with causation, and determinism with Laplacian determinism. I have tried to clarify further these points in two articles that may be regarded as supplements to this book: "Causality, Chance, and Law," in *American Scientist*, Vol. 49 (1961), p. 432, and "*Causality: A Rejoinder*," in *Philosophy of Science*, Vol. 29 (1962), p. 306. The latter is reproduced in the present edition as an Appendix, and the former is included in my new book, *The Myth of Simplicity*. The differences between lawfulness and determination are dealt with in my "Kinds and Criteria of Scientific Law," *Philosophy of Science*, Vol. 28 (1961), p. 260, and the incompatibility of the "law" of continual creation of matter out of nothing with the genetic principle (a component of determinism *lato sensu*) in my "Cosmology and Magic," to appear in the first issue of the new *Monist*.

I have taken the opportunity of this new edition to correct a few misprints in the original edition and to add a couple of footnotes.

Buenos Aires
October 1962.

MARIO BUNGE

Preface to the First Edition

This is an essay on determinism, with special emphasis on causal determinism—or causality, for short. To some, causation and determination—and consequently causalism and determinism—are synonymous. But to most people, determinism is a special, extreme form of causality—and even a particularly displeasing one, for it is wrongly supposed to deny man the possibility of changing the course of events. I take sides with the minority that regards causal determinism as a special form of determinism, namely, that kind of theory that holds the unrestricted validity of the causal principle to the exclusion of every other principle of determination. The rational ground for regarding causality as a form of determinism, and not conversely, is that modern science employs many noncausal categories of determination or lawful production, such as statistical, structural, and dialectical, though they are often couched in causal language.

In this book, the causal principle is neither entirely accepted nor altogether rejected. My aim has been to analyze the meaning of the law of causation, and to make a critical examination of the extreme claims that it applies without restriction (causalism), and that it is an outmoded fetish (acausalism). I have tried to do this by studying how the causal principle actually works in various departments of modern science. However, I hope I have succeeded in avoiding technicalities—save in a few isolable passages. The book is, in fact, addressed to the general scientific and philosophic reader.

The chief result of the above-mentioned examination is that the causal principle is neither a panacea nor a superstition, that the law of causation is a philosophical hypothesis employed in science and enjoying an approximate validity in certain fields, where it applies in competition with other principles of determination. A by-product of this analysis is a fresh examination of various topics in metascience, ranging from the status of mathematical objects to the nature and functions of scientific law, explanation, and prediction.

A systematic rather than a historical order has been adopted. However, owing to the fact that probably all of the ideas on causation that have been thought are still extant, and assuming that genuine interest in ideas always leads to inquiry into their historical development, I have usually mentioned some of the historical roots of the most important ideas dealt with; more exactly, I have recalled some of their Western roots, a limitation that it would be difficult to exculpate. A further, and more severe, restriction concerns the kind of illustrations that have been chosen; I fear that physics has here got the lion's share. I might try to apologize for this preference by claiming that, after all, physics is the lion of the so-called empirical sciences—which is probably true; but it will be more honest to grant that the one-sidedness in question, which I hope is not exaggerated, is just a sign of a professional deformation.

Error seekers should find this book particularly rewarding. I myself regard everything in it as controvertible and consequently as replaceable or improvable. In contrast to dogmatic commenting and arbitrary speculation, scientific philosophy is essentially critical and self-correcting, requiring that its assertions be put to the test. Philosophy, I believe in common with an increasing number of people, deserves to be called 'scientific' solely to the extent to which its hypotheses are somehow testable—whether directly (by their logical compatibility with a given set of principles) or indirectly by the verifiable consequences such ideas may have on practical human activity and on scientific research.

This book should satisfy no orthodoxy; if it did, I would regard it as a failure. Not that I have tried to seek novelty for the sake of innovation; but I claim that no single philosophic school has ever afforded a satisfactory provisional solution of the problem of determinism. Moreover, it might be argued that we owe no single contemporary philosopher an analysis and systematization of the categories of determination comparable to Aristotle's. On the other hand, valuable elements for a modern scientific approach to the problem of determination can be found scattered in various modern philosophies of science and ontologies influenced by science.

I should be satisfied if this book were to add some amount of fruitful controversy over the much-debated question of causality, stimulating—if only as a reaction—some fresh thought on the inexhaustible problem of determinism. Moreover, I hope that it may give rise to new ideas resulting in the improvement of what is said in it.

Buenos Aires
December 1956.

M. B.

Acknowledgments

I have discussed the substance of this book in the Círculo Filosófico de Buenos Aires (1954), which was attended at that time by Dr. Enrique Mathov, Mr. Hernán Rodríguez, Prof. Manuel Sadosky, and Prof. José F. Westerkamp, to all of whom I owe illuminating critical remarks. I have also profited from discussions with colleagues and students of the Instituto Pedagógico of the Universidad de Chile, where most of the material of this book was delivered in lecture form (1955). My thanks are also due, for various criticisms and suggestions, to Dr. Irving Horowitz (New York), Prof. Henry Margenau (Yale), Prof. Ralph Schiller (Stevens Institute of Technology), and Prof. Gerold Stahl (Santiago, Chile), as well as to an anonymous critic of Cornell University. I have been stimulated in various ways by Professors Félix Cernuschi (Montevideo), Georg Klaus (Berlin), and Willard V. Quine (Harvard), no less than by the Conselho Nacional de Pesquisas (São Paulo, 1953) and the Fundación Ernesto Santamarina (Buenos Aires, 1954), both of which supported my studies in the so-called causal interpretations of quantum mechanics. But above all I have been stirred by Prof. David Bohm—the David who faced the Goliath of quantum indeterminacy—with whom I had the privilege of arguing over determinism during a couple of years, both verbally and by letter. Finally, it is a pleasure to thank the staff of the Harvard University Press for their care and kindness, and especially Mr. Joseph D. Elder, who performed a difficult and delicate piece of editorial work on the manuscript. Without the advice, stimulus, and criticism of the above-mentioned persons, and without the support of the above-mentioned institutions—particularly appreciated at a time of hardship—this book would have been definitely worse.

But, of course, the mistakes in it are my own, and all I hope is that some of them may be interesting enough to deserve being refuted.

M. B.

CAUSALITY AND MODERN SCIENCE

PART I

A Clarification of Meaning

NOTE TO THE READER

The reader is asked not to attach to the words 'causality' and 'determinism' in the text the meanings he has customarily assigned to them. Definitions of these and other key terms are worked out in Part I of the book.

The few elementary mathematical and logical signs occurring in the text should not put anybody off, since they are usually explained in plain words.

Unless explicitly stated otherwise, italics in quotations belong to the authors quoted.

References to books are given in the text in a sketchy form. Details will be found in the Bibliography.

1

Causation and Determination, Causalism and Determinism

The bewildering confusion prevailing in contemporary philosophic and scientific literature with regard to the meanings of the words 'causation', 'determination', 'causality', and 'determinism', obliges us to begin by fixing the terminology. A certain nomenclature will be proposed in this chapter, and the place of causal determination in the frame of general determinism will be sketched.

1.1. Causation, Causal Principle, and Causal Determinism

1.1.1. *The Threefold Meaning of the Word 'Causality'*

The word 'causality' has, unfortunately enough, no fewer than three principal meanings—a clear symptom of the long and twisted history of the causal problem. The single word 'causality' is in fact used to designate: (*a*) a *category* (corresponding to the causal bond); (*b*) a *principle* (the general law of causation), and (*c*) a *doctrine*, namely, that which holds the universal validity of the causal principle, to the exclusion of other principles of determination.

In order to minimize the danger of confusion, it will be convenient to try to keep to a definite nomenclature in correspondence with such a semantic variety. Let us designate by:

(*a*) *causation*, the causal connection in general, as well as any particular causal nexus (such as the one obtaining between flames in general and burns produced by them in general, or between any particular flame and any particular burn produced by it);

(*b*) *causal principle*, or principle of causality, the statement of the law of causation, in a form such as *The same cause always produces the same effect*, or any similar (and preferably refined) one. It will be convenient to restrict the term '*causal law*' to particular statements of causal determination, such as "Flames invariably cause burns in the human skin"; finally,

(*c*) *causal determinism*, or *causalism*, and often simply *causality*, the doctrine asserting the universal validity of the causal principle. A few formulations of the kernel of this theory are: "Everything has a cause", "Nothing on earth is done without a cause", "Nothing can exist or cease to exist without a cause", "Whatever comes to be is necessarily born by the action of some cause", and "Everything that has a beginning must have a cause".

In short, while the causal principle states the form of the causal bond (causation), causal determinism asserts that *everything* happens according to the causal law.

1.1.2. *Causation: A Purely Epistemological Category of Relation, or an Ontological Category?*

As here understood, causation is synonymous with causal nexus, that connection among events which Galileo[1] described as "a firm and constant connection", and which we shall try to define more accurately in Chapter 2. But what is the status of the category of causation? Is it a form of interdependence, and has it consequently an ontological status? Or is it a purely epistemological category belonging solely, if at all, to our description of experience?

[1] Galileo (1632), *Dialogo sopra i due massimi sistemi del mondo*, giornata 4a, in *Opere*, vol. 7, p. 471: "If it is true that an effect has a single primary cause [*cagione* = cause, reason, or ground], and that between the cause and the effect there be a firm and constant connection [*ferma e costante connessione*], then it necessarily follows that whenever a firm and constant alteration is perceived in the effect, there be a firm and constant alteration in the cause". See also p. 444.

The modern controversy over this question began with the skeptic and empiricist criticism. According to modern empiricism, the status of the causation category is purely epistemological, that is, causation concerns solely our experience with and knowledge of things, without being a trait of the things themselves—whence every discourse on causation should be couched in the formal, not in the material, mode of speech.[2] Thus Locke[3] proposed the following definitions: "That which produces any simple or complex idea, we denote by the general name 'cause', and that which is produced, 'effect'. Thus finding that in that substance which we call 'wax' fluidity, which is a simple idea that was not in it before, is constantly produced by the application of a certain degree of heat, we call the simple idea of heat, in relation to fluidity in wax, *the cause* of it, and fluidity *the effect*". Moreover, like Kant after him, Locke held the causal principle to be "a true principle of reason": a proposition with a factual content but not established with the help of the external senses.

The conception of causation as a mental construct, as a purely subjective phenomenon, was emphasized by Locke's followers Berkeley[4] and Hume,[5] as well as by Kant.[6] But, whereas Locke had regarded causation as a *connection*, acknowledging production as its distinctive mark, his followers

[2] Lenzen (1954), *Causality in Natural Science*, p. 6: "Causality is a relation within the realm of conceptual objects. The relation of cause and effect refers to conceptual events regardless of the relation of the latter to reality". See also Frank (1937), *Le principe de causalité et ses limites*, *passim*, where the causal relation is restricted to data of experience.

[3] Locke (1690), *An Essay Concerning Human Understanding*, book II, chap. xxvi, sec. 1.

[4] Berkeley (1721), *De motu* (*Concerning motion*), secs. 41, 71, and *passim*, in *Works* (Fraser ed.) vol. 1, pointed out that, owing to the purely mental character of causation, the real "efficient causes of motion" fall entirely outside the scope of mechanics and are instead the concern of metaphysics.

[5] Hume (1740), *A Treatise of Human Nature*, book I, part III, secs. ii–iv; (1748) *An Enquiry Concerning Human Understanding*, sec. vii. Hume uses indifferently 'relation' and 'connection', but means always a relation emerging from a comparison between perceptions or "ideas".

[6] Kant (1781, 1787), *Kritik der reinen Vernunft* (B), pp. 163 ff. According to Kant the causal law does not apply to things but to experience alone; that is, it applies to the phenomenal, not to the noumenal, world, being nothing but a direction enabling us to order or to label phenomena, so as to read them *qua* experiences.

have held that causation is only a *relation*,[7] and moreover one relating experiences rather than facts in general. Hume emphasized this point particularly, on the ground that it would not be empirically verifiable that the cause *produces* or engenders the effect, but only that the (experienced) event called 'cause' is invariably associated with or followed by the (experienced) event called 'effect'—an argument which is of course based on the assumption that only empirical entities shall enter any discourse concerning nature or society.

The empiricist doctrine of causality will be examined rather closely in the course of this book, especially as it is a widespread belief that Hume gave the final, or almost final, solution of the causal problem. For the time being we shall merely state the opposite thesis, namely, that causation is not a category of *relation* among *ideas*, but a *category of connection and determination* corresponding to an actual trait of the factual (external and internal) world, so that it has an ontological status—although, like every other ontological category, it raises epistemological problems.[8] Causation, as here understood, is not only a component of experience but also an objective form of interdependence obtaining, though only approximately, among real events, i.e., among happenings in nature and society.

But before we can ascertain whether causation is a category of determination, and long before we can hope to show that, while being so far used in both science and philosophy, it is not the sole category of determination, we shall have to examine what is meant by 'determination'.

1.2. Toward a General Concept of Determination

1.2.1. *Two Meanings of 'Determination': Property, and Constant Connection*

Let us take a glance at the different uses of 'causation' and 'determination', two concepts that are frequently regarded as

[7] Connections should be regarded as a subclass of relations. If we say that "John is taller than Peter", we state a *relation* between John and Peter; if, on the other hand, we say that "John is Peter's partner", we state a connective relation between two individuals.

[8] Ontology is here understood as the theory of the most general traits of reality, involving both the study of categories (such as space) and the analysis of generic laws (such as the causal principle).

equivalent[9] although some philosophers have acknowledged their difference.[10] In actual use the word 'determination' designates various concepts, among which the following are particularly relevant to our discussion: (*a*) *property* or characteristic; (*b*) *necessary connection*, and (*c*) *process* whereby an object has become what it is—or way in which an object acquires its determinations in sense (*a*).

In its first acceptation, 'determination' is synonymous with 'characteristic', whether qualitative or quantitative; this is what *determinatio* meant in post-Roman Latin, and in this way it is used in various European languages, notably in German (*Determination* and *Bestimmung*). In this sense, that is determinate which has definite characteristics and can consequently be characterized unambiguously; when applied to descriptions and definitions, 'determinate' is used as an equivalent of precise or definite, in contradistinction to vague. Thus Locke[11] called determinate or determined the ideas which Descartes had described as clear and distinct; and Claude Bernard[12] termed *indéterminés* the facts collected without precision, those ill-defined facts "constituting real obstacles to science".

But in science the most frequent use of the word 'determination' that is relevant to our concern seems to be that of *constant and unique connection* among things or events, or among states or qualities of things, as well as among ideal objects. (Thus, for instance, machines that run regular and reproducible—hence fully predictable—courses have been called *determinate*. Their successive states follow one another in a constant and unique way to the exclusion of new, unexpected states—contrarily to *indeterminate* machines, whose states are only statistically

[9] See University of California Associates (1938), "The Freedom of the Will", in Feigl and Sellars, eds., *Readings in Philosophical Analysis*, p. 602: "Determination and causation are, indeed, identical concepts. '*A* determines *B*' and '*A* causes *B*' are identical propositions". See also Ayer (1940), *The Foundations of Empirical Knowledge*, chap. iv, esp. pp. 179 ff.

[10] For instance, Hartmann (1949), *Neue Wege der Ontologie*, p. 56: "Cause and effect constitute only one among many forms of determination".

[11] Locke (1690), *An Essay Concerning Human Understanding*, Epistle to the Reader, p. xv.

[12] Bernard (1865), *Introduction à l'étude de la médecine expérimentale*, part I, chap. ii, sec. vii.

determined and constitute moreover an open set, that is, one admitting new elements.[13]) If by *necessary* is meant that which is constant and unique in a connection,[14] then in sense (*b*) the word 'determination' amounts to necessary connection.

If the form of a necessary connection is known, then some features of certain relata will be inferable from the knowledge of certain other relata. Thus, for example, the relation 'twice', that is, $y = 2x$, enables us to derive unambiguously the value of y when that of x is specified. Analogously, an equation such as

$$x - 1 = 0 \tag{1.1}$$

is called *determinate* because it has a unique solution ($x = 1$). On the other hand, the equation

$$x + y - 1 = 0 \tag{1.2}$$

is called *indeterminate* because, although it expresses a rigid interdependence between x and y, a further datum (that is, a further equation) is needed in order to solve the problem. As it stands, equation (1.2) admits infinite solutions; in other words, the problem is not *uniquely determined*, although every particular answer (such as $x = \frac{1}{3}$, $y = \frac{2}{3}$), far from being lawless, will fit the law $x + y - 1 = 0$. Therefore, probably a better name for such equations would be that of partially determined, or incompletely determinate, in contrast to those of the former type, which are fully determined or completely determinate.

1.2.2. *Constant Unique Connections Need Not Be Causal*

When turning to the sciences of nature and of society, the variables $x, y, \ldots$ may symbolize properties of concrete objects, and functions like $y = f(x)$ may consequently designate relations among qualities or among intensities of qualities. Thus thermal expansion of a metal rod occurs (approximately) in accordance with the law

$$L(t) = L(0)(1 + \alpha t), \tag{1.3}$$

[13] Ashby (1956), *An Introduction to Cybernetics*, p. 24.

[14] Margenau (1950), *The Nature of Physical Reality*, p. 407.

where $L(0)$ stands for the length of the rod at the temperature $t = 0$, α being the coefficient of thermal expansion. This law is usually regarded as an instance of the causal principle: namely, as the mathematical transcription of the causal law "Heating causes the expansion of metals". To test this causal interpretation of (1.3), let us write it in the following, mathematically equivalent form:

$$\alpha t = \frac{L(t) - L(0)}{L(0)}. \tag{1.3'}$$

The right-hand member certainly represents the relative value of the effect, that is, the relative expansion. But the left-hand member, αt, contains not only the intensity of a quality belonging to *both* the "agent" and the "patient"—namely, the temperature t—but also α, which denotes a thermoelastic property (a disposition) characterizing the piece of metal and having nothing to do with the specific nature of the extrinsic agent (fire, warm water, radiation) producing the modification (expansion) concerned. In short, while the second member of our relation does represent the effect (expansion), the first member does not represent the cause (heating), at least not in an equally satisfactory way. Hence, (1.3) is a fully-determined (constant and unique) relation, that is, it stands for a necessary connection among the qualities length and temperature, and the disposition expansibility; but it does not reflect the causal connection between the heating agent and the effect it produces. This should suffice, for the time being, to suggest the inadequacy of the functional view of causation, which will be dealt with at length in Sec. 4.1.

In case the foregoing example was not found modern enough, we might recall Einstein's famous law of the numerical equivalence of the mass m of a body and its energy E:

$$E = mc^2, \tag{1.4}$$

where c symbolizes the velocity of light in vacuum. This, too, is a relation expressing a necessary connection among properties of a physical object. One can, and usually does, say that the numerical value of one of the properties is *determined* by the value of the related property. But, clearly, when used in this way, the word 'determination' does not convey the activity and productivity inherent in causation. And it could not be otherwise, as neither temperature nor mass are physical agents, events, or phenomena, but are instead qualities of physical objects. In other words, since qualities and dispositions have by themselves no productive virtue, the laws (1.3) and (1.4) *are not causal laws* although they express necessary connections. Hence the word "determination" is used in connection with them in a sense which is more restricted than that of causation, namely, in the sense of necessary (constant and unique) relation between properties.

This point is seen even more clearly if, in the function $y = f(x)$, the independent variable is made to denote time. For example, the velocity $v(t)$ acquired by a freely falling body after t seconds have elapsed is

$$v(t) = gt + v_0, \quad (1.5)$$

g and v_0 being constants. One can say that $v(t)$ is *determined* by t, but not, of course, in the sense of being *produced* by it, since time has no productive virtue of its own—except in a highly metaphorical sense. Both the unique connection between $v(t)$ and t and the hypothesis that such a connection is constant (invariable) enable us in principle to predict the value of the velocity after any prescribed time has elapsed. In general, in this restricted sense, 'determination' means possibility of calculation and consequently of quantitative forecast. The same applies to the law (1.3) of the thermal expansion of a metal rod: although it does not contain time, it enables us to predict the value $L(t)$ when $L(0)$, α, and the temperature t are known; analogously, Einstein's law (1.4) enables us to predict, for example, the energy required to dissociate a molecule if its mass defect is known. The apparent paradox of prediction with the help of laws that do not contain time is dissolved as soon as it is recalled that time enters *our* use of such laws.

But the meaning of the laws of nature is not restricted to the possibility of calculation on their basis—contrary to what some empiricists have maintained.[15] Indeed, if determination has an epistemological status, it is because it has an ontological support; thus, if we can predict the intensity of the property y from the value of the associated property x through the relation $y = f(x)$, it is only because, and to the extent to which, such a mathematical relation has an objective counterpart. In the sciences dealing with concrete objects we are not interested in calculating for the sake of calculation, but in calculating what we assume to have some relation, even if remote, with real events and processes. That is, meaningful and useful calculation follows in these sciences (as contrasted to mathematics) from the assumed factual validity of the law statements with the help of which calculation is carried out.

1.2.3. *A Third Meaning of 'Determination': Way of Becoming*

From the foregoing we conclude that the word 'determination', as used *in science*, has a very restricted meaning: it means less than causal determination, for though it denotes constant unique connection (which is certainly a characteristic of the causal bond), it lacks the essential ingredient of productivity. In other words, the current scientific meaning of the word 'determination' does not coincide with the third acceptation of it which we have recorded, namely, the way (act or process) whereby an object acquires a property. In this sense, not only that which is fully qualified (hence susceptible to unambiguous description) is 'determinate', not only that which has definite characteristics, but also that which has acquired them in one or more definite ways. This addition may at first sight seem otiose; but it is not, for, according to indeterminism, there may be determination in senses (*a*) and (*b*) but not in sense (*c*).

[15] Schlick (1932), "Causality in Everyday Life and in Recent Science", in Feigl and Sellars, *Readings in Philosophical Analysis*, p. 525: "Let us see how the scientist uses the word determination—then we shall find out what he *means* by it. When he says that the state E at the time t is determined by the state C at the time t_0, he means that his differential equations (his Laws) enable him to *calculate* E, if C and the boundary conditions are known to him. Determination therefore means Possibility of Calculation, *and nothing else*". See also Guido Castelnuovo, "Il principio di causalità", *Scientia 60*, 61 (1936), where determinism is defined as possibility of precise prediction.

Indeed, according to the propounders of accidentalism—or tychism, as C. S. Peirce called it—things do have definite characteristics, but they acquire them in a lawless (hence unpredictable) way: nothing determines what they are, events popping up here and there without any antecedent or contemporaneous condition.

Among the simplest forms of determination in sense (*c*) stands the acquisition, by concrete objects, of quantitative characteristics without the emergence of new qualities but, instead, through continuous evolution. This is the type of determination envisaged by mechanical determinism, which regards qualities as fixed and consequently takes only quantitative changes into account. Poincaré's description[16] of this particular type of determinism (which he called *l'hypothèse déterministe*) is worth recalling owing to its conciseness, clearness, and wide acceptance.[17] According to Poincaré, "On the deterministic hypothesis the state of the universe is determined by an excessively large number, n, of parameters which I shall call $x_1, x_2, \ldots, x_n$. If the values of these n parameters are known at any given instant, and their time derivatives are also known, then the values of the same parameters at a previous or at a later time can be calculated. In other words, these n parameters satisfy n differential equations of the form

$$\frac{dx_i}{dt} = \psi_i(x_1, x_2, \ldots, x_n), \quad i = 1, 2, \ldots, n\text{''}.$$

But not every pattern of change fits the foregoing scheme: things may acquire their determinations in ways other than the one envisaged by the science of mechanics and by mechanistic ontologies. There are in the world qualitative changes besides quantitative variations, so that mechanical determinism is but a subclass of what I shall call *general determinism*, or determinism *lato sensu*.

Determination need not be effected through quantitative variations only, as mechanical determinism holds; and it need

[16] Poincaré (1908), *Thermodynamique*, p. ix.

[17] For a similar definition of "deterministic system", see Russell (1912), "On the Notion of Cause", in *Mysticism and Logic*, p. 188.

not be brought forth by external compulsion alone, as causal determinism claims; nor need it be unique or well defined, as both causal and mechanical determinism hold. All that is needed in order to maintain determinism in a general sense is to hold the hypothesis that events happen in one or more definite (determinate) ways, that such ways of becoming are not arbitrary but lawful, and that the processes whereby every object acquires its characteristics develop out of preëxisting conditions. (The essential components of general determinism will be described more precisely in Sec. 1.5.)

1.2.4. *Chance: Alien to Determinism?*

Even chance, which at first sight is the very negation of determination, has its laws, and accidents emerge from preëxisting conditions; consequently chance has a place in what we are calling general determinism. Thus the appearance of a "head" in coin throwing, far from being a lawless event, and far from emerging out of nothing, is the determinate result of a determinate operation. Only, it is not the sole possible result, it is not a unique outcome of a given process; or, as may also be said, it is not a well-defined result. Coin throwing is a *determinate* process because: (*a*) far from being unconditional, it requires the meeting of definite conditions, such as the existence of a coin and of a throwing person (or machine), a horizontal surface on which the coin may fall, a gravitational field, and so forth; (*b*) far from yielding completely indeterminate, arbitrary, lawless results, coin throwing yields just "heads" and "tails" and this, moreover, in accordance with the definite statistical law stating that, in the long run, the number of "heads" will nearly equal that of "tails" if the coin is well balanced—whereas if tossing a coin produced at times "heads" and other times elephants, newspapers, dreams, or any other objects in an arbitrary, lawless, way, with no connection at all with the antecedent conditions, then it would be an indeterminate process.

In games of chance, the final results arise out of definite conditions in accordance with definite laws; they cannot consequently be used to illustrate or to support indeterminism, as is so often done. What happens is that games of chance do not follow a certain customary type of law, namely, the Newtonian; they follow instead statistical laws, they are *statistically determined.*

This brings to the fore the question whether the quantum theory has led to the bankruptcy of determinism, as is so often heard.

1.2.5. *The Quantum Theory: A Restriction on Determinism or on Causality?*

The answer to the question whether the quantum theory entails the failure of determinism depends not only on the definition of determinism but also on the interpretation of the quantum theory that is chosen—and there are by now quite a number of consistent and empirically equivalent interpretations of quantum mechanics.[18]

The usual presentation of the quantum theory, as advanced by Bohr and Heisenberg, does eliminate causality as regards the results of observation, in the sense that the "same" physical situation may be followed in an unpredictable way by a large (usually infinite) number of different states. But this restriction on *causality* does not entail a breakdown of *determinism*, since statistical determinacy is definitely retained in that interpretation—not to speak of the obviously nonstatistical laws of quantum mechanics, such as the conservation laws, the selection rules, or the exclusion principle.

Moreover, even the orthodox interpretation of quantum mechanics restricts the scope of causality without rejecting it entirely. Thus when we write down the probability for a transition of a physical system from state 1 to state 2, we often assign this transition to some force (cause), usually represented by an interaction potential (Fig. 1). Only, the cause and the effect are here not tied in the constant and unique (that is, necessary) way asserted by the causal principle. In other words, the usual interpretation of quantum mechanics does not sweep out

[18] See Mario Bunge, "A Survey of the Interpretations of Quantum Mechanics," *American Journal of Physics 24*, 272 (1956).

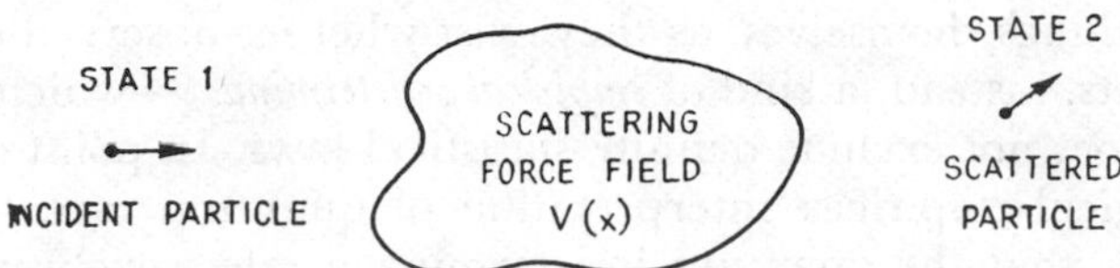

Fig. 1. Scattering of a particle by a field of force. The deflection of the particle is not uniquely determined, for several paths are possible, but at any rate it is *caused* by the force deriving from the potential $V(x)$.

causes and effects, but rather the rigid causal nexus among them.

By calculating and by interpreting such transition probabilities p_{12}, we imply (*a*) that the transitions are not arbitrary and do not arise out of nothing, but follow from definite states under the action of definite forces, so that, for example, we can ascertain with certainty that an electron will not radiate unless it is accelerated by an external field; (*b*) that the transitions are not necessary, in the sense that state 2 does not develop regularly and unambiguously from state 1: there is only a certain probability p_{12} for this occurrence among many other possible transitions, which means that in a large number, N, of cases, the transition $1 \rightarrow 2$ is almost certain to occur about Np_{12} times.

In short, the usual interpretation of the quantum theory actually does not eliminate determinism in the general sense; moreover, it retains a certain dose of causality. But it does drastically restrict the Newtonian form of determinism, according to which all physical processes boil down to changes of position determined by both the previous state of motion and externally impressed forces, the paths of the point masses concerned being precisely defined (fully determined) trajectories in space-time. It may also be said that Newtonian determinism is sublated in quantum mechanics in its orthodox interpretation, since it is found to obtain on the average.

On the other hand, the positivistic philosophy built on (and partially built into) the usual interpretation of the quantum theory, eliminates determinism—but, then, also indeterminism —in the *ontological* sense, that is, in connection with the behavior of the things themselves, as they exist whether observed or not. It asserts, instead, a kind of *empirical indeterminacy*—which, however, does not exclude definite statistical laws. In point of fact, the logical-empiricist interpretation of quantum mechanics[19] declares that the quantum indeterminacy refers exclusively to results of observation, not to matter itself—a term which it regards as a meaningless metaphysical fiction.

The root of this empirical indeterminacy is easy to see: the bond between two successive states of an atomic system is taken to be the observer, who acts freely with respect to the system. In other words, the behavior of the observer is regarded, in this interpretation, as independent of the object of observation, but not vice versa: the properties of the observed system are those which the observer decides to prepare or to "conjure up"—to employ the term introduced by a renowned physicist. In the absence of direct objective connections between the consecutive states of a physical system it is no wonder that every form of scientific determinism in the ontological sense is lost, with the sole exception of statistical determinacy. This sort of indeterminism is clearly a consequence of the subjectivistic doctrine about the almost arbitrary intervention of the observer, who is regarded as the very conjurer of atomic-scale phenomena. Quantum indeterminacy is, then, a consequence of the idealistic hypothesis inherent in modern positivism.[20] This point will concern us again in Sec. 12.5.

In some recent interpretations of quantum mechanics, not even the causal principle is entirely forsaken. Thus in Bohm's presentation[21] of the quantum theory Newtonian determinism, with its causal ingredient, is partially restored by the assignment of a physical meaning to the wave function ψ, which in the usual interpretation performs only a mathematical function. A force equation is derived, according to which the acceleration of a "particle" is the effect of the joint action of the external forces and of a new, internal, force depending on the

19 See Frank (1946), *Foundations of Physics*, pp. 48 ff.

20 Mario Bunge, "Strife about Complementarity", *British Journal for the Philosophy of Science 6*, 1, 141 (1955).

21 David Bohm, *Physical Review 85*, 166, 180 (1952).

ψ-field surrounding the "particle". And the statistical distribution of the results of a measurement is not regarded as an unintelligible *ultima ratio*, but is explained as the outcome of a definite interaction between the physical system in question and the measuring device (a further physical system). In this way quantum jumps cease being elementary (unanalyzable) *Urphänomene*; chance at the quantum level ceases to be an ultimate and is instead analyzed into further categories of determination. But the statistical determinacy peculiar to quantum mechanics is not thereby eliminated or explained away; it is shown to be the outcome of lower-level processes. (More on the recent revaluation of the causal problem in connection with quantum mechanics will be found in Secs. 12.5.2 and 13.4.2.)

Whether we accept some causal interpretation of quantum mechanics or not, we see that this physical theory, in either of its interpretations, does not dispense with determinism in general but at most forsakes the Newtonian type of determinism. We also realize that, whether chance is regarded as a radical ultimate (as Peirce and Eddington thought) or not, statistical determinacy has to be accounted for by every philosophy of modern science; it is no longer possible to state dogmatically that chance is but a name for human ignorance, or to declare the hope that it will ultimately be shown to be reduced to causation. Chance is a peculiar type of determination, and its relations to other categories of determination are worth investigating.

1.3. The Spectrum of Categories of Determination

Since we wish to characterize causal determination as neatly as possible, we have to point out its place in general determinism. As here conceived, causation is only one among several categories of determination. A surely incomplete list of such categories appearing in the ontology of modern science is the following:

Quantitative self-determination: determination of the consequent by the antecedent. Illustrations: (*a*) The successive positions of a freely moving macroscopic body are uniquely determined by its position and velocity at any prescribed instant of time. (*b*) The spontaneous transformations of an isolated thermodynamic system are such as take it to states of increasing entropy. Quantitative self-determination is the category of determination prevailing in the continuous unfolding of states that differ from one another in quantitative respects only; in some cases, notably in thermodynamics, quantitative self-determination may be shown to emerge from processes characterized by other categories of determination, among them causation.

Causal determination, or *causation*: determination of the effect by the efficient (external) cause. Illustrations: (*a*) If a bullet is fired against a window, the glass is broken. (*b*) If an electromotive force is applied to the ends of a piece of metal, an electric current is set up in the metal in accordance with Ohm's law. The causation category is particularly conspicuous when the main changes are produced by external factors.

Interaction (or reciprocal causation, or functional interdependence): determination of the consequent by mutual action. Illustrations: (*a*) The orbits of the components of a double star are determined by their gravitational interaction. (*b*) The functioning of every gland in the human body depends on that of the remaining glands.

Mechanical determination: of the consequent by the antecedent, usually with the addition of efficient causes and mutual actions. Illustrations: (*a*) Forces modify the state of motion of bodies (but motion may exist before the application of the forces). (*b*) The streamlines in a fluid are determined by the latter's previous state, by the external forces acting upon it, by internal friction (viscosity), and by internal pressure differences.

Statistical determination: of the end result by the joint action of independent or quasi-independent entities. Illustrations: (*a*) In the game of dice, the long-run frequency of the event "throwing two aces in succession" is 1:36. (*b*) About one-half of the newborn children are females. As in the case of other categories of determination, statistical determinacy may emerge from processes on deeper levels, in which still other categories of determination are involved.

Structural (or wholistic) determination: of the parts by the whole. Illustrations: (*a*) The behavior of an individual (a molecule in a fluid, a person in a social group) is determined by the over-all structure of the collection to which it belongs. (*b*) The functioning of an organ is partially determined by the needs of the whole organism. But, of course, the whole, far from being prior to its members, is in turn determined by them.

Teleological determination: of the means by the ends, or goals. Illustrations: (*a*) Birds build their nests "in order to" safeguard their young. (*b*) Standardization is adopted in industry in order to lower production costs. Needless to say, goal-directed structures, functions, and behaviors need not be purposefully planned by anybody.

Dialectical determination (or qualitative self-determination): of the whole process by the inner "strife" and eventual subsequent synthesis of its essential opposite components. Illustrations: (*a*) Changes of state in matter in bulk are produced by the interplay and final predominance of one of the two opposite trends: thermal agitation and molecular attraction. (*b*) The contrasting economic interests of social groups determine changes in the very social structure of such groups. In opposition to quantitative self-determination, internal dialectics involves qualitative changes. And, needless to say, it has nothing to do with logical contradiction.

Further types of determination would certainly be recognized as a result of deeper analysis. At any rate, causal determination appears as only one among various ways of determination, as one of the types of lawful production (or nomogenesis). Hence causal determinism, or causality, which is placed between the two extremes of fortuitism and fatalism, is but one variety of determinism; along with other types of determinism, causality is subsumed under general determinism.

1.4. Connections Among Different Types of Determination

Any clear-cut distinction among types of determination involves the assumption that they are *irreducible* to one another, that is, that none of them can be regarded as a mere mixture of other forms of lawful production, every one of them being characterized by a peculiar newness of its own. That this is the case can be seen from the respective definitions of the categories of determination, and will be shown with the help of more detailed analysis as occasion arises.

At this point I only wish to point out, first, that the above-mentioned determination categories, however mutually irreducible they are, are also *connected* with one another, constituting a hierarchy of types of determination and second, that, as a consequence of this connection, no type of determination is found to operate in all purity, to the exclusion of all others, save in ideal cases. To illustrate my first contention, take mechanical determination, which is a peculiar combination of purely quantitative self-determination (in this case, inertial motion) and reciprocal action, which can often be polarized into cause and effect. Or take statistical determination, which emerges, with characteristics of its own, as a result of the interplay of a large number of elements that are individually determined in accordance with other types of determination (mechanical, or teleological). The various types of determination are genetically connected with one another, and the higher types are *dependent* on the lower without being entirely *reducible* to them.

No type of determination can be assigned a territory where it operates to the complete exclusion of other types of lawful production. True, one might try to characterize different sectors of reality according to the relative *predominance* of the various categories of determination; but no clear-cut stratification would result from this. Take, for instance, the domain of life processes, which vitalists claim is unambiguously and exclusively characterized by final causes. Biology seems to use a certain amount of teleology—which is harmless as long as it is cleansed of its traditional unscientific associations; but it also needs the category of determination by virtue of inner processes, that self-determination or freedom which is found in all the corners of the material universe, and which living beings possess to a supreme extent; biology also needs, in addition, wholistic determination (the German *Ganzheitsbestimmung*), since one of the characteristics of life phenomena is the organismic partial subordination of the parts to the whole; it further needs statistical determination, as actually the adequacy of means to ends is never necessary but only statistical; besides, biology cannot dispense with reciprocal action, which within the organism is characteristically synthetic or integrative, and which plays such an eminent role in evolution through selection; and, of course, biology cannot forget efficient extrinsic causes, as living beings are never entirely free from the compulsive and often almost one-sided actions of their environment. Thus the entire spectrum of categories of determination seems to be used in biology—and, a fortiori, in the cultural sciences, which ontologically (not historically) presuppose biology even if they are not reducible to it.

A detailed investigation of the interconnections among the various forms of determination would show more clearly that they can be ordered according to their increasing complexity, in such a way that all but the first two (quantitative self-determination and causation) can be rooted to the lower types. We would thus have a scale of degrees or levels of determination, each characterized by a peculiar trait of its own, but grounded on the lower levels.

Hence it is not a question of choosing one type of determination at the expense of others, by decreeing that the chosen category shall reign undisputed in all sectors of reality; like all monistic solutions, this one is too simple to be adequate. Unlike dogmatic philosophies, the philosophical examination of modern science requires us to realize that a rich assortment of types of lawful production, or determination, is actually employed in the description and explanation of the world, that they all have an ontological counterpart, though not necessarily in the same sectors of reality or to the same extent in all sectors.

This rather proofless sketch of a theory of levels of determination was only meant to point out the modest but as yet indispensable place of causation in the wider context of general determinism.

1.5. The Essential Components of All Types of Determinacy: Productivity and Lawfulness

1.5.1. *The Principle of Lawfulness or Orderliness*

If the philosophic concept of determination is wider than both the usual acceptation of this word in science and the causation concept, then causal determinism can only be a special type of determinism in the broad sense—although most authors[22] identify them.

But what are the essential marks of determinism in the broad sense? According to some philosophers, determinism only states that "everything depends on conditions and happens only when these conditions are met".[23] But the dependence upon conditions might be regular or not. If conditionalness is *regular*, i.e., if it fits definite (not necessarily unchanging) patterns, then it seems preferable to call it *lawfulness*, that is, conformity to law.[24] And *this* is the type of conditionalness we are interested in, because it is what science seeks to establish.

Now, the principle of lawfulness may be worded as follows: *There are laws.*[25] And the principle of universal lawfulness, a stronger postulate, may be taken to read thus: *Every single event is lawful*, i.e., is determined in accordance with a set of objective laws—whether we know the laws or not. Or, again, *Every single fact is the locus of a set of laws.*[26] Note that, in this wording, the principle of universal lawfulness does not assert that facts are determined *by* laws, but *in accordance with* laws, or simply *lawfully*. Thereby the idealistic doctrine is avoided, according to which natural and social laws are not the immanent form of facts, but prescribe them *ab extrinseco*.[27] There is no Rule of

[22] For example, Bergson (1888), *Essai sur les données immédiates de la conscience*, p. 151.

[23] Hartmann (1949), *Neue Wege der Ontologie*, p. 57. Unfortunately Hartmann, who met science late in life, disowned statistical lawfulness.

[24] Lawfulness is sometimes called *orderliness*: see Wisdom (1952), *Foundations of Inference in Natural Science*, chap. viii.

[25] According to Russell (1948), *Human Knowledge: Its Scope and Limits*, p. 496, "in any verifiable form, such a postulate would be either false or a tautology".

[26] This statement will be explained in sec. 10.4.3.

[27] The idealistic notion that laws tower above the objects to which they apply, that the objects are somehow created by the laws, is common to both Kantianism and logical positivism. See Frank (1938), "Determinism and Indeterminism in

Law, laws do not determine anything: they are the *forms or patterns of determination*—and this is one of the reasons why determinacy is not synonymous with lawfulness. Thus, for example, the constraints to which a dynamical system is subjected contribute to the determination of its motion; but Gauss and Hertz's principle of least constraint does not *determine* the motion along the corresponding least-curvature path: it is just the form of the action of the constraints on motion.

The principle of lawfulness, however, does not require that every individual phenomenon should always occur in the same way whenever certain conditions are fulfilled; universal lawfulness is consistent with individual *exceptions*, with occurrences in a given low percentage of cases. Individual irregularity in some respects is consistent with collective regularity, i.e., with the individuals concerned fitting a collection having laws of its own, *qua* collection. Statements of statistical law apply to situations in which there are several alternatives, and exceptions are nothing but the least frequent alternatives. Moreover, the property of being exceptionless (or, rather, almost exceptionless) belongs not so much to facts as to *statements* about facts; indeed, statements of statistical law are often made with a high accuracy and the probability of their being true can be increased almost indefinitely.

Finally, the principle of lawfulness is not committed to a specific form of determinism, such as causal determinism, or mechanical determinism—contrary to what Peirce[28] stated in a famous paper which is still held in awe in some quarters.

In short, since the chief aim of science is the search for, explanation, and application of laws, if we wish to build an ontological theory of determination having a scientific ground,

Modern Physics", in *Modern Science and its Philosophy*, p. 178. It is an inescapable consequence of the empiricist principle according to which things can be said to exist solely to the extent to which they can be experienced, for instance, observed, or measured.

[28] Peirce (1892), "The Doctrine of Necessity Examined", in *Philosophical writings*, p. 324. Peirce gave the name "necessitarianism" to the belief in the principle of universal lawfulness. The whole of his celebrated criticism of legality relied on the erroneous identification of scientific law with *mechanical* law. Since obviously not everything can be made to fit mechanical laws, it was rather easy for Peirce to "disprove" the principle of universal legality.

we have to qualify bare conditionalness as the peculiar mark of determinism, we have to admit that not mere conditionalness but regular conditionalness, that is, *lawfulness*, is an essential component of general determinism. Still, lawfulness is insufficient.

1.5.2. *The Genetic Principle*

Although it is commonly held that 'determinism' is nothing but "the belief in the absolute necessity of laws",[29] 'determinism' should mean more than lawfulness. In fact, despite Schopenhauer's naïve belief to the contrary, laws are conceivable which could "govern" the appearance of things, or the emergence of properties, out of nothing. Moreover, statements of such laws, and even their precise mathematical formulation, have been propounded by the inventors of the so-called "new cosmology", or steady-state theory of the homogeneous universe; this theory involves the hypothesis of the continuous spontaneous creation of matter *ex nihilo*.[30] And this is not precisely what is usually meant by respecting scientific determinism, even in its widest sense, for the concept of emergence out of nothing is characteristically theological or magical—even if clothed in mathematical form.

In order to build a definition of determinism both elastic enough to admit new categories of determination and strict enough to exclude irrational and untestable notions (such as creation out of nothing), I propose to combine the principle of lawfulness with that of productivity, that is, the ancient principle according to which *nothing comes out of nothing or passes into nothing*.[31] An alternative formulation of this principle is: *There are neither absolute beginnings nor absolute terminations*, but everything is rooted to something else and leaves in turn a track

[29] Le Chatelier (1936), *De la méthode dans les sciences expérimentales*, p. 24.

[30] See Hoyle (1952), *The Nature of the Universe*, pp. 97 ff. For a criticism of this theory, see Herbert Dingle, *Monthly Notices of the Royal Astronomical Society 113*, 393. (1953). Curiously enough, Dingle objects to *continuous* creation rather than *creation*.

[31] See Lucretius (c. 58 B.C.), *Of the Nature of Things*. The very first principle that nature teaches us is, that not even divinity can produce something out of nothing: "*Nullam rem e nihilo gigni divinitus unquam*" (I, 150), "*Nil. . .fieri de nilo*" (I, 206). And nothing goes over into naught (I, 249–250).

in something else. For the sake of brevity, this old materialistic principle will be called the *genetic principle.*

The genetic principle is *consistent* with the principle of lawfulness, on which it puts a restriction, but is independent of it. In fact, everything might be the outcome of a process and might in turn give rise to other events, yet not in a lawful way. Thus, if we were to believe intuitionists, the processes of artistic and intellectual creation are lawless, even though every stage in them is the outcome of a preceding state of the individual and has in turn definite consequences.[32]

Needless to say, the genetic principle is rejected not only by theologians but also by subjectivist philosophers, whether idealistic[33] or sensationistic;[34] it is essential for them that the world, even if fluent, be *barren* without the assistance of some psychic entity. That is, both transcendentalism and subjectivism may grant that events *follow* one another, but not that they are *produced* by one another. They may deny lawfulness, as Plato did, or grant it, as Hume did; what they cannot accept is the hypothesis that there is, in the external world, a power of originating anything. Transcendentalists and subjectivists tend therefore to reduce determination to succession, whether uniform or not, without production.

1.5.3. The Principle of Determinacy

Our definition of *general determinism* will then be the following: Determinism in the large sense is that ontological theory whose necessary and sufficient components are

the genetic principle, or principle of productivity, according to

[32] Many who would not dispute that the chief aim of scientific research is the discovery and application of *laws*, would, on the other hand, be prepared to admit the intuitionistic thesis that research itself, and particularly the process of scientific creation, is at least partially lawless—and this on the bare grounds that (*a*) we still know very little about the psychology of intellectual work, (*b*) there are no golden (infallible) rules for scientific discovery, and (*c*) the principles of scientific methodology that assist (or hinder) discovery and invention are far from sufficient to ensure success (or failure). See Bunge (1959), *Metascientific Queries*, chap. 3.

[33] Hegel (1817), *Encyklopädie der philosophischen Wissenschaften im Grundrisse, Logik*, sec. 88, 5, regarded the genetic principle as inconsistent with change.

[34] Mach (1883), *The Science of Mechanics*, p. 609, called this principle an "empty maxim".

which nothing can arise out of nothing or pass into nothing; and

the principle of lawfulness, stating that nothing happens in an unconditional and altogether irregular way—in short, in a lawless, arbitrary manner.

The two principles can be conjoined to yield the following sentence: *Everything is determined in accordance with laws by something else*, this something else being the external as well as the internal conditions of the object in question. This statement may be termed the *principle of determinacy*. It is a philosophical assumption of science confirmed by the results of scientific research; it clearly cannot be refuted, since future investigation is expected to confirm it wherever it may now seem to be falsified. Any theory of structure or change or both respecting the principle of determinacy will hereafter be called *deterministic*.

The causal principle is a particular case of the principle of determinacy; it essentially obtains when determination is effected in a *unique* or unambiguous way by *external* conditions. General determinism, as here conceived, allows for causal, mechanical, statistical, teleological, and other kinds of determination, such as the qualitative change brought about by the increase or decrease in quantity, the so-called "struggle" of opposites, etc. The sole kind of determinism excluded from our definition is fatalism, because it violates the principle of lawfulness, since the alleged determination by virtue of some *fatum* is presumed to occur no matter what the prevailing conditions are, and entails moreover a transcendent agent (see Sec. 4.3). The richer kind of determinism outlined in the foregoing description requires only productivity (or genetic connection) and legality (or conditionalness and regularity). General determinism does not restrict a priori the various forms that change and the scientific laws of change may assume.

1.6. Causation and Determination: Main Views

As is well known, the most diverse stands have been taken with regard to the causal problem, from the flat rejection of the causation category to the assertion of its coinciding with determination. All of them belong to one of the following classes: *causalism*, or panaitism, *semicausalism*, or hemiaitism, and *acausalism*, or anaitism. Although we shall examine them all in detail in the course of this book, it will be convenient to give here a brief characterization of each of these groups of theories and of their main varieties.

Causalism. (*a*) According to the *traditional* theory (often wrongly credited to the founders of modern mechanism), causation is the sole category of determination, so that science is coextensive with causality; on this view, no scientific law or explanation is possible that does not hinge on the causation category. This is probably the most widespread belief, from Aristotle, who wrote that "what is called Wisdom is concerned with the primary causes and principles",[35] to Claude Bernard, who gave "the name *determinism* to the proximate or determinant cause of phenomena".[36]

(*b*) According to the *rationalistic* doctrine, the causal principle is a necessity of thought (*Denknotwendigkeit*), an a priori regulative principle, hence a presupposition rather than a result of science. This is the opinion of Leibnizians, to whom the causal principle is nothing but a form of the principle of sufficient reason. (Schopenhauer took this belief over from Leibnizianism.) It is also the belief of Kantians,[37] who assert that the

[35] Aristotle, *Metaphysics*, book I, chap. i, 981b.

[36] Bernard (1865), *Introduction à l'étude de la médecine expérimentale*, 2nd ed., p. 348 and *passim*. Along with other French authors, Bernard employed the word 'determinism' to designate determination, and identified it in turn with causation. Another characteristic statement of Bernard is this: Empirical or conjectural medicine should be replaced by a medicine of certainties (*médecine certaine*), "which I call experimental medicine because it is based on the experimental *determinism* of the cause of illness" (p. 339).

[37] Kant (1787), *Kritik der reinen Vernunft*, (B), p. 232: "All changes take place in accordance with the law of the connection [*Vernküpfung*] between cause and effect". But (p. 233) this connection is "the product of a synthetic faculty of the imagination [*Einbildungskraft*]". The causal principle is then not a result but a presupposition of experience—it renders experience possible (B, p. 765 and *passim*). See also Cassirer (1937), *Determinismus und Indeterminismus in der modernen Physik*, pp. 25 ff, 73 ff, and *passim*: the essential meaning of the causal principle is that it is not a statement about things but about experience; it is not an assertion regarding things or processes, but a statement about the systematic interconnection of *Erknenntnissen*. Also Nagel (1956), *Logic Without Metaphysics*, p. 124, holds that the causal principle states no law of nature but "functions as a maxim, as a somewhat vague rule for directing the course of inquiry".

causal bond is synthetic, in the sense of being verifiable *in* experience but not derivable from it nor further analyzable; and that the causal principle "could never be refuted by any possible experience—It is nothing but the demand of understanding everything".[38]

Semicausalism. (*a*) The *eclectic* theory recognizes the validity of causation in certain domains (for example, in macrophysics) along with the unrestricted validity of other categories of lawful production (such as the statistical or the teleological ones) in other domains, without, however, establishing links among the diverse categories of determination and without acknowledging the possibility that several of them may concur in one and the same process. This nomic pluralism is quite widespread among physicists.[39]

(*b*) According to the *functionalist* or interactionist theory, the causation category is a particular case of the interaction or interdependence category; on this doctrine, it is always a sheer abstraction to disentangle simple, linear, cause-effect bonds from the universal interconnection or interdependence (*Zusammenhang*), which has an organic character. This opinion is typical of romantics and is shared by most dialectical materialists.

[38] Helmholtz, *Physiologische Optik*, vol. III, p. 30, quoted in Weyl (1949), *Philosophy of Mathematics and Natural Science*, p. 192. See also Ostwald (1908), *Grundriss der Naturphilosophie*, p. 41. Einstein himself adopted some features of the Kantian doctrine of causality when he stated that, even if the causal principle is not found to hold *de facto* in the domain of experience, it does hold *de jure* in the realm of ideas, so that it should be possible to build a presentation of quantum mechanics in which the initial state of a system determines entirely its later states. See the debate on determinism and causality in contemporary physics in the *Bulletin de la Société Française de Philosophie*, No. 5 (Oct.–Dec. 1929). Likewise Grete Hermann, "Die Naturphilosophischen Grundlagen der Quantenmechanik", *Die Naturwissenschaften 23*, 718 (1935), held that the difficulties encountered by the partisans of causality in connection with the quantum theory have nothing to do with the causal principle, but derive instead from the assumption that natural phenomena are independent of the observer.

[39] Reichenbach (1929), "Ziele und Wege der physikalischen Erkenntnis", in Geiger and Scheel, ed., *Handbuch der Physik*, vol. IV, pp. 69, defended a dualism of causality and "probability" (*sic*) as independent principles intervening in the description of all phenomena. See also Born (1951), *Natural Philosophy of Cause and Chance*, where a similar dualistic doctrine is expounded, according to which "nature is ruled by laws of cause and laws of chance in a certain mixture" (p. 3).

(*c*) *General determinism*, or neodeterminism—the theory advocated in this book—asserts in this connection that causation is only one among several interrelated categories concurring in real processes; on this view, the causal principle has a limited range of validity, being no more and no less than a first-order approximation.

Acausalism (*a*) The *empiricist* theory reduces causation to external conjunction or succession of events—or, rather, to the concomitance or the temporal sequence of experiences; it may grant the lawfulness of phenomena but asserts the contingency of qualities and of the laws themselves, regarding the latter as nothing but rules of scientific procedure. Empiricism holds that the notion of causation is an "episode in the history of ideas", an outdated fetish that is gradually being replaced by functional laws (Mach) or by empirically found statistical correlations (Pearson) or, in general, by probability laws (Reichenbach). According to Russell,[40] "the law of causality . . . like much that passes muster among philosophers, is a relic of a bygone age, surviving, like the monarchy, only because it is erroneously supposed to do no harm".

(*b*) The *indeterminist* doctrine denies every lawful link among events and qualities; in particular, it does not recognize the existence of causal bonds, and asserts that events just happen, and that qualities are mere idiosyncrasies, or characteristics that, being loose, might have been different. This extreme development of empiricism does not seem to have been systematically defended by anybody.

1.7. Conclusions

As may be gathered from what has been said in the foregoing, the target of my arrows will not be the *causal principle* but only the claim that causation is the sole category of determination and that, as a consequence, the causal principle enjoys an unlimited validity. In fewer words: I will not argue against the notion of causation but against causalism.

The theory to be worked out in the following pages belongs

[40] Russell (1912), "On the Notion of Cause", in *Mysticism and Logic*, p. 171.

therefore to the group of theories named in Sec. 1.6 semi-causalism. It is specifically characterized by the following theses, which it will be my purpose to substantiate in the course of the present work:

Causation (efficient and extrinsic) is only one among several categories of determination; there are other types of lawful production, other levels of interconnection, such as statistical, teleological, and dialectical determinacy.

In real processes, several categories of determination concur. Purity in types of determination (such as purity of causation) is as ideal as any other kind of purity.

The causation category, far from being external to other categories of determination, is connected with them. Thus multiple causation leads to statistical determinacy, the latter may in turn lead to quantitative self-determination, and reciprocal causation is interaction or interdependence.

The causal principle holds approximately in certain domains. The degree of approximation is satisfactory in connection with certain phenomena and very poor with regard to others.

Causal determinism, without being altogether erroneous, is a very special, elementary, and rough version of general determinism.

Before examining in more detail the pros and cons of the doctrine of causality, and before suggesting how to repair it, we shall have to recall and analyze some of the statements of the causal principle.

2

Formulations of the Causal Principle

In the present chapter we shall be concerned with listing and analyzing some typical definitions of causation, that is, we shall examine various statements of the causal principle. Such an examination of definitions is indispensable in every rigorous approach to the causal problem—although it is entirely overlooked in many works on the subject [1]—and, indeed, in every scientific treatment of philosophic questions. For, although philosophizing is not restricted to framing and discussing definitions, the lack of precise—that is, sufficiently precise—definitions facilitates wild speculation.

2.1. Definitions of Cause

2.1.1. *The Aristotelian Teaching of Causes*

Almost every philosopher and scientist uses his own definition of cause, even if he has not succeeded in formulating it clearly. The earliest and most systematic codification of the meanings of this thorniest of words we owe to Aristotle, who elaborated Plato's scattered ideas on causality. According to the Stagirite [2] and to the peripatetic schoolmen, a single kind

[1] For example, Cassirer (1937), *Determinismus und Indeterminismus in der modernen Physik*, an otherwise highly stimulating treatise.

[2] Aristotle, *Metaphysics*, book I, chap. iii, 983a, b: "There are four recognized kinds of causes. Of these we hold that one [the formal cause] is the essence or essential nature of the thing (since the 'reason why' of a thing is ultimately reducible to its formula, and the ultimate 'reason why' is a cause and principle); another [the material cause] is the matter or substrate; the third [the efficient cause] is the source of motion; and the fourth [the final cause] is the cause which is opposite to this, namely the purpose or 'good'; for this is the end of every

of cause was not sufficient for the production of an effect, whether in nature or in art (industry). Four kinds of cause were instead needed, namely: the material cause (the scholastic *causa materialis*), which provided the passive receptacle on which the remaining causes act—and which is anything except the matter of modern science; the formal cause (*causa formalis*), which contributed the essence, idea, or quality of the thing concerned;[3] the motive force or efficient cause (*causa efficiens*), that is, the external compulsion that bodies had to obey; and the final cause (*causa finalis*) was the goal to which everything strove and which everything served. The first two were causes of being; the efficient and the final causes were causes of becoming.

The Aristotelian teaching of causes lasted in the official Western culture until the Renaissance. When modern science was born, formal and final causes were left aside as standing beyond the reach of experiment; and material causes were taken for granted in connection with all natural happenings—though with a definitely non-Aristotelian meaning, since in the modern world view matter is essentially the subject of change, not "that out of which a thing comes to be and which persists".[4] Hence, of the four Aristotelian causes only the efficient cause was regarded as worthy of scientific research.

Some of the grounds for the Renaissance reduction of causes to the *causa efficiens* were the following: (*a*) it was, of all the four, the sole clearly conceived one; (*b*) hence it was mathematically expressible; (*c*) it could be assigned an empirical correlate, namely, an event (usually a motion) producing another event (usually another motion) in accordance with fixed rules; the remaining causes, on the other hand, were not definable in empirical terms, hence they were not empirically testable; (*d*) as a consequence, the efficient cause was controllable; moreover, its control was regarded as leading to the harnessing

---

generative or motive process". See also book V, chap. ii, and *Physics*, book II, chaps. iii and vii.

[3] Shape, or geometric form, was but a particular kind of form, formal cause, or essence (*eidos*). Shape played a dominant role in the Aristotelian doctrine of artificial bodies, a subordinate one in the Philosopher's teaching on natural bodies.

[4] Aristotle, *Physics*, book II, chap. iii, 194b.

of nature, which was the sole aim of the instrumental (pragmatic) conception of science advocated by Bacon and his followers.

As has been usual since the beginnings of modern science, we shall hereafter restrict the meaning of the term 'cause' to *efficient cause*, or extrinsic motive agent, or external influence producing change—in contrast to other kinds of cause, such as final, internal (or *causa sui*), etc. The classical definition of efficient cause was, of course, put forward by Aristotle: it is "the primary source of the change or coming to rest; e.g., the man who gave advice is a cause, the father is cause of the child, and generally what makes of what is made and what causes change of what is changed".[5] The efficient cause is, in short, the *agent* producing some change in what is (erroneously) imagined to be a *patient*, upon which the cause acts *ab extrinseco*, from the outside.

2.1.2. *Galileo's Definition of Cause*

Modern thought, while retaining the externality of causation, has preferred other definitions of the efficient cause. One of the clearest of them all was given by Galileo, who defined the efficient cause as the *necessary and sufficient condition for the appearance of something*: "that and no other is to be called cause, at the presence of which the effect always follows, and at whose removal the effect disappears".[6] Hobbes,[7] in many ways a follower of the great Florentine, distinguished carefully

[5] Aristotle, *Physics*, book II, chap. iii, 194b.

[6] Galileo (1623), *Il Saggiatore*, in *Opere*, vol. 6, p. 265: "se è vero che quella, e non altra, si debba propriamente stimar causa, la quale posta segue sempre l'effetto, e rimossa si rimuove; solo l'allungamento del telescopio si potrà dir causa del maggior ricrescimento: avvengachè, sia pur l'oggetto in qualsivoglia lontananza, ad ogni minimo allungamento ne seguita manifesto ingrandimento".

[7] Hobbes (1655), *Elements of Philosophy* (*Elementa philosophiae, Sectio prima, De corpore*), chap. ix, 3, in Woodbridge, *Selections*, p. 94: "The cause, therefore, of all the effects consists in certain accidents [properties] both in the agents and in the patients; which when they are all present, the effect is produced; but if any one of them be wanting, it is not produced; and that accident either of the agent or patient, without which the effect cannot be produced, is called *causa sine qua non*, or *cause necessary by supposition*, as also the *cause requisite* for the production of the effect". Note the modern conception of causation couched in scholastic terminology (agent, patient, accident).

between the *causa sine qua non*, or necessary cause, and the complex of sufficient causes, which may alternatively produce the same effect.

Galileo's definition is at first sight satisfactory, especially as it has not only an ontological but also a methodological meaning; indeed, it states a practical criterion for deciding whether a factor is a necessary cause or not, namely, its removal. But a closer inspection shows its inadequacy in the following respects. First, the definition involves an indefinite number of factors, as it includes in the cause any thing or event that could make some difference to the result or effect; and, since indeterminateness or haziness is inconsistent with causal determinacy, the above definition of cause does not look adequate. Suppose we used it in the context of the theory of the universal causal interconnection (too often erroneously called determinism); then, since every event would be regarded as the effect of an infinity of events of the most diverse kinds, Galileo's definition of cause would lead to its identification with the state of the whole universe immediately preceding the event in question—which is actually how Laplace conceived of causes. But this would render the concept of cause practically worthless, for causal analyses would then be impossible owing to the infinity of factors (all of about the same importance) presumably constituting the cause. And the empirical test of causal hypotheses would be equally impossible, since the removal of any of the infinite factors would make some difference, hence would require the control of an infinity of parameters.

The second reason why Galileo's definition of cause is no longer entirely correct is that in a sense it is too general—so much so that it may apply to statistical, dialectical, and other processes, since what it states is the set of *conditions*, both necessary and sufficient, for the occurrence of an event of whatever kind, produced by a process of any sort, whether causal or not. Galileo's definition is, in fact, a statement of regular conditionalness—which in Sec. 1.5.1 was recognized as a necessary component of all kinds of determinacy.

The vagueness of the definitions of causal bond current until about one century ago has prompted the framing of more precise, and consequently more schematic and abstract, formulations of the causal principle. Let us examine some of them; but let us begin by stating the general conditions that any adequate formulation should meet.

2.2. General Features of Any Formulation of the Causal Principle

The following sentences are sometimes regarded as correct formulations of the causal principle:

$$C, \textit{ therefore } E, \tag{1}$$

or

$$E \textit{ because } C. \tag{1'}$$

These forms are not, however, adequate to pour causation into. First, they have the forms of explanatory statements: the terms 'therefore' and 'because' in them may suggest that a *reason* rather than an "agent" is involved in the causal bond. In the second place, they assert not only a bond between *C* and *E* but suggest also the *existence* of both: as soon as the values of the symbols '*C*' and '*E*' are specified, the forms (1) and (1′) become propositions that may be understood as singular factual statements. Now scientific laws—and, a fortiori, principles of scientific ontology—are not singular factual statements in the sense that they refer to matters of fact; they are, on the other hand, general *hypotheses*, and, moreover, of the conditional form. It was recognized long ago that a scientific law "does not express what happens but what would happen if certain conditions were met".[8]

The hypothetical status of law statements, that is, of the conceptual reconstructions of the patterns of being and becoming, is apparent from the fact that they *apply* to more or less idealized models of reality, even though they are supposed to *refer* to concrete pieces of reality. In other words, the referent of a nomological statement may be real, but its range of exact validity consists of a set of ideal cases only approximately

[8] Meyerson (1908), *Identité et réalité*, p. 19. For an analysis of this problem, see Braithwaite (1953), *Scientific Explanation*, pp. 314 ff.

coincident with real situations. Consider, for instance, the most general physical law, namely, the principle of conservation of energy, which can be stated in the following way: "If a system is isolated, its total energy is conserved". Or, in the subjunctive, "If a system were isolated, its total energy would be conserved". These propositions are deemed to be universally true (except perhaps in the case of quantum virtual transitions) even if perfect isolation in every respect is nowhere to be found; the principle does not assert the existence of isolated systems, and consequently it does not contradict the assertion of their factual inexistence. In general, while statements of matters of fact (singular factual statements, like "The sun is shining") can be categorical, law statements are hypothetical.

It follows, then, that an adequate statement of the causal principle should not involve the assumption that *C* actually exists but should instead say that, *if C* is the case, then *E* will also be the case; in short, the statement must be a *conditional*. The emphasis should be on the *relation* rather than on the relata—as Russell[9] has untiringly insisted—and on the *conditions* for the occurrence of facts of a certain class, rather than on the facts themselves. And the conditionalness peculiar to scientific lawfulness renders the use of hypotheticals desirable—whether in the indicative or in the subjunctive moods. In short, law statements in the nonformal sciences should begin with *if*.

Let us then try the following form:

$$\textit{If } C, \textit{ then } E, \tag{2}$$

or simply: *If C, E.*[10] In either of these equivalent hypothetical sentences, '*C*' and '*E*' can be read as designating *singulars belonging to any classes of concrete objects*—events, processes, conditions, and so on, to be specified as occasion arises. Every specification of the class of facts will, of course, lead from the empty scheme (2) to a specific law statement. Needless to say,

[9] Russell (1914), *Our Knowledge of the External World*, pp. 219 ff.

[10] Ostwald (1908), *Grundriss der Naturphilosophie*, p. 44, regarded (2) as the structure of the "purified causal law"—with the qualification that the antecedent should denote the complex of necessary and sufficient factors. But only the relation of logical equivalence (reciprocal implication) can be used to denote necessary and sufficient conditions; in this case we would have: *If, and only if, C, then E.*

C and *E*—that is, the referents of the symbols '*C*' and '*E*'—must differ from one another in at least one respect.

Let me insist on the warning that '*C*' and '*E*' should be taken as representing different specific objects (or, rather, particular features of concrete objects) insofar as these objects belong to definite *classes*. But '*C*' and '*E*' should not refer to particular things or properties, if the causal principle is to be concerned with *differences* among existents rather than with existents, with becoming rather than with being. And '*C*' and '*E*' should refer to a limited number of features, and not to the unlimited richness of real events, if the principle is to take on a verifiable form. That is to say, '*C*' and '*E*' stand for *kinds* of individual, specific, concrete events regarded in definite respects; if preferred, '*C*' and '*E*' designate specific events insofar as they belong to classes—else we would not be dealing with the skeleton of a general law, but with a bare statement of fact concerning singulars.

Now, from a logical point of view, '*C*' and '*E*' may be regarded either as individual or as propositional variables. In the former case, (2) is an incomplete framework, whereas in the latter it is just a compound proposition. It may be convenient to regard '*C*' and '*E*' as kinds of free variables; but then, in order to obtain from (2) a proposition (a statement with a definite truth value), hence a sentence capable of being verified or refuted, we have to specify the range of the terms involved—i.e., we have to affix a quantifier to (2). This will now be done.

2.3. The Constant-Conjunction Formula of Causation

Let us then regard the formulations of the causal principle as stating relations among variables '*C*' and '*E*' symbolizing kinds (classes) of facts, or rather definite features of facts. Now, with regard to a relation there are the following possibilities: that it holds *sometimes* (whether in a fixed or in a variable percentage of cases), or that it holds *always*, i.e., for all the values of the variables. The usual interpretation of the causal principle is obviously inconsistent with the former alternative; the causal connection is supposed to hold universally, hence the causal principle must assert the exceptionless repetition of *E* whenever *C* is the case. Consequently the word 'always' (the all-operator) must be added to (2) if it is to become a statement of the causal principle. This leads us to the sentence

If C, then E always, (3)

or, what is equivalent,

For all C and E, if C is the case, then E is the case. (3′)

This general conditional is often regarded as the precise formulation of the vulgar statements *Same cause, same effect,* or *Every event has a cause, and this cause is always the same*—which have the shortcoming of emphasizing the sameness of causes and effects instead of the constancy of their relation. Needless to say, the word 'always' in (3) must not be taken as meaning forever, but in all cases, or without exception in the given universe of discourse, or generally. The time concept is excluded from (3): we are still in the realm of being (or, if preferred, in the domain of timeless logical relations); we are not yet in the domain of process.

An analysis of (3) must now be performed, in order to ascertain whether, and to what extent, it does convey the core meaning usually attached to the causal principle. It turns out that the following concepts are contained in (3): conditionalness, what may be called existential succession, and constancy. Let us treat them separately.

Conditionalness. From an ontological point of view, the terms *If C* state the clause(s) or condition(s) for the occurrence of *E*; in other words, they assert that *E* occurs provided *C* happens. But conditionalness is not peculiar to *causal* lawfulness; it is a minimum requirement to be fulfilled by *any* kind of law, whether of the causal type or not. Lawfulness may, indeed, be defined as regular conditionalness (see Sec. 1.5.1); the unlawful, on the other hand, is by definition the unconditional or arbitrary, that which exists or happens no matter what the conditions or the circumstances are.[11]

[11] Mill (1843, 1872), *A System of Logic,* book III, chap. v, sec. 6, regarded *unconditionalness* as a characteristic of the causal connection; but he thereby meant that, given *C*, *E* will follow no matter what the *remaining external* circumstances may be.

Asymmetry, or existential succession. The effect E will appear only provided the conditions summarized by C have been fulfilled—not however necessarily *after* C. To employ a term of which traditional philosophers are fond, the cause is existentially prior to the effect—but need not *precede* it in time. Indeed, so far nothing has been said of a time delay between C and E: formula (3) does not contain time—a point that will be dealt with in detail in 3.2. The priority of C over E, the unsymmetrical dependence of the effect upon the cause,[12] together with the uniqueness pointed out above, is conveniently symbolized in the form $C \to E$—which should not, however, be understood as logical implication, since the causal connection is a synthetic, not an analytic, connective, and has consequently nothing to do with logical (analytic) necessity.[13]

Constancy. If C is the case, E will ensue *invariably*: this is what the all-operator 'always' in (3) means. Our statement does not assert that the existence of C *may* entail that of E or that, given C, E will ensue in a certain percentage of cases: it asserts, on the contrary, that the connection obtains invariably. Now, constancy and uniqueness make up necessity as defined in Sec. 1.2.1; hence, (3) asserts that the causal bond is conditional, unsymmetrical, and necessary.

Yet—should the concept of necessity be included in an adequate formulation of the causal principle? The traditional answer is, of course, in the affirmative; still, nowadays we would

[12] Russell (1927), *An Outline of Philosophy*, p. 121, has referred, on the other hand, to "the reversibility of causal laws", in the sense that they enable us to *infer* backward as well as forward, to retrodict as well as to predict. This statement involves the identification—unavoidable in orthodox logical empiricism—of *laws* with their *statements*. Beyond doubt, we can make law *statements* function in a "reversible" way, when using them for either postdictive or explanatory purposes; but such a reversibility at the epistemological level has nothing to do with the processes to which such statements refer, and it applies not only to causal laws but to other kinds of laws as well. For a distinction between objective laws of nature or society, and law statements (the conceptual reconstructions of the former), see sec. 10.1.

[13] Logicians sometimes assume that the meaning of the causal principle consists in the logical structure of the statements that are formulated in order to reconstruct the causal bond in thought. This would be correct if, as Hegel believed, the structure of the world were logical. For a treatment of the formalizations of the causal connection, see sec. 9.6.

hardly accept any of the traditional meanings attached to the word 'necessity', which are repugnant to both modern science and modern philosophy. Indeed, the traditional—and also the popular—notion of necessity identifies it with either (*a*) *unconditionalness*, that is, the quality of that which occurs under all circumstances, whether or not certain conditions are met, or (*b*) *passive obedience* to an external or even transcendent power that does not lie in the very nature of things, or, finally, (*c*) *want* not too different from human needs. Empiricism duly reacted against these anthropomorphic views, reducing the valid use of 'necessity' to mean lack of exception,[14] or regularity in the sense of prestatistical science. But there is no point in employing two words for designating a single concept. If the notion of necessity is stripped of its anthropomorphic and fatalistic associations, it reduces either to efficient causation—as Spinoza[15] had implicitly proposed—or to constancy and uniqueness, as agreed before.

So far, so good; the statement *If C, then E always* is shown, upon analysis, to contain three notions that are usually associated with causality: the conditionalness peculiar to lawfulness, the existential priority of the cause over the effect, and lack of exception. But are these traits enough to describe causation unambiguously?

2.4. Criticism of the Constant-Conjunction Formula of Causation

2.4.1. *The Uniqueness of the Causal Bond: Neglected in the Previous Formula*

The first objection that can be raised against (3) as an adequate formula of causation is that it does not account for the uniqueness of the causal bond, it does not say that there is

[14] See, for example, Schlick (1932), "Causality in Everyday Life and in Recent Science", reprinted in Feigl and Sellars, *Readings in Philosophical Analysis*, p. 523.

[15] Spinoza (1677), *Ethics*, part I, Def. VII: that thing "is called necessary, or rather compelled (*coacta*), which by another is determined to existence and action in a fixed and prescribed manner". This definition of necessary determination overlaps with the notion of efficient causation, whence it does not seem advisable to adopt it. A further meaning of necessity is that of lawfulness, or strict accordance to law: this is how it was conceived by Boole (1851), "The Claims of Science", in *Studies in Logic and Probability* (1952), pp. 192 ff.

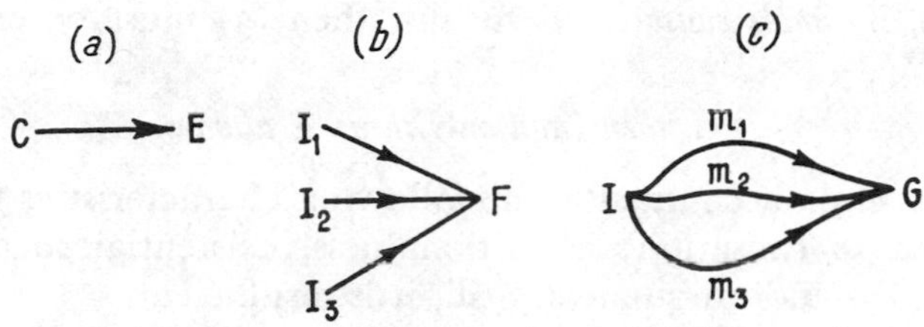

Fig. 2. (*a*) The causal nexus is unique. (*b*) The same final statistical distribution F can be attained from a number of initial states. (*c*) Several means m_i may be employed by an organism to attain a given goal G that is invariant under changing conditions (within limits).

a one-to-one correspondence between C and E. Indeed, in *If C, then E always*, the term 'C' may denote any of the *sufficient* causes, hence the formula makes room for multiple causation (see Sec. 5.1). On the other hand, simple causation is characterized by the 1 : 1 reciprocal correspondence between the cause and the effect, that is, the relation between C and E is such that there is a *single E* for every C and vice versa. The existence of E follows (not necessarily in time) in a *unique* or unambiguous way from the existence of C; or, again, E is a single-valued function of C.

Unlike conditionalness, which is a generic trait of scientific law, uniqueness, or lack of ambiguity, is absent from certain kinds of law, such as statistical regularities, which establish many-to-one connections between causes and effects. Uniqueness is a characteristic, though not an exclusive, trait of causation. The uniqueness of causation, often described as the *rigidity* of causal determination, is to be contrasted to the *souplesse* of statistical determination, largely insensitive as it is to precise details (such as the initial states of the individual components of a mass of gas). This unambiguous character of the (simple) causal bond is also to be contrasted to the so-called *plasticity* of teleological determination, in which a given goal may be attained through a whole range of alternative means (see Fig. 2).

If only necessary and sufficient (instead of just sufficient) factors are to be regarded as antecedents in a causal connection, that is, if *simple causation* is meant, then (3) must be changed into

If C, then (and only then) E always. (4)

This formula accounts for the following characteristics usually assigned to causation: conditionalness, existential succession, constancy, and uniqueness. Still, it is insufficient.

2.4.2. *The Efficacy of Causation: Denied by the Humean Doctrine of Causation*

The sentence *If C, then (and only then) E always* is a universal conditional statement expressing the *constant conjunction*[16] of two kinds of terms. It does not state a *genetic* connection but an external association, an *invariable coincidence*. It says nothing about the active and productive nature that causal agents are usually supposed to possess, nothing of a process out of which *E* emerges. The sentences (3) and (4) are typically phenomenalist maxims in which the cause concept is not regarded as a category of determination through change, but only as an *antecedent*.

The reduction of causation to constant conjunction—and, in particular, to constant association of successives, that is, to regular succession—must have been common in antiquity, as even an amateur philosopher like Cicero[17] criticized it. This view was in modern times adopted by Joseph Glanvill in his *Scepsis scientifica* (1665), and later on by Malebranche in his *Recherche de la vérité* (1675); but it only became popular with Hume's *Treatise* (1739–40).[18] Since Hume, the statement *If C,*

[16] Constant conjunction is here meant in an ontological, not in a logical sense; that is, it does not mean the joint assertion of two propositions, but the concomitant occurrence of two events.

[17] Cicero (c. 44 B.C.), *De fato*, 34–35.

[18] French historians of philosophy often take a pleasure in calling Hume's criticism of causality a popularization of Malebranche's analysis. See Brunschvicg (1922), *L'expérience humaine et la causalité physique*, 3rd ed., p. 13. French philosophers, on the other hand, forget too frequently that the agnostic contention of the unknowability of the true and ultimate causes of things has been characteristic of empiricism since antiquity, and was stated explicitly by Glanvill (1665) and by Locke (1690). See Richard H. Popkin, "Joseph Glanvill: Precursor of Hume", *Journal of the History of Ideas 14*, 292 (1953).

then E always has usually been regarded by empiricists as exhausting the meaning of causation, hence as the correct statement of the causal principle. Thus Ayer[19] wrote that "every general proposition of the form '*C* causes *E*' is equivalent to a proposition of the form 'whenever *C*, then *E*', where the symbol 'whenever' must be taken to refer, not to a finite number of actual instances of *C*, but to the infinite number of possible instances". And Reichenbach,[20] in another influential book, stated that "by a causal law the scientist understands a relation of the form *if–then*, with the addition that the same relation holds at all times". In agreement with the empiricist tradition, he held that "the meaning of the causal relation consists in the statement of an exceptionless repetition".[21] As may be seen, contemporary empiricists have not improved Hume's definition of the causal relation as constant conjunction, as the "constant union betwixt the cause and effect",[22] which would give us what James[23] called "a world of mere *withness*, of which the parts were only strung together by the conjunction 'and'".

2.4.3. *Inadequacy of the Constant-Conjunction Formula*

It is high time to put the constant-conjunction view of causation to the test. This can be done by giving values to the (bound) variables '*C*' and '*E*', seeing whether the resulting singular propositions are actually causal or not. Take "Wars cause worries", on the one hand, and "Red apples are sweet", on the other. The former is an openly causal proposition, whereas the latter asserts a correlation, for nobody would be prepared to regard a quality, like redness, as the cause of another quality, like sweetness. Still, both propositions fit the

[19] Ayer (1936), *Language, Truth and Logic*, 2nd ed., p. 55.

[20] Reichenbach (1951), *The Rise of Scientific Philosophy*, p. 157.

[21] *Ibid.*, p. 158. Note, in the first place, that causation is defined by Reichenbach as a *relation*, not as a connection—as if it could be placed in the same bag with statements like "B lies between A and C". Secondly, Reichenbach seems to have abandoned here the positivist (pragmatist) verifiability criterion of meaning, since the property of *exceptionlessness* is anything but empirically verifiable.

[22] Hume (1739–1740), *Treatise of Human Nature*, book I, part III, sec. xv.

[23] James (1907), *Pragmatism*, p. 105.

general formula *If C, then E always*. This counter example should suffice to indicate that the Humean formula of causation is not specific enough to be regarded as an adequate conceptual reconstruction of the causal bond.

The chief reason of the inadequacy of the Humean formula is that it is too weak: it expresses a *relation*, not a connection. By the assertion that *If C, then E always* only a constant relation between two terms is meant, an invariable correlation between events that are "loose and separate", "conjoined but not connected"—as the Humean description runs. The Humean formula expresses, indeed, the conditionalness, unsymmetry, and lack of exception that characterize the causal link; but it accounts neither for the uniqueness nor for the *genetic* character of the relation between *C* and *E*. It does not convey the productivity or efficacy of causation: it does not, in short, say that the effect happens to be *produced* by the cause, but only that it is regularly conjoined to it.

The correlated variations of the elements in a structure fit the Humean formula of causation. Yet, a law of correlation is not a causal law, because it does not state that a given entity (or a change in it) is *produced* by another entity (or by a change in it), but just that the two are regularly associated. Thus hair on the one hand and teeth or horns on the other are often correlated with one another.[24] The two arms of a lever are rigidly associated, but are not causally connected with each other. Similarly, the sides and angles of a triangle determine each other in a rigid (unique) way, yet not in a causal way. In either case the genetic, productive element is absent—and this productivity is chiefly what renders the cause-effect connection essentially unsymmetrical.

There is more to it: most statements in pure mathematics fit the *If–then always* formula. And it would be hard to assign to

[24] See, for example, Darwin (1859), *The Origin of Species*, 6th ed., pp. 11–12: "Hairless dogs have imperfect teeth; long-haired and coarse-haired animals are apt to have, as is asserted, long or many horns; pigeons with feathered feet have skin between their outer toes; pigeons with short beaks have small feet, and those with long beaks large feet. Hence if man goes on selecting, and thus augmenting, any peculiarity, he will almost certainly modify unintentionally other parts of the structure, owing to the mysterious laws of correlation." See also pp. 149 ff.

mathematics the job of accounting for causal connections: uninterpreted mathematical formulas do not refer to the external world, but constitute a world by themselves. Mathematical objects are certainly a product of human activity and may be *correlated* with material processes by way of rules of correspondence—but in themselves mathematical objects lie outside change: they are immutable, and in particular they have not the power of changing on their own account.[25] Productivity is as absent from mathematics as from the Humean formula of causation.

Besides being ontologically defective, Hume's celebrated criticism of the efficacy of causation can be shown to be circular.[26] In fact, it rests on the following assumptions: (*a*) *sense impressions* are the sole relata to be considered—as fits an empiricist theory; (*b*) sense impressions are momentary, that is, unrelated to either past or future; (*c*) since the past is no longer actual, it cannot act upon the present, so that *every event is a fresh entity* having no connection with entities existing in the past—a consequence which, if consistently developed, should lead to the conclusion that the world is created afresh at every instant in some mysterious way. Consequently, Hume presupposed what he set out to prove, namely, that there is no connection between past and future; in other words, his disproof of productive causation is circular.

The empiricist reduction of causation to regularity is grounded in the original sin of empiricism, namely, the identification of *truth* with its *criterion*, the reduction of the meaning of a proposition to the mode of its verification. The road is short: it is first stated that the sole test of causation is

[25] There is neither a *mathematics of rest* nor a *mathematics of change*, any more than there is a *logic of rest* or a *logic of change*. Ideal objects, like those handled by logic and mathematics, are characterized, in contrast to material objects, by their timelessness and lack of self-movement. The ability of scientists in connection with the use of mathematics as a language for expressing facts consists in establishing the right correspondences between symbols and material referents, particularly among the static abstract structures and the concrete processes in the restless external world.

[26] Abner Shimony, "An Ontological Examination of Causation", *Review of Metaphysics 1*, 52 (1947).

the observation of constant conjunction;[27] then, on Wittgenstein's verifiability doctrine of meaning, it is concluded that the *meaning* of causation is exhausted by regular association. Now, it is indisputable that, in order to *test* hypotheses concerning any kind of facts, the examination of a large number of similar instances is very convenient: that is, whenever possible, it is desirable to study large collections or series of (almost) repetitive phenomena (that is, repetitive in some respect). But, apart from the fact that this is not always possible in practice, such a sound methodological prescription does not entail that, *in re*, Causation = Regularity. Such a reduction involves a confusion of epistemology (and, more particularly, of scientific methodology) with the theory of the most general traits of reality, namely, ontology. Likewise, the recognition of the fact that every criterion of material existence involves sensing, acting, and judging subjects does not imply that every existent is dependent upon some sensing, acting, or judging being.

In summary, the reduction of causation to regular association, as propounded by Humeans, amounts to mistaking causation for one of its tests; and such a reduction of an ontological category to a methodological criterion is the consequence of epistemological tenets of empiricism, rather than a result of an unprejudiced analysis of the laws of nature.

2.5. Causation as Necessary (Constant and Unique) Production

What we need is a statement expressing the idea—common to both the ordinary and the scientific usage of the word—that causation, far more than a relation, is a category of genetic connection, hence of change, that is, a way of *producing* things, new if only in number, out of other things. This efficacy or productivity of the efficient cause,[28] this dynamic character of

[27] See Schlick (1932), "Causality in Everyday Life and in Recent Science", reprinted in Feigl and Sellars, *Readings in Philosophical Analysis*. Production and reproduction play a role in the pragmatic variety of empiricism—but, then, in an exaggerated way detrimental to theory.

[28] Plato went so far as to identify the cause with "that which creates". See *Greater Hippias*, 296 E, and *Cratylus*, 413 A. Cicero, in *Concerning Destiny*, 34, insisted that causes are productive, and advanced an obviously circular definition: "The

the causal connection left aside in the Humean formula, is what we shall try to express in a third approximation.

Let us try the following proposition:

If C happens, then (and only then) E is always produced by it. (5)

Translated into the categorical mode, this statement may be worded as follows: *Every event of a certain class C produces an event of a certain class E.* These sentences are refinements of the vulgar maxim *The same cause always produces the same effect*, or of the even shorter *C always brings forth E.*

Unlike (4), the last statements go beyond asserting a mere constant and unique (that is, a necessary) conjunction: they say that the effect is not merely *accompanied* by the cause, but is *engendered* by it—in conformity with both the ordinary and the scientific usage. Contrary to what Hume[29] and his followers have maintained, I am here rejecting the identity of production and causation: for, as science shows us—and as is abundantly argued at other places of this study—some things are produced or originated in a noncausal way. Causation is a particular case of production; the latter need not be restricted to causal production or to any other special form of generation.[30]

According to (5), something, *E*, is brought forth by something else, *C*, in a necessary (constant and unique) manner.

---

cause is that which effectively produces what it causes". Sextus Empiricus, *Outlines of Pyrrhonism*, book III, chap. 4, stated that most "dogmatic" (i.e., nonskeptic) philosophers agreed to call cause that by whose action an effect is born. Among our contemporaries, Meyerson (1908), *Identité et réalité*, p. 37, saw in productivity the kernel of causation: "Cause is that which produces, that which must produce the effect."

29 Hume (1739–40), *A Treatise of Human Nature*, book I, part III, sec. ii. Later on, in sec. xiv, Hume asserts that "the terms of *efficacy*, *agency*, *power*, *force*, *energy*, *necessity*, *connexion* and *productive quality*, are all nearly synonymous; and therefore it is an absurdity to employ any of them in defining the rest."

30 Needless to say, production (the Greek generation, or *genesis*) need not be goal-striving nor, particularly, purposeful. Incidentally, Neoplatonism holds just the opposite. See Inge (1928), *The Philosophy of Plotinus*, 3rd ed., vol. 1, pp. 179–180: "Causation implies creative action; it is a teleological category, and belongs to the processes of nature only as determined once for all by a 'First Cause', or as directed by an immanent will."

The following concepts, usually regarded as essential components of the causation category, are included in the proposition (5): *conditionalness*; *uniqueness*; *one-sided dependence* of the effect upon the cause; *invariability* of the connection, and *productivity*, or genetic nature of the link. Other formulations of the causal principle have been proposed, and there may be better ones than (5). But presumably most of the alternative formulations of the principle of causation either are reducible to (5), or are not rigorous enough, or contain some extracausal concept—as will be made plausible in what follows. Therefore I propose to employ the statement *If C happens, then (and only then) E is always produced by it* as an adequate formulation of the causal principle—until new notice.

Let us now investigate whether further specifications are likely to lead to better formulations.

2.6. Supposed Further Refinements of the Necessary-Production Formula of Causation

Among the statements that are at first sight more precise than (5), the following is noteworthy:

If C happens under the same conditions, then (and only then) E is always produced by it. (6)

Or, in the categorical form, *The same cause, under the same conditions, always produces the same effect.* Or, again, *Other things being equal, the same cause always produces the same effect.*

Is (6) really better than (5)? In order to answer this question we must distinguish two cases: the "conditions" and the cause are interdependent, or they are independent from each other. If the conditions or circumstances are connected with *C*, if together with *C* they constitute *a single causal complex C′*, then not much has been gained with the qualification—except that it is now explicitly stated that the cause is not simple but complex. Thus in the almost classical example "If a match is struck, it lights", at least the following conditions must be assumed for the link to hold: that the match be dry, and that there be enough oxygen. But these two conditions, subsidiary or background causes, can be regarded as accompanying the primary cause, which is the striking of the match. In fact, the above statement can be remodeled to read "If a *dry* match is struck *in an oxygen-containing atmosphere*, it lights". In short, if the conditions are linked to the cause, then *C* may be taken as the *chief or primary cause* (necessary but not sufficient), the conditions being the subsidiary causes. It is then the whole complex of determiners which constitutes the necessary and sufficient cause.

If, on the other hand, the conditions required for a causal link to obtain are *contingent* upon the cause, then it may be wondered whether (6) is a strictly causal proposition. Think, for example, of auxin, the vegetal hormone which is growth stimulating when located in the tips of a plant but inhibits growth when placed in the roots; in this case no *direct* bond exists between the cause (hormone action) and the effect (growth). And whenever a third component is added, which may be as important as the cause while remaining external to it, a *non*-causal type of determination is at stake.

The conditions referred to in (6) may include, not only the environment of the object on which the cause acts, but also the specific properties and the state of the object, as well as its intrinsic processes—and the latter are noncausal determinants, since efficient causes are by definition (see Sec. 2.1.1) extrinsic. (Self-movement and the corresponding self-determination are just the opposite of causal determination; indeed, whatever is *spontaneous* in some respect and to some extent is not caused—though it is not therefore indeterminate.) Hence if the accompanying conditions are contingent upon the cause, formula (6) expresses *more* than a simple, direct causal bond: the efficient cause involved in it acts in combination with the processes inherent in the object concerned, and the cause must then be regarded as the *unchainer* or triggerer of a process rather than as the necessary and sufficient producer of the effect.

In conclusion, if the accompanying conditions are contingent upon the cause, the statement (6) cannot be reduced to (5)—but, then, it definitely *departs* from causality, as it implies that *C* is necessary but not sufficient for the production of its effect. If, on the other hand, the accompanying conditions are functionally bound with the cause, then (6) reduces to (5). Consequently, (6) cannot be regarded as a better formulation of the causal principle.

Undoubtedly, all events are immersed in complex situations, every cause is embedded in a complex of determiners out of which it cannot be extricated except by abstraction—an abstraction that may or may not work as a first approximation. Moreover, no cause succeeds in effecting a result if it remains, so to speak, indifferent to the proper nature of the object concerned and consequently outside its self-movement—this being the reason why science works and magic does not. Hence formula (6) does fit the *nature of things* better than the former scheme; only, to the extent to which it actually says more than (5), it is not a faithful reconstruction of *causation*, since an element of spontaneity or self-determination is then introduced in it.

Let us now consider a further supposed improvement over the statement of necessary production. Ever since Leibniz, in a Heraclitean mood, formulated his celebrated maxim that no two things in the material world are identical (that is, the principle of the identity of indiscernibles), both philosophers of change and thoughtful scientists have become more and more convinced that, as Maxwell put it, "it is manifest that no event ever happens more than once, so that the causes and effects cannot be the same in *all* respects".[31] This is why the uniqueness condition is sometimes relinquished and the causal principle is stated under the form *Like results under like circumstances*, or, more explicitly, *Similar causes, under similar conditions, always produce similar effects*. A translation of the last sentence into the conditional mood would read

If similar causes happen under similar conditions,
then similar effects are always produced by them. (7)

Does such a replacement of 'sameness' by 'likeness' constitute an improvement over (6)? Not at all, as from the start we have pointed out (see Sec. 2.2) that the causal connection, being lawlike, does not refer to isolated facts, but to facts be-

[31] Maxwell (1877), *Matter and Motion*, chap. i, sec. 19, p. 13.

longing to certain *classes* or kinds—which takes care of the variations in the individual instances. Russell has explained this point clearly: "The law of causation docs not state merely that, if the *same* cause is repeated, the *same* effect will result. It states rather that there is a constant relation between causes of certain kinds and effects of certain kinds. For example, if a body falls freely, there is a constant relation between the height through which it falls and the time it takes in falling. It is not necessary to have a body fall through the *same* height which has been previously observed, in order to be able to foretell the length of time occupied in falling . . . In fact, what is found to be repeated is always the *relation* of cause and effect, not the cause itself; all that is necessary as regards the cause is that it should be of the same *kind* (in the relevant aspect) as earlier causes whose effects have been observed".[32] The same holds for all law statements, and their lack of reference to specific events is just what lends them a universal extension. In conclusion, if in all the previous formulations of the causal principle we have understood under C a *class* of events and under E another *class*, then the statement (7) does not constitute any improvement on (5), as it just renders manifest what had been implicit in it.

Finally, cautious scientists might object to the all-operator in (5). It is not that the word 'always' (or, conversely, 'never') should never be used in connection with events in the material world, as fortuitism claims. But the legitimate extension of these concepts is really very restricted: in factual (material or empirical) matters more frequently than not *almost always* replaces *always*.[33] Consequently the following statement may be preferred to the preceding ones:

If similar causes happen under similar conditions, then similar effects are produced by them in most cases. (8)

[32] Russell (1914), *Our Knowledge of the External World*, pp. 234–235.

[33] The statistical interpretation of the probability value $p = 1$ for a class of events is not 100 percent (certainty) but "almost always" (i.e., with the eventual exception of a set of zero measure). Analogously, the interpretation of $p = 0$ in factual terms is not 0 percent (impossibility, hence certainty that the concerned event will not happen), but "almost never".

Who doubts that this principle fits facts more closely than the preceding ones? But what a far cry from the candid and superficial *if–then always* relation! Actually, (8) belongs to the domain of statistical determinacy: in fact, the qualification 'in most cases' designates maximum frequency of occurrence, that is, a value around which all values tend to pack. Thus, a closer approximation to facts has taken us further away from causality. Consequently, the category of causation cannot be regarded as exhausting determination—a conclusion we anticipated in Chapter 1; and we would proceed wisely retaining the formula (5) of necessary production as a faithful schematic reconstruction of causation.

2.7. Retrospect and Conclusion

We started by recalling the Aristotelian teaching of causes, finding that, of all four (formal, material, final, and efficient), modern thought had retained only the efficient cause regarded as an agent acting extrinsically. Then we examined Galileo's clear but not altogether distinct definition of cause as the necessary and sufficient condition for the appearance of an event: we found that, far from specifying the causal connection, it was general enough to apply to all types of lawful determination. After that, a more precise and abstract statement was tried, namely, the *if–then always* relation, that is, the constant-conjunction formula of causation. This Humean enunciation turned out to cover various marks usually attributed to causal determination, save the uniqueness and the genetic nature of the causal link. This purposeful limitation of the Humean formula was criticized for blurring the distinction between correlation and production.

Thereafter we tried to remedy the shortcomings of the constant-conjunction formula of causation by introducing the concept of uniqueness (1 : 1 correspondence between C and E) and, above all, that of production as distinct from that of causation. This took us to a refined version of the common maxim "Same causes same effects", to wit, *If C happens, then (and only then) E is always produced by it.* This proposition happened to mark a crucial stage in our search for an adequate statement of the causal principle, as successive attempts to refine it either did not add anything new or took us away from causality into other types of determinacy (self-movement and statistical determination). This was regarded as adding strength to the claim that proposition (5), expressing necessary (constant unique) production, is the adequate formulation of the principle of causation—even though it falls short of reconstructing the full richness of determination.

This does not mean, however, that the statement *If C happens, then (and only then) E is always produced by it*, or any other single proposition, can be regarded as covering the whole richness of causation or all the implications of the doctrine (causalism) holding the coincidence of determinacy and causality: except in formal matters, adequacy does not entail completeness.

Thus far I have concentrated on a few verbalizations of the causal connection. In what follows I propose to examine the doctrine of causality, or causal determinism; in the course of this analysis further formulations of the causal principle will be met.

PART II

What Causal Determinism Does Not Assert

3

An Examination of the Empiricist Critique of Causality

Before attempting to discover the valid kernel of causalism we should endeavor to clear up several misunderstandings about it—if only because it is unfair to attribute to a doctrine sins that it has been unable to commit. Hence in this and in the following chapter we shall point out what causality is not—thus imitating Thomas Aquinas's method of remotion to acquire a knowledge of God.

In the present chapter we shall analyze some theses on causation that are characteristic of empiricism; we shall do it partly because of their intrinsic philosophic importance, partly because they are shared by many contemporary scientists—which is in turn owing, at least in part, to the simplicity and precision with which such theses can be stated.

In contrast to most empiricist analyses of causation, the following treatment will neither rely on a study of the psychological origin of the causation concept, nor restrict itself to a linguistic analysis of it or to a methodological examination of the empirical verifiability of causal hypotheses, a subject on which masterly studies are available.[1] In what follows we shall of course touch on these subjects, but we shall concentrate on causation regarded as an ontological category.

Furthermore, it will be taken for granted that countless hypotheses concerning causal connections have been empirically verified, to within experimental error, millions and

[1] See Wisdom (1952), *Foundations of Inference in Natural Science.*

millions of times. For an assumption of this work is that, to the extent to which the causal principle works, it reflects not only a feature of our cognitive relation with reality, but a trait of reality itself. This accounts for our focusing on the ontological status of causation.

3.1. Does Causality Involve Contiguity?

3.1.1. *Contiguity: An Essential Component of Causation According to Humeans*

The causal principle involves what may be called *continuity of action* between the cause and the effect, as every discontinuity or interruption in the causal thread would have to be assigned a further cause if noncausal events, such as gaps in a causal chain, are to be avoided. But such a continuity of action, which excludes noncausal events, should not be confused, as it so often is, with *spatial contiguity*, that is, with the continuous transmission of actions through space (action by contact).

It was Hume who not only held that the idea of contiguity in space is one of the essential components of the idea of causation, but went so far as to assert that it was nearly the first of the component parts of the idea of causation appearing upon an analysis of the latter: "I find in the first place, that whatever objects are considered as causes or effects, are *contiguous*; and that nothing can operate in a time or place, which is ever so little removed from those of its existence. Though distant objects may sometimes seem productive of each other, they are commonly found upon examination to be linked by a chain of causes, which are contiguous among themselves, and to the distant objects; and when in any particular instance we cannot discover this connection, we still presume it to exist. We may therefore consider the relation of *contiguity* as essential to that of causation".[2]

It is strange that Hume should hold this opinion. For one thing, he was writing precisely in the midst of an era when the idea of action at a distance, doubtless contradictory to contiguity, was triumphant in his own country and beginning to

[2] Hume (1739–1740), *Treatise*, book I, part III, sec. ii.

conquer the European continent,[3] gradually overcoming the Cartesian and Leibnizian doctrines of nearby action effected by contact forces through a subtle medium, as well as the Democritean principle that the interaction of bodies can occur only through collision.[4] This is one of the pieces of evidence in favor of the suspicion that Hume's philosophy of science was behind his own times, being in fact pre-Newtonian.

3.1.2. *Contiguity: A Hypothesis Inconsistent with Empiricism*

A second, and more important, reason why it is difficult to understand that an empiricist should cling to the belief that contiguity is not only an essential component of causation, but even an indispensable ingredient of a scientific world-view, is that contiguity is not a fact directly ascertainable by experience, but rather (as Hume himself recognized[5]) a *hypothesis*. Moreover, it is not a hypothesis the consequences of which can be tested with any degree of accuracy: the experimental test of contiguity would require an infinity of observations between any two events at different points in space. The hypothesis of contiguity entails, in short, an empirically unprovable interpolation—which, of course, does not diminish its value. In other words, it is the principle of contiguity, rather

[3] A decade before Hume wrote his *Treatise*, Voltaire had written in his *Lettres philosophiques* (1728) that in London "*on n'accorde rien au Français et on donne tout à l'Anglais*"—meaning Descartes and Newton respectively. See Letter xiv, in *Oeuvres*, vol. XVII, p. 79. Ten years later the situation had altered substantially on the continent, partly owing to Voltaire himself, whose *Éléments de la philosophie de Newton* (1738) had an enormous success. Moreover, the popularizations of Newton's physics (including Voltaire's own book) were practically restricted to the theory of gravitation, conceived as an instantaneous *actio ad distans*; they were almost unconcerned with the principles of dynamics—which Voltaire himself misunderstood.

[4] Prefield physics acknowledged two kinds of action by contact: (1) collision, or direct contact among particles, and (2) propagation of pressure differences (density waves) in continuous media, such as fluids. The principle of spatial contiguity was consequently consistent with both the Democritean *vacuum* and the Cartesian and Leibnizian *plenum*. In the former case, *physical* causes were collisions of atoms in a void; in the latter they were pressure differences in continuous media. Therefore Hobbes, in his *Elements of Philosophy* (*Concerning Body*), chap. ix, 7, wrote: "There can be no cause of motion, except in a body contiguous and moved." But this was in 1655, not in 1739, when Hume wrote.

[5] Hume, *Treatise*, book I, part III, sec. ii.

than the hypothesis of *actio ad distans* (which Leibniz regarded as "barbarous"), that is a "metaphysical" assumption about physical reality.

Let us dwell on this point, which has an important bearing on field physics, as the principle of nearby action is embodied in it. The validity of the very concept of field cannot be demonstrated empirically just by measuring field strengths, and this for the following reasons. First, what is measured is not the field strength itself but the force exerted by the field on a test body; for example, in the case of the electric field, the ponderomotive force $F = eE$ is measured, from which the value of the field intensity E is *inferred* on the basis of the knowledge of the electric charge e, obtained from an independent experiment. Second, field strengths $E(x, t)$ are *defined* but not *measured* at a given *point* in space and at a given *instant* of time: what experiment affords is not a point function $E(x, t)$ but a collection of *averages* over both a small but nonvanishing field region and a small but nonvanishing time interval; moreover, owing to the atomic structure of matter, it does not seem possible at present to diminish that volume without limit.[6] In short, the field concept is not an operationally defined one.

The validity and usefulness of the field concept are demonstrated, at least in the macroscopic domain, in an *indirect* way, as is usual with elaborate theoretical concepts, namely, by the far-reaching testable consequences to which field theories lead. And this is not peculiar to field physics, but is a pervasive feature of science. If scientific concepts had to be restricted to operationally defined "observables" having an immediate empirical content, only a few unintelligible empirical rules would remain in every branch of science—and Mach's ideal of an economical description of experimental facts would be attained. If the hypothesis of contiguity were abandoned in accordance with the requirements of consistent empiricism, all field theories (gravitation, electromagnetism, and mesodynamics), as well as the whole physics of continuous media (classical fluids and elastic solids) would have to be ruled out. Such a surgical policy was actually demanded by opera-

[6] See Heitler (1954), *The Quantum Theory of Radiation*, pp. 81 ff.

tionalism several years ago,[7] and is still being suggested by pragmatists who feel that Maxwell's theory of the electromagnetic field is "overdescriptive" because it overflows the operationist cup.

3.1.3. *Explicit Definitions of Causation Do Not Involve Contiguity*

But our primary concern is not to show that the principle of action by contact is definitely incompatible with empiricism, as in fact it is. Our business is to weigh the claim that contiguity is essential to causation—a Humean idea that has modern followers.[8] If causation is *defined* as involving contiguity, then the causal principle will of course imply the principle of nearby action—as was the case with Hume's allegedly "precise definition of cause and effect".[9] But such definitions, specifying a restricted type of causation, common as they were during the

[7] Bridgman (1927), *The Logic of Modern Physics*, p. 150: since it is operationally meaningless "to burrow inside the electron", which has to be regarded as a whole (and why not as a complex whole too?), we should "try to get along without the field concept", reducing the whole of electromagnetic theory "to the ultimate elements that have physical meaning, namely, a dual action between pairs of electrical charges, with no implications about physical action where the charges are not", that is, where no measurements can be made. Bridgman went so far as to deny the existence of light in empty space, on the ground that light can be detected only at emitters and absorbers: "from the point of view of operations it is meaningless or trivial to ascribe physical reality to light in intermediate space, and light as a thing travelling must be recognized to be a pure invention" (p. 153). Bridgman's operationalist program was in part carried through by J. A. Wheeler and R. P. Feynman, *Reviews of Modern Physics 21*, 425 (1949), and by R. P. Feynman, *Physical Review 76*, 749, 769 (1949). It is at least doubtful whether explicit reference to the field concept could be avoided if electrodynamics turned out to require nonlinear equations, as is likely.

[8] Carnap (1926), *Physikalische Begriffsbildung*, p. 57; Reichenbach (1929), "Ziele und Wege der physikalischen Erkenntnis", in Geiger and Scheel, ed., *Handbuch der Physik*, vol. IV, p. 60; Reichenbach (1944), *Philosophic Foundations of Quantum Mechanics*, p. 117 and *passim*; Ernest H. Hutten, "On the Principle of Action by Contact", *British Journal for the Philosophy of Science 2*, 45 (1951). Born (1951), *Natural Philosophy of Cause and Chance*, chaps. ii and iv, owns that contiguity is the subject of a separate principle, but regards it all the same as an attribute of causality.

[9] Hume, *Treatise*, book I, part III, sec. xiv: "A *cause* is an object precedent and contiguous to another, and so united with it that the idea of the one determines in the mind to form the idea of the other, and the impression of the one to form a more lively idea of the other".

Cartesian period, are now mostly pieces of historical interest. A mere inspection of any of the more or less adequate and general formulations of the causal principle advanced in the last two centuries (see Chapter 2) shows that causation and nearby action are two logically *independent* categories, hence the subjects of separate principles. Causation is consistent with contiguity but does not entail it; moreover, contiguity is consistent with a reversal of the causal nexus, in the sense that theories are conceivable in which everything proceeds *de proche en proche*, as demanded by the principle of contiguity, but in which effects appear before their causes.[10]

Finally—and this is usually overlooked by Humeans—any reference to spatial contiguity in the formulation of the causal principle would render it almost inapplicable outside physics.

To sum up, causation is consistent with contiguity but does not entail it.

3.2. Does Causality Involve Antecedence?

3.2.1. *Causality Is Consistent with Instantaneous Links*

The principle of retarded action, or of antecedence, states that there is always a time delay between the cause and the effect, the former being prior in time to the latter, so that (relatively to a given physical system, such as a reference system), C and E cannot be both distant in space and simultaneous. Hume,[11] Schopenhauer,[12] and many other students of the causal problem have stated that the time priority of the cause over the effect is essential to causality; Russell asserted that "if there are causes and effects, they must be separated by

[10] Differential equations containing even-order derivatives with respect to time, and an arbitrary number of derivatives with respect to space coördinates, can be read backward in time. Thus the equations for the electromagnetic potentials have not only retarded but also advanced solutions; the latter describe waves (probably without any physical meaning) in which the result (acceleration of a charge or absorption by it) appears earlier than the condition (emission).

[11] Hume, *Treatise*, book I, part III, sec. ii.

[12] Schopenhauer (1813), *Ueber die vierfache Wurzel des Satzes vom zureichenden Grunde*, sec. 20.

a finite time-interval";[13] and N. Hartmann held that "causality means only that, in the course of happenings, the later are determined by the earlier".[14] Finally, the philosophic notion of causality that has come to be accepted among scientists in the last hundred years actually entails the idea of a time lag of the effect relative to the cause.

However, the principle of antecedence and the causal principle are independent of each other. To verify this assertion, it suffices to recall any of the formulations of the causal principle that were examined in the preceding chapter, particularly the empiricist definition of causation as constant-conjunction (Sec. 2.3) and the definition of causation as necessary production (Sec. 2.5); none of them contains the idea of temporal priority. What they all do state is what we called *existential* priority of the cause over the effect (Sec. 2.3); that is, they all require that the cause be there if the effect is to occur, but they do not entail succession in time.

Causality is consequently *compatible with instantaneous links*, whether at a given point in space (as in the case of the dynamical law "Force causes acceleration") or among systems placed at different regions in space. The latter case is important, as actions at a distance involving no time delay have actually been imagined by physicists not only before field theories were born but also in some present theories, independently of considerations about causation[15]—although scientists who affect to

[13] Russell (1912), "On the Notion of Cause", in *Mysticism and Logic*, p. 175. In his contribution to the article "Cause" in Lalande's *Vocabulaire technique et critique de la philosophie*, vol. I, p. 101, Russell wrote: "A causal proposition can be stated in the following way: A exists at time $t \supset$ B will exist at time $t + \Delta t$". But in *Our Knowledge of the External World* (1914), p. 219, he asserted that "It is not essential to a causal law that the object inferred should be later than some or all of the data. It may equally well be earlier or at the same time."

[14] N. Hartmann (1949), *Einführung in die Philosophie*, p. 23.

[15] Instantaneous propagation of physical actions appears in connection with both electrodynamics (classical and quantal) and quantum mechanics. In electrodynamic theory, the instantaneous propagation of the field is obtained by superposing ordinary (retarded) and advanced waves; since the former propagate from past to future and the latter from future to past, the net result is a zero time lag between the cause and the effect. But not many scientists take instantaneous electrodynamic actions seriously. In quantum mechanics, instantaneous actions appear in connection with the so-called instantaneous collapse of the wave packet

ignore philosophy altogether tend to mix up questions of antecedence with the causal problem.

The principle of retarded action, or antecedence, was not generally adopted until about the middle of the 19th century. Before that time the hypothesis of instantaneous physical actions (both at a distance and through a medium) prevailed in gravitation theory and in electromagnetic theory. Even the Cartesians, who remained faithful to the principle of contiguity, stuck to the belief that the propagation of light was instantaneous. It was not until the deformations of continuous media were studied by Cauchy, and the concept of electromagnetic field was introduced by Faraday, that the principle of retarded action began to be regarded as a necessary restriction upon the principle of causation in the domain of physics.[16] The recognition of the principle of antecedence was essentially linked with the successes of field theories and entailed a restriction upon the causal principle in the domain of physics—not, however, in domains where no velocities of propagation enter. Special relativity (1905) finally banished instantaneous links from mechanics in connection with events taking place at different points in space.[17]

---

associated with a physical system upon measurement of any dynamical attribute of it. On the orthodox (positivist) interpretation of quantum mechanics, this collapse is not regarded as a physical process, but merely as a contraction of a "field of knowledge" defined in an abstract Hilbert space, since the wave function is not deemed to represent any physical situation, but only our information about it. A reasonable explanation of such a sudden contraction is a task for non-positivistic theories that regard the ψ-wave as representing a real field in ordinary space, but this would presumably involve a reinterpretation of second quantization.

[16] The recognition of the principle of antecedence meant also, *de jure* if not *de facto*, a radical shrinkage of the program of geometrization of physics which was begun by Descartes and which Lagrange had tried to fulfil in analytical mechanics with the derivation of forces from functions (potentials) depending solely upon the relative distances among mass-points (that is, upon geometric configuration). The finite velocity of propagation of interactions entails that the interaction potential cannot be solely a function of the space coördinates of the interacting particles, but must also contain time—which, like mass, is not a geometric concept.

[17] The principle of antecedence in mechanics was introduced in two ways: first, by implicitly stating that all mechanical systems are or can be connected with one another through the electromagnetic field, as suggested by the fundamental role played in relativistic mechanics by the velocity of light in empty space; second, by showing that two phenomena that are simultaneous in a given frame of

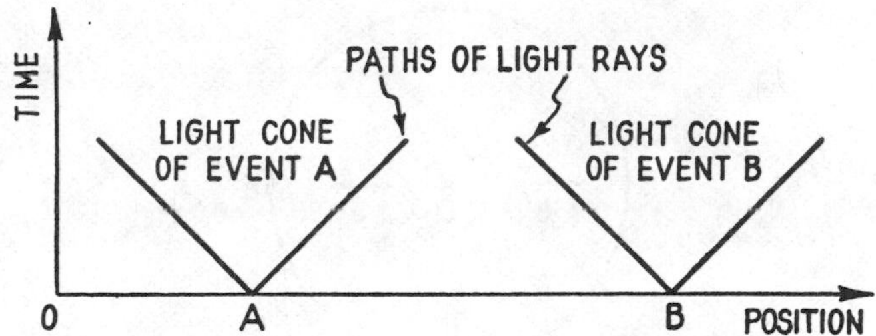

Fig. 3. Event *B* is to the right of event *A*, and the two are simultaneous relative to this particular frame of reference. They are *causally unrelated*, because not even the fastest known signal (a light ray) could bind them. Light rays emerging from points *A* and *B*, both to the left and to the right, propagate in space-time along the oblique straight lines shown in the figure; these lines constitute the forward light cones of events *A* and *B*. Event *A* can only act on the events lying *inside* its own forward light cone, and so for *B*.

3.2.2. *The Principle of Retarded Action in Special Relativity*

We have asserted that the principle of retarded action, rather than being involved by the causal principle, imposes a *restriction* upon physical causation whenever the former is accepted. This means that the principle of antecedence entails the assertion that *there are systems that cannot interact with one another*—and that, a fortiori, cannot be *causally* tied to one another. In other words, the joint assertion of the principles of causation and of antecedence restricts the number of possibilities of physical connectedness. In relativity physics, this statement takes on the following special form: as a consequence of the principle according to which all (known) physical actions are propagated with finite velocities, any two points in space-time that cannot be tied to each other by means of the fastest (known) chain of events (light) are *causally unrelated* (see Fig. 3). Contrary to what has been claimed,[18] this does not mean a violation of determinism; it just excludes as physically impossible certain

reference may cease to be simultaneous relative to other reference systems, owing to the finite velocity of propagation of electromagnetic interactions.

[18] H. Bondi, *Nature 169*, 660 (1952).

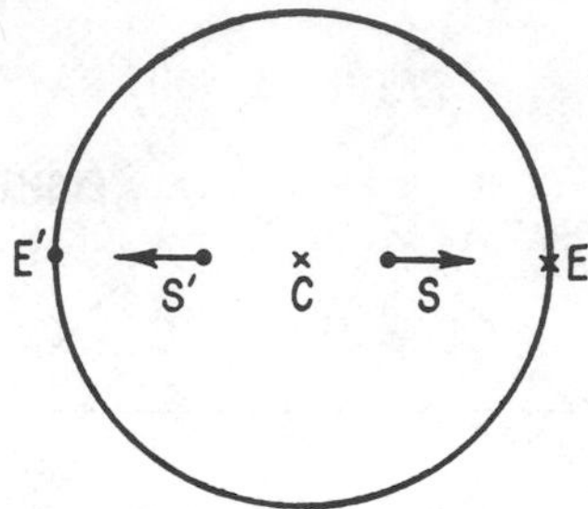

Fig. 4. Relativity of the time series. At a given initial instant a spherical light wave is emitted by the source *C* (the cause), where two reference systems are initially superposed. At the same time one of these systems, say *S*, starts moving to the right, while the other system, *S′*, goes in the opposite direction. Relative to the source *C*, the wave front arrives simultaneously at *E* and *E′* (the effects); but relative to *S* the wave front arrives earlier at *E* than at *E′*, and with respect to the reference system *S′* the time order of the events *E* and *E′* will be reversed. But in no case will the events *E* and *E′* happen before light has been produced at *C*. To sum up, the order of *E* and *E′* can be reversed, but both *E* and *E′* happen *after C*.

conceivable causal lines, but it retains all causal chains within light cones.

A further radical change in classical ideas about succession in time was introduced by relativity: it was shown that time series are relative, that is, they depend upon the reference system, though not necessarily upon the observer, which is one among an infinity of possible reference systems. Relativity has moreover shown that the time order of *certain* events can be reversed. Thus two light signals arriving at a certain reference frame (rigid material system) in the order (1,2) may arrive in the reverse order (2,1) at a different system, in motion relative to the former (see Fig. 4). On the other hand, relativity does not hold that light (or any other physical object) can arrive at a given point before it has been produced: it asserts that the *order of production* of the successive light flashes coming from a given distant source is unique: only their time differences (the time intervals) are relative, that is, differ numerically from one reference system to another. Time reversal is possible only for those pairs of events that, being separated by a distance, do *not* stand in a causal relation to each other.

In other words, relativity admits the reversal of time series of physically disconnected events but *excludes the reversal of causal connections*, that is, it denies that effects can arise before they have been produced, and consequently does not assert that the past can be changed. If preferred, according to relativity the time order of *causes* is relative (to the reference system) while *causal connections* are invariant. That is, duration is relative, and the time order of any two genetically disconnected events E and E' can be reversed. As a particular case, E and E' may both stem from a single cause, C, as in the example of Fig. 4, but in no case will C happen after either E or E'. Moreover, the cause cannot even occur (in relativity physics) at the same time as E and E', for such a simultaneity would entail that a light signal could travel in no time, thereby contradicting the principle of retarded action embodied in relativity. In short, events whose order of succession is reversible cannot be causally connected with one another; at most they may have a common origin. That is, the concepts 'earlier' and 'later' have an *absolute* meaning for causally connected events —and for them alone. Hence, sequences of the Humean type do not involve a unique arrow of time; but genetic chains, whether causal or not, are on the other hand no less than the very substratum of time.[19]

To conclude, a condition for causality to hold is that C be previous to or at most simultaneous with E (relative to a given reference system). Antecedence is relative in connection with pairs of causally unconnected events, and absolute with regard to events that are causally connected and, in general, genetically linked to one another. The principle of retarded action is independent of the causal principle; and, whenever it is

[19] Our remarks on the irreversibility of causal series are also valid with regard to the general theory of relativity, the gist of which is the invariance of the laws of nature (that is, their independence relative to the choice of reference systems, in particular, observers) and, as a special case, the invariance of causal laws (or of the causal aspects of natural laws).

postulated, it entails a restriction on the possible genetic connectivity of the physical level of reality. Moreover, the mere statement of retarded action, far from being essentially committed to causality, is consistent with noncausal categories of determination.[20]

3.3. Is Causation Identical with Invariable Succession in Time?

3.3.1. *The Interpretation of Causal Process as Succession of States*

Most positivists have held that the concept of causation should be replaced by or actually reduced to invariable (uniform) succession in time. Thus Comte asserted that in the "positive" (scientific) stage the human mind does not seek to know the intimate causes of phenomena, but only "their effective laws, that is, their invariable relations of succession and similitude".[21] Mill stated that "the Law of Causation, the recognition of which is the main pillar of inductive science, is but the familiar truth that invariability of succession is found by observation to obtain between every fact in nature and some other fact which has preceded it".[22] The ground for this claim has been well known since Hume: only invariable succession and conjunction are shown *in experience* (a curious sort of experience that ignores active production and reproduction, acknowledging only passive observation). Hence, if philosophy is to be empirical, no other meaning than that of uniform succession can be attached to the causal principle: "What experience makes known, is the fact of an invariable sequence between any event and some special combination of antecedent conditions, in such sort that wherever and whenever that union of antecedents exists, the event does not fail to

[20] Essentially the same result concerning the independence of causality and antecedence is obtained, along different lines, by J. S. Wilkie, "The Problem of the Temporal Relation of Cause and Effect", *British Journal for the Philosophy of Science* *1*, 211 (1950).

[21] Comte (1830), *Cours de philosophie positive*, 1e leçon, vol. I, p. 3. However, Comte did not exclude necessity from his own *loi des trois états*, which he described as a great and fundamental law to which human intelligence "is subjected by an invariable necessity".

[22] Mill (1843, 1875), *A System of Logic*, book III, chap. v, sec. 2.

occur".[23] A similar reduction of causation to invariable succession was defended by Pearson[24] and by more recent representatives of positivism.[25]

The identification of cause with constant antecedent, of causation with invariable (uniform) succession, has been criticized many times by philosophers. It has been pointed out, for instance, that the reduction of determination to precedence would prevent us from distinguishing the few relevant antecedents in the mass of precedents. However, scientists, concerned as they are in the most interesting part of their work with laws of change, have too often ignored such criticisms; many scientists still believe, in fact, that every statement of regular and unique succession of events, or every continuous series of states of a system, constitutes a causal law. Thus the following sentence is frequently regarded as a rigorous formulation of the causal principle:

The states of a (closed) system unfold themselves
in time in a unique and continuous way. (9)

Other times, this or some equivalent proposition is considered as the particular form that the causal principle takes in physics.[26] Since it is difficult to understand how a law of succession of states could be described as causal without further ado, sometimes the *ad hoc* definitions Cause = Initial State, and Effect = Final State are added.[27] Thereupon the vulgar maxim "Same cause, same effect" is sometimes replaced by the following formula, which has become quite popular among physicists and philosophers of science:

The same initial state is always followed by the same final state. (10)

[23] Mill (1865), *An Examination of Sir William Hamilton's Philosophy*, vol. II, p. 279.

[24] Pearson (1892), *The Grammar of Science*, 3rd ed., chap. iv, sec. 8.

[25] Lenzen (1954), *Causality in Natural Science, passim.*

[26] Weyl (1927, 1949), *Philosophy of Mathematics and Natural Science*, p. 191.

[27] The identification of *cause* with *initial state* has been in the air since the end of the 19th century, owing to both the (impossible) attempt to frame thermodynamics in a causal language, and the influence of phenomenalism (Mach, Ostwald, Duhem). An early explicit statement of that identity is found in Le Dantec (1904), *Les lois naturelles*, p. 97. More recent equivalent statements are those of Northrop (1947), *The Logic of the Sciences and the Humanities*, p. 219, and Popper (1950), *The Open Society and its Enemies*, rev. ed., pp. 445–446.

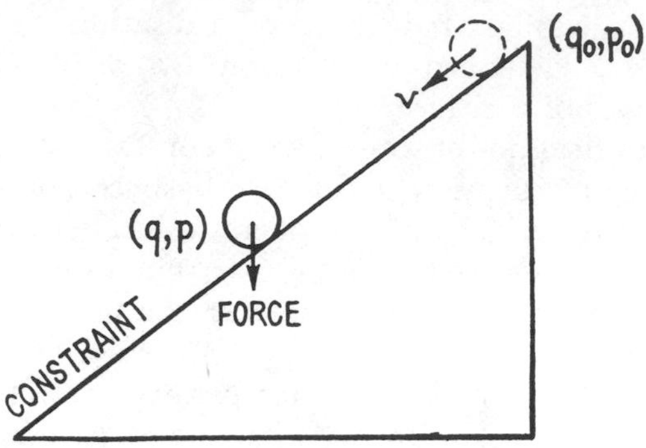

Fig. 5. In Newtonian mechanics the final state of motion (q, p) follows uniquely from the initial state (q_0, p_0) by the combined action of self-motion (inertia), external forces, and constraints.

A much-exploited illustration of this sentence is Newton's law of motion, namely, "Force equals mass times acceleration". In this case the initial state consists of the set of values of the positions q_0 and momenta p_0 of the intervening mass-points at a prescribed time. By Newton's law of motion, subsequent states (q, p) follow in a regular, unique, and continuous way from the initial state. But, save in the extreme case of inertial motion, this evolution is *not*, as required by (10), exclusively dependent on the initial state (q_0, p_0); instead, usually both constraints and external forces contribute to determining the subsequent states (see Fig. 5)—and the constraints are not specified in the initial state. Since genuine causes are supposed to be necessary *and sufficient* for the production of their effects, it is not correct to call initial states causes whenever additional determiners (such as constraints and external forces) are present. The simple case of analytical mechanics shows, then, that *causal lines are more than patterns of uniform succession.* Likewise, the thermodynamic theory of isolated systems, concerned as it is with successive states that unfold themselves spontaneously (that is, in the absence of external determiners) shows that *intrinsic patterns of uniform succession need not be causal.*

It is easy to see why states cannot have a productive virtue of their own. The state of a material system is a system of qualities, not an event or a string of events. Every state is the *outcome* of a set of determiners (usually causal and noncausal); if preferred, the values of the parameters that specify a state are the result of both inner processes and external connections. Consequently, there can be no action of one state upon another state of a given system; in particular, *there can be no causal links among states*, nor among any other systems of qualities. States are not *causes*, but simply *antecedents* of later states. To regard states as causes amounts consequently to committing the fallacy of the *post hoc, ergo propter hoc*.

3.3.2. *The Interpretation of Causation as Predictive Ability*

Sometimes a more sophisticated sentence is regarded as the correct formulation of the causal principle, namely,

The knowledge of the initial state of a (closed) system is sufficient for the prediction of its state at any other later time. (11)

This is how many contemporary physicists understand causality.[28] However, (11) is first of all incomplete, as the law of motion (and eventually the force function) as well as the constraints and the specification of the surroundings (boundary conditions) are needed besides the initial state in order to perform some prediction. But even if these qualifications are added, the uniqueness that characterizes the causal bond is lost in (11), since empirical information is never quantitatively exact; instead of an unambiguous set of values specifying the initial state of the system concerned, observation and measurement usually yield statistical distributions of the relevant variables. This uncertainty in the initial information—which always shows a spread around average values—spoils the one-to-one correspondence among neatly defined states even if, as in classical physics, the theoretical values are supposed to be sharply defined. This is one of the reasons for regarding (11) as an inadequate formulation of the causal principle.[29] (More-

[28] See Weizsäcker (1943), *Zum Weltbild der Physik*, 5th ed., p. 73.

[29] See Nagel (1951), "The Causal Character of Modern Physical Theory", in Feigl and Brodbeck, ed., *Readings in the Philosophy of Science*, p. 425.

over, *all* laws, whether causal or not, when framed in observational terms acquire statistical features—but this has nothing to do with the laws at the ontic level, which are not couched in operational terms. The transition from the level of theory to the level of experiment, from the conceptual to the empirical level, takes place when it is a question of *verifying* a theory, not when its *formulation* is at stake; see 12.1.2.)

Besides, (11) is not an ontological statement, not a proposition concerning the world, but a sentence concerning our knowledge and prognosis of events. If it were correct (but we have seen that it is incomplete), since it refers to initial information and to prediction it might be regarded as an *epistemological partner* of an ontological principle (the one about the unique and continuous unfolding of states of an isolated system). Certainly, the knowledge of a certain type of law of nature, alongside a set of items of information (such as initial conditions), makes an almost unique prediction of future states often possible—at least as long as no qualitative changes occur, nor unexpected external perturbations arise; provided we know how to solve the mathematical problems eventually involved in the prognosis, our prediction may be adequate for limited periods of time. But the causal problem, far from having solely an epistemological side, is chiefly an ontological question. Hence, considerations about knowledge and foreknowledge are out of place in connection with statements of the causal principle; they belong, on the other hand, in statements concerning the verification or the application of law statements, whether causal or not.

The empiricist identification of the *meaning* of an empirical statement with either its predictive usage or the technique by means of which it is put to the test (hence confirmed or infirmed) leads to confusing *criteria* of determination with *kinds* of determination, and particularly criteria of causal connection with kinds of causal bond. Thus Rapoport distinguishes three "kinds of causality": (1) *observational causality*, the meaning of which would be, "Watch for the occurrence of A, and you will observe the occurrence of B"; (2) *manipulative causality*, which is taken to mean, "Make A occur, and you will observe B—or prevent A from occurring, and B will not occur"; finally, (3) *postulational causality*, which is said to obtain if "B appears to us surprising when we are ignorant of the occurrence of A, and ceases to be surprising when we learn of the occurrence of A".[30] Of course, if the verifiability theory of meaning—which underlies the preceding classification—is not accepted, then the foregoing kinds of causality are seen not as different types of causal connection but as *criteria* of causal connection or as *techniques* for verifying causal hypotheses. To summarize, causation is not identical with any of the tests for it.

3.3.3. *Descriptions of Change as Sequence of States Need Not Be Causal*

Descriptions of change as time sequences of states need not be causal, that is, need not point to a process whereby certain effects are *produced* by certain causes according to a given pattern. Moreover, predominantly phenomenological accounts of change need not be intelligible and verifiable descriptions, that is, need not be scientific; they may be mythical or magical, as is the case with the archaic and ancient myths about the successive metamorphoses of fabulous beings.[31] The phenomenalist program of accounting for invariable succession in a purely descriptive way, without enquiring into the "mechanism" of change, is in fact characteristic of a poorly developed culture rather than peculiar to the "positive" stage of mankind.

Of course, science contains assertions of invariable succession in time. But, first, it does not seem legitimate to call them 'causal laws' whenever they lack the essential component of a productive, genetic, connection, if they just describe uniform

[30] Rapoport (1954), *Operational Philosophy*, pp. 57 ff.

[31] See Frankfort *et al.* (1946), *Before Philosophy: The Intellectual Adventure of Ancient Man*, p. 27: "Changes can be explained [by primitive and ancient man] very simply as two different states, one of which is said to come forth from the other without any insistence on an intelligible process—in other words, as a transformation, a metamorphosis. We find that, time and again, this device is used to account for changes and that no further explanation is then required. One myth explains why the sun, which counted as the first King of Egypt, should now be in the sky. It recounts that the sun-god Rē became tired of humanity, so he seated himself upon the sky-goddess Nūt, who changed herself into a huge cow standing foursquare over the earth. Since then the sun has been in the sky."

sequences without pointing out the *causes* of the process. Second, science seldom remains satisfied with mere statements of regular succession in time: these are usually the first, not the last, word—unless the change concerned actually happens to be purely quantitative; further laws, underlying the bare *fluxus formae* and explaining it, are always sought and sometimes even found. Thus the hypothesis of the electromagnetic nature of light explains geometric optics, and the hypothesis of the class struggle is intended to explain certain historical events.

Finally, if causation were nothing but uniform succession, relativity could not hold. Indeed, relativity establishes a deep difference between the succession of genetically unconnected events (the time order of which is relative) and causal succession, whose time order is irreversible (Sec. 3.2.2). The relativity of the time order of causally unconnected events, as contrasted to the absolute character of the "existential", and eventually also temporal, character of the precedence of the cause over the effect, shows that the empiricist reduction of causation to regular succession in time is at least outdated.

In short, causation is not exhausted by regular association in time, although causal chains may of course develop in time (or, rather, they engender time) in accordance with fixed patterns.

3.4. Is Causation Mirrored by Differential Equations?

3.4.1. *Differential Equations as Mirror Images of Uniform Sequences: A Confusion of Dimensions of Language*

If causation were exhausted by uniform, unique, and continuous succession, then differential equations with time as the "independent" variable could be regarded as the mirror images of causal laws, just as (partial) differential equations with space coördinates as "independent" variables are the adequate mathematical forms for expressing the contiguity characteristic of fields and continuous media. Such is, in fact, a very widespread belief.[32] Russell once went so far as to state that "Scien-

[32] Carnap (1926), *Physikalische Begriffsbildung*, p. 57. Frank (1937), *Le principe de causalité et ses limites*, pp. 144 ff. Lenzen (1938), *Procedures of Empirical Science*, in *International Encyclopedia of Unified Science*, vol. I, no. 5, p. 36: "The differential analysis gives an exact formulation of the popular concept of causality"; (1954)

tific laws can only be expressed in differential equations".[33] Quite a deal of boldness was required to utter this sweeping generalization at a time when integro-differential equations, finite-difference equations, matrix equations, and other mathematical forms were being introduced everywhere in physics and technology—not to speak of older acquaintances: arithmetic laws such as Mendel's, algebraic laws like Ohm's, or integral principles like Fermat's and Hamilton's.

If the empiricist reduction of becoming to uniform, unique, and continuous succession in time is accepted, then the reduction of the mathematics of "empirical" science to differential equations may follow (on the additional assumption that science should not be concerned with laws of structure). But the stubborn fact is that science uses other mathematical forms as well—which shows that research would have been seriously impaired had it followed the positivistic doctrine (or any other limitationistic policy) in this regard. Every a priori limitation on the possible mathematical forms of scientific laws is at best ineffective and at worst crippling. Experienced scientists know that the mathematical instrument that is chosen in each case to express a given objective pattern depends not only on the sort of phenomena themselves that are concerned but also on the scientist's own mathematical equipment and ability. It is furthermore a commonplace truth that the advances of science elicit the invention of fresh mathematical tools, and vice versa, the advances of mathematics afford in turn a richer and richer assortment of forms that sooner or later are "filled" with (that is, correlated to) the most varied contents—or at least take part in some scientific theory. To imagine that this two-way flow could some day cease is a mark of unwarranted

---

Causality in Natural Science, p. 54: "Causality has become functional relationship which is most adequately expressed by the differential equation". Mises (1939, 1951), *Positivism*, pp. 158–159. Weizsäcker (1943), "Naturgesetz und Theodizee", in *Zum Weltbild der Physik* 5th ed., p. 165. Margenau (1950), *The Nature of Physical Reality*, pp. 404 ff. Hutten (1956), *The Language of Modern Physics*, p. 214. Bergson (1907), *L'évolution créatrice*, p. 19, regarded the expressibility of scientific laws in terms of differential equations as a characteristic of inorganic matter.

[33] Russell (1927), *An Outline of Philosophy*, p. 122; in *Human Knowledge* (1948), p. 334, the statement was restricted to the laws of physics.

pessimism. The dogmatic dicta of philosophers have sometimes a large stopping power in connection with the progress of science, but the latter ends always by finding its way.

Not only are differential equations not the sole mathematical instruments of science, but they do not even reflect causation, one of the simplest forms of determination. In fact, differential equations, even if duly interpreted in material terms, do not state that changes are *produced* by anything, but only that they are either *accompanied* or *followed* by certain other changes. Just as functions may be used to express constant associations, so differential equations may be employed to express regularly *associated* (concomitant) changes.

Consider the simplest of differential equations with time as the "independent" variable:

$$\frac{dx}{dt} = f(t),$$

which can also be read

$$dx = f(t)dt.$$

These formulas [34] do not state that the change dx is *produced* by dt; they merely assert that the variation dx undergone by the property designated by 'x', during the time interval dt, equals $f(t)dt$. They should not be read as asserting that the "flow of time" dt has *caused* or produced by itself the change dx—since time, far from having productive virtues and far from "flowing" *per se* apart from the succession of events, is just the rhythm of process, the order of successives (Leibniz).

When a causal interpretation of a mathematical form (whether equation or inequality, differential or not) is reasonable, it has to be *added* to the mathematical entity. Such a semantic rule of correspondence, attached to the syntactic

[34] It is often stated that differential equations expressing causal relations should not, unlike that of the text, contain the time variable in explicit form. Thus Russell (1912), "On the Notion of Cause", in *Mysticism and Logic*, pp. 193–194, asserted that "no scientific law involves the time as an argument"; and others have regarded such a restriction as the very essence of the principle of causality. If such an interpretation were accepted, everyday phenomena like damped oscillations or motion in a fluid should not be regarded as "obeying" causal laws, as in either case the force function may depend explicitly on time.

form concerned, is usually expressed by means of words; in other words, such an interpretation does not belong to the mathematical symbols themselves but to the (semantic) system of relations linking the signs with the physical, chemical, biological, . . . entities in question. Sometimes such an interpretation is not made explicitly but is taken for granted; but it is precisely a job of philosophers of science to unearth hidden assumptions and to distinguish the various dimensions of language.

Ampère, who was both a scientist and a metascientist, knew this very well, when he argued that the law of the mutual and pairwise action of electric currents, which he had established, was consistent with two different explanatory hypotheses; he was convinced that, whatever the process by means of which electrodynamic phenomena were explained, his formula accounted for the facts. And de Broglie said once that equations remain while their interpretations pass away. (This may be, by the way, the chief justification for most present meson-field theories: their equations may finally turn out to have empirical correlates in some section of reality.)

One and the same mathematical form, such as a differential equation, is then susceptible of different interpretations (contents or meanings); if it is not possible to attach to it an explanation in terms of material processes, it remains a simple matter-of-fact statement—although it may serve to explain further facts. Consider Fourier's equation of heat diffusion,

$$\frac{\partial U}{\partial t} = D\frac{\partial^2 U}{\partial x^2}.$$

This equation was highly praised by Comte[35] and Mach[36] precisely because, according to them, it does not tell *what* heat is or *why* it flows, but is instead limited to telling *how* it flows. Fourier, on the other hand, had clearly stated that his own

[35] Comte (1830), *Cours de philosphie positive*, 31e leçon.

[36] Mach (1900), *Die Principien der Wärmelehre*, 2nd ed., p. 115: "Fourier's theory of thermal conduction may be termed a *model* physical theory. It is not based on a *hypothesis* but on an observable *fact*, namely, that the rate of equalisation of small temperature differences is proportional to the differences themselves".

work on heat conduction was consistent with different explanations of the mechanism of heat propagation and with different hypotheses regarding the nature of heat.[37] Moreover, as is now well known, the same equation can be used to describe other phenomena as well, such as the transport of materials by diffusion. The moral is simple enough, to wit: syntactic forms may be invariant under the most varied semantic transformations.

3.4.2. *Noncausal Laws Formulated with the Help of Differential Equations*

If a more fashionable illustration is required, Schrödinger's wave equation,

$$\frac{ih}{2\pi}\frac{\partial\psi}{\partial t} = H_{op}\psi,$$

will do, as it probably is the most reinterpreted equation that has ever been written. According to the logical empiricist interpretation of quantum mechanics,[38] the theory based on Schrödinger's equation is ontologically neutral, and in particular it supports neither determinism nor indeterminism, as it asserts nothing about the real world—but everything about measurements on the atomic scale; the theory would, on the other hand, be decidedly indeterministic with regard to the results of such measurements, which are not deemed to be determined by anything. Now, such empirical information is summarized in the ψ-function (often named "wave of knowledge") that solves the Schrödinger equation and that, according to the interpretation espoused by most positivists, has only a probability meaning. In other words, the usual, positivistic, interpretation of the quantum theory entails an *empirical indeterminacy*, although probabilities evolve in a strictly prescribed and unique way according with Schrödinger's equation —which is precisely a differential equation with time as the "independent" variable. As a consequence, it seems plain that

[37] Fourier (1822), *Théorie analytique de la chaleur*, in *Oeuvres*, vol. I, p. 538.

[38] Philipp Frank, "Philosophische Deutungen und Missdeutungen der Quantentheorie", *Erkenntnis 6*, 303 (1936); this issue of *Erkenntnis* was entirely devoted to the causal problem. See also Frank (1946), *Foundations of Physics*.

positivists should either drop the orthodox interpretation of the quantum theory or abandon their contention that differential equations of a certain type mirror causation, as the two are incompatible with each other—not to speak of their separate disputability.

The identification of causal laws with differential equations that can be made to *represent* an unlimited variety of types of continuous evolution in time (if the adequate semantic "rules" are attached to them) has led to a widespread naïve solution of the puzzle of quantum mechanics. According to this interpretation, the theory, while renouncing causality as regards things themselves (for example, atomic systems), retains it in connection with the evolution of what is usually regarded as *our information*, as probabilities themselves (or, rather, probability amplitudes) evolve in accordance with an allegedly causal law, namely, Schrödinger's equation.[39]

Mathematical forms say *by themselves* nothing about material reality, and this is just why they may be used (in combination with semantic "rules") to say so much about the external world. The eventual objective content that can be poured into mathematical forms lies entirely in the factual (physical, biological, . . .) meaning attached *ad hoc* to the symbols appearing in them, that is, in the semantic "rules". Consequently the causal problem—like the problem of the mechanical nature of the objects referred to by a given theory—cannot arise in either the formulations or the representations of a given scientific theory, but only in its *interpretations*.[40] Whoever accepts the doctrine of the "voidness" of mathematical structures (and logical positivists are among those who have contributed to clear this point up) cannot consistently maintain that differential equations are by themselves the mirror image of causation —nor of any other form of determination.[41]

[39] The germ of this interpretation was proposed by Max Born in the paper in which he advanced the statistical interpretation of the quantum theory (1926). It is also the nucleus of Margenau's interpretation.

[40] See Mario Bunge, "Lagrangian Formulation and Mechanical Interpretation", *American Journal of Physics* 25, 211 (1957).

[41] That mathematics is formal, in the sense that it does not speak about material reality (although it can be given an almost arbitrary variety of interpretations in

That many features of the material world can be described *with the help* of mathematical tools is true, and is another matter. But, to restrict ourselves to differential equations, it must be realized that they may be of help in the formulation of both causal and *non*-causal laws. Thus differential equations of the type

$$\frac{\partial \rho}{\partial t} = f(q, p, t)$$

(where q and p may in turn be functions of t) appear not only in theories having a strong causal component—such as classical fluid dynamics—but also in statistical theories of nonstationary processes, such as transport phenomena or stochastic chains. In these theories, ρ usually denotes a probability density, so that the meaning of the foregoing equation, in the context of such statistical theories, is not precisely causal. The simplest and best-known illustration of that type of equation is provided by the law of the (average) rate of radioactive decay, namely,

$$\frac{dN}{dt} = -\lambda N,$$

which can be derived on the assumption of the mutual independence or randomness of the successive individual disintegrations. Integration of this equation yields $N = N_0 e^{-\lambda t}$, which is plotted in Fig. 6. But the same mathematical form is employed to describe the absorption of light in homogeneous media, the decay of the intensity of excitation in neural circuits, the death rate of bacteria under the action of a disinfectant, Daniel Bernoulli's law of marginal utility, and a host

---

material terms) has been known, of course, to all non-Pythagoreans from Aristotle on, and to all those who do not accept the traditional empiricist and vulgar materialist thesis that every mental object is the transcription or reflection of some empirical fact. Even Kepler, strongly influenced as he was by Pythagoreanism, Platonism, and Neoplatonism, realized finally that mere games with signs are neutral, that is, prove nothing about the external world, so that no secret of nature can ever be revealed with their sole help. The thesis of the emptiness of mathematical and logical formulas was defended by Leibniz (1696) and adopted by Wittgenstein (1922) and by the Vienna Circle, whose followers often describe these formulas as tautological or noninformative. See Carnap (1939), *Foundations of Logic and Mathematics*.

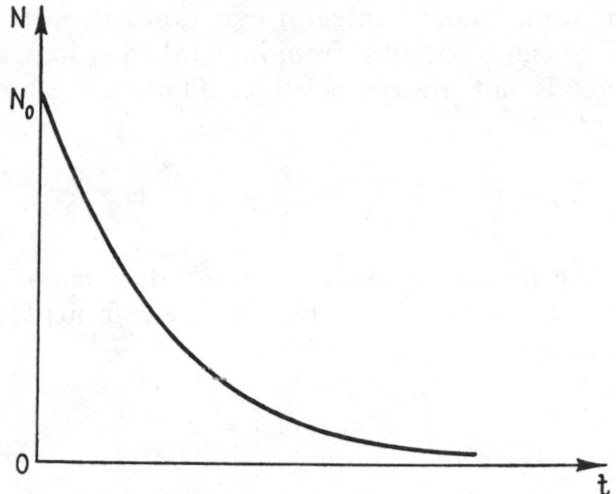

Fig. 6. Exponential decay, $N = N_0e^{-\lambda t}$, a conspicuous pattern of real processes in all known levels.

of other processes, whether causal or not. What matters in this respect is not so much the mathematical form as the *meaning* attached to the symbols appearing in it, that is, their interpretation. In the case of statistical theories, initial states are initial probability densities, or frequency distributions, or average numbers, out of which other distributions follow in time (in a unique way, as is the case of the so-called stochastically definite processes, or *not*, as in the general case). To call 'cause' the initial value of a distribution function specifying, for example, the probability that a system shall be in one or other of a number of possible states, sounds paradoxical; it is, in fact, begging the question. We therefore refuse to identify causes with initial states and causation with regular, unique, and continuous succession of states.

3.4.3. *Integral Equations and Teleology*

Since the opinion we are examining is so widespread, a further argument may be found useful. Another fact showing the untenability of the contention that all that can be said about causation is stated by means of differential equations of a certain type is that these ideal objects are contained in still wider structures, namely, integral equations. In fact, differential equations can be derived from integral equations, although the converse is not always possible. Thus, for example, the Lagrange equation,

$$\frac{d}{dt}\frac{\partial L}{\partial \dot{q}} - \frac{\partial L}{\partial q} = 0,$$

(probably the most frequently recurring differential equation of theoretical physics) can be deduced from Hamilton's integral principle,

$$\delta \int_{t_1}^{t_2} L(q, \dot{q}, t)dt = 0.$$

If one insists on assigning every mathematical form a specific ontological content, why not maintain, then, that integral instead of differential equations are the mirror image of causation?

Actually, the opposite has been claimed, namely, that integral equations (and, more specifically, extremum principles, such as Fermat's or Hamilton's) are the mathematical translation of—purpose.[42] Is this any better than attributing life to self-regulating material systems like gyroscopes or automatic furnaces, just because they respond to small perturbations in such a way that some property of them (angular momentum and temperature, respectively) is conserved? Moreover, since every differential equation can be rewritten in an integral way, and since from extremum principles certain differential equations are deducible (namely, the Euler-Lagrange equations), that strange opinion might lead to identifying efficient and final causation, or to asserting their interconvertibility—a thesis that has actually been defended,[43] and which essentially boils down to confusing final state with goal.

[42] See Planck (1937), "Religion und Naturwissenschaft", in *Vorträge und Erinnerungen*, p. 330. Two decades before, Planck had explicitly stated that integral principles have nothing to do with teleology; see "Das Prinzip der kleinsten Wirkung", in Hinneberg, ed. (1915), *Die Kultur der Gegenwart, Physik*, pp. 692 ff.

[43] See D'Abro (1939), *The Decline of Mechanism*, pp. 265–266; Weizsäcker (1943), "Naturgesetz und Theodizee", in *Zum Weltbild der Physik*, p. 166.

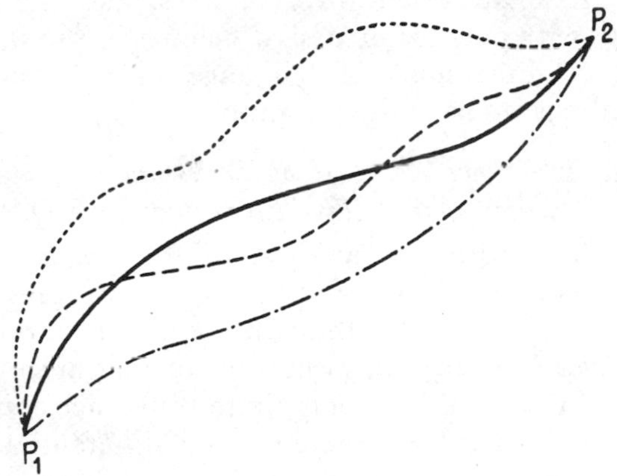

Fig. 7. Hamilton's stationary-action principle. Of all conceivable (fictitious!) paths between points P_1 and P_2 (corresponding to the instants t_1 and t_2 respectively), the system *actually* goes along that one (full line) for which the total action is either the smallest or the largest relative to the action corresponding to the remaining conceivable paths.

Those who, from Aristotle on, insist on attempting to introduce teleology in physics would do well to read Voltaire's *Histoire du Docteur Akakia*, where he made fun of Maupertuis' teleological interpretation of the least-action principle. They might also recall that the multiplicity of means characterizing end-seeking behavior (see Sec. 2.4.2) is just the opposite of the *uniqueness* of processes asserted by extremum principles (provided the integrand satisfies certain conditions that are usually required in physics just to ensure the uniqueness of the solution). This is precisely why extremum principles have sometimes been regarded as formulations of the postulate of the uniqueness of natural processes (see Fig. 7). To say that in behaving the way they do physical objects move "with the purpose" of minimizing or conserving the intensity of a given quality is not too different from asserting that things happen as they do "in order that" the laws of nature may be satisfied. Extremum principles are no more indicative of end-seeking behavior than any other physical laws, and the association of integrals with purpose or design belongs in the same class of confusion of dimensions of language as the association of differential equations with causation.

3.4.4. *The Empirical Test of Differential Equations and the Question of the "True Elementary Laws of Nature"*

To conclude with our analysis of the claim that certain differential equations mirror causal processes, a last argument will be set forth, namely, that differential equations are not *directly testable by experiment*, even after having been interpreted in empirical terms. In fact, only finite-difference equations and integral relations (such as the solutions of differential equations) can be correlated with empirical data. Think of the very elementary notion of instantaneous velocity, defined as dx/dt; only mean velocities $\Delta x/\Delta t$ are afforded by measurement. Or think of the term curl $\mathbf{E}$ appearing, for instance, in one of Maxwell's equations; only its integral equivalent, namely, the line integral $\oint \mathbf{E} \cdot d\mathbf{s}$, can be assigned an operational meaning (to wit, the electromotive force).

It might be argued that an arbitrarily close approximation to the elementary (differential) expression can be obtained, so that after all the problem of the empirical verification of a scientific hypothesis couched in differential terms resembles the replacement of geometric lines by fine threads. This is actually so in many cases, notably in connection with formulas involving simple symbols like dx/dt. But in some cases not a mere quantitative difference is at stake but a *qualitative* difference between the theoretical law and the experimentally testable expression; more exactly, sometimes no *unique* correspondence can be established between the differential laws and experimental data, so that the choice among the various data-fitting hypotheses is arbitrary, unless one of them is backed by theoretical reasons (or by ontological tenets, such as simplicity). Such a lack of unique correspondence between experimental data and the theoretical model is particularly acute whenever closed lines of flow or of force appear, as in fluid dynamics, electromagnetic theory, and other classical field theories. In such cases no less than an *infinity* of different "elementary" (differential) laws is consistent with a single integral law (the one accessible to experiment).[44] In other words, largely *arbitrary* modifications of the differential laws have no empirically testable consequences; hence, it is empirically meaningless to ask which of these infinite *differential* formulas is the true one. The failure to understand this point was the source, a century ago, of a long and rather sterile controversy among the partisans of Ampère and those of Grassmann about the "true elementary laws" of the mutual actions of electric currents. The argument was over two different "elementary" laws which, integrated along a circuit, gave the same result and were thus empirically equivalent. It is now understood—though not made explicit enough—that in such cases only the *integral* law statements can be true or false, while the choice among the various differential laws consistent with the integral law is largely a matter of convenience or even of taste—to the despair of the upholders of the reflection theory of knowledge.

Moreover, the same set of experimental data and the same reasoning suggesting the choice of a given differential equation as "the" reflection of a certain group of phenomena suggests also a finite-difference equation. Only further consequences of the two alternative hypotheses (equations) may eventually be tested; while no different consequences have been tested, such alternative hypotheses are of course empirically equivalent and our preference for differential over finite-difference equations is dictated chiefly by the greater simplicity of the former or by the acceptance of the hypothesis of continuity.

All this has the important philosophical consequence that the theory according to which differential equations are the ideal or even the sole form of scientific laws is (*a*) consistent with the *empiricist* reduction of determination to uniform succession and (*b*) inconsistent with the radical *empiricist* requirement that scientific laws should contain solely parameters that are accessible to experiment, that is, having a

[44] An infinity of exact (total) differentials can be added to the given differential expression, owing to the vanishing of the circulation (integral taken along a closed path) of exact differentials.

direct operational significance. The empiricist choosing the second alternative should propose calling only integrated expressions and finite-difference equations scientific laws, and should consequently abstain from calling Newton's law of motion or Maxwell's laws of the electromagnetic field physical laws, since they obviously are not laboratory equations having a direct operational meaning.[45] However, few scientists would be prepared to receive this idea, which lends itself too easily to subjectivism, for the solutions of a differential equation, unlike the equation itself, depend upon the choice of initial or boundary values, reference systems, and so on, and these are in turn dependent, to a large extent, upon the information, mathematical equipment, and needs of the scientist. Besides, those following the operational course of identifying physical laws with laboratory equations should part company with the propounders of the positivist teaching on causality (see Sec. 3.4.1).

To sum up, differential equations are not the mirror image of becoming in general, nor of causation in particular; on occasion they can be assigned a causal meaning by means of *ad hoc* semantic rules, but this can be done with other types of mathematical objects as well. In short, the causal problem is not a syntactic but a semantic one; it has to do with the interpretation rather than with the formulation and representation of theories.

3.5. Summary and Conclusions

Causality involves neither contiguity nor antecedence, although it is consistent with both; indeed, causality is compatible with action at a distance—and instantaneous action at that, as gravitation was once supposed to be. Since Hume regarded both contiguity and antecedence as the essential components of causation, it follows that his criticism of causality was not valid, however psychologically effective it may have

[45] See Lindsay (1941), *Introduction to Physical Statistics*, p. 1: "Physical laws are equations containing quantities which have a direct operational significance in the laboratory and to which numbers can be assigned by experiment". Hence, only the solutions of a differential equation, not the equation itself, can be named physical laws, since by the foregoing definition physical laws are "laboratory equations" (p. 2).

been. The principle of action by contact is moreover inconsistent with empiricism, as it is not directly testable by experiment; though reasonable and fruitful in the domain of physics, the hypothesis of nearby action may lose any significance in other fields of research—which is a further ground for not regarding contiguity as essential to causation. Contiguity and antecedence, on the other hand, each impose a restriction upon causation.

States of physical objects are not agents but systems of qualities; consequently they have no causal efficacy. Initial states cannot therefore be causes but may be produced by previously acting causes. Hence causation is not reducible to invariable, unique, and continuous successions of states. Predictability by means of laws of succession is a criterion not of causal connection but of the validity of nomological hypotheses about time sequences. Causation is not mirrored by differential equations, which by themselves have no more factual content than any other mathematical form. Such equations do not state that changes are produced by anything, but can be read as stating that certain changes are accompanied by certain other changes. One and the same differential equation can be assigned a causal interpretation alongside other, noncausal, interpretations. Differential equations are as little the carriers of efficient causation as integral equations are the carriers of final causation. The opinion that differential equations are the paradigm of scientific law—an 18th-century belief—is not confirmed by modern science and is moreover inconsistent with the (equally untenable) operationalistic requirement that scientific laws should contain solely parameters having a direct empirical significance. In conclusion, the empiricist theory that determination is nothing but regular, unique, and continuous succession not only is unwarranted by modern science but leads empiricism into contradictions that do not seem surmountable without a retreat from fundamental positions.

One of the severest blows suffered by the empiricist reduction of causation to uniform succession came from relativity physics. In fact, relativity introduces a radical difference between time series of genetically unconnected events (which are reversible), and causal series (which are irreversible). Moreover, it suggests that it is the flow of time (relative to every reference system) that can be regarded as rooted to genetic sequences or, if preferred, as measuring their tempo. Change is thereby regarded as primordial and time as derivative. The time theory of causation, defended by Hume and his followers, is thereby reversed and a *causal theory of time* is established. As soon as this important step is taken, a traditional cornerstone of empiricism is left aside; and when such a step is taken by eminent empiricists,[46] it may be a sign of the deep crisis in contemporary empiricism—though not much deeper, it is true, than the crisis in any other living philosophic school.

All in all, the empiricist critique of causality has been as erroneous as famous and rigor-provoking.

[46] See Reichenbach (1951), *The Rise of Scientific Philosophy*, pp. 148 ff; (1956) *The Direction of Time*, p. 24: "time order is *reducible* to causal order". What, we may wonder, will remain of empiricism after the following tenets have been abandoned: the Humean theories of causation, induction *qua* a truth-establishing procedure, the wholesale reduction to sense data, and the analytic-synthetic dichotomy?

4

An Examination of the Romantic Critique of Causality

The starting point of the empiricist critique of causality was the obvious remark that experience, by itself, said nothing about any causal bonds—whence it was concluded that, experience being the sole source of legitimate knowledge (a clearly trans-empirical statement), the category of causation in the sense of genetic link was a dispensable figment of abstract thought. The net result of the empiricist critique of causality was consequently a *restriction* of determination to regular association or invariable succession.

The romantics, from Schelling to James and Bergson—including Peirce, the romantic empiricist—have on the other hand criticized causality on an ontological plane, and not just because they deemed the claims of causalism to be exaggerated—moderation not being precisely a romantic characteristic—but because they regarded causality as too poor a theory of change. Their colorful vitalistic *Weltanschauung* was embroidered on a canvas of multiple subtle organic interconnections. Compared with the richness of this universal interdependence or *Zusammenhang*, the one-sided causal nexus, decried by empiricists for its alleged anthropomorphic associations—decried, in short, for containing more than experience could warrant—causation looked rough, miserable, dry, lifeless, and impersonal.

The target of the romantic critique of causality was, to sum up, the *enlargement* instead of the restriction of causality, which empiricism had preached. It is likely that the romantic criticism would have remained unnoticed had it not been for the fact that empiricists had already done a great deal in the way of discrediting causalism, and that 19th-century science, in all of its newest branches, was rapidly disclosing richer, noncausal, kinds of connection, mainly mutual actions and "organic" connections. Above all, science was realizing that the picture of change in terms of separate events, into which the causation category can fit, had to be replaced by a picture in terms of processes.

In its turn, the increasing awareness of the complexity of the categorical outfit actually used by science, combined with the romantic criticism of causality, exerted a beneficial influence on the positivistic approach to the causal problem—although at first sight no two approaches could seem as polarly opposed as the neopositivistic and that of the *Naturphilosophen*. In point of fact, the traditional empiricist efforts to reduce causation to regular association, to the external juxtaposition of concomitant events, were replaced, by some of the followers of Hume, by the attempt to substitute functional interdependence for causal dependence. This was especially the case with Mach, who, unlike most of his forerunners and followers, had a keen and somewhat romantic feeling for the diversity of nature and for the intimate interconnections existing among its different members and aspects. Like the romantics—whom he probably despised and certainly ignored—Mach decried causation as an abstraction and as an insufficient concept which becomes superfluous as soon as interconnections are taken into account. We shall see below, however, that Mach failed to grasp, or at least to acknowledge, the notion of genetic interconnection, so that the affinity of the two approaches, which I am suggesting, should not be exaggerated.

In the present chapter we shall analyze some of the criticisms raised against causality by romanticism. Actually, we shall here be concerned solely with the main critiques that I consider unjustified, namely, the identification of causality with mechanical determinism, the attempt to replace causation by functional interdependence, and the claim that causality makes no room for freedom. The kernel of the positive contribution of the romantic philosophers to the problem of determinism will be mentioned in later chapters.

4.1. Should Causation Be Replaced by Interdependence?

4.1.1. *The Functional View of Causation*

As was pointed out above, both romantics and neopositivists have demanded the replacement of causation by interdependence. But modern positivists have usually lacked the romantic sensitivity to process and newness; and, contrary to the *Naturphilosophen*, they accepted mathematics enthusiastically, though just as a shorthand enabling us to perform an economical description of experience. Further, while the romantics decried mathematics on account of its alleged dryness and tautologous character, and opposed universal organic interconnectedness to the one-sided and unidirectional causal bond, Mach proposed on the other hand that the mathematical concept of function be used as the precise scientific tool for reflecting interdependence. Mach showed thereby that by interdependence he did not mean *genetic* interrelation but rather mutual dependence among existents, a static net of reciprocal dependences like that among the parts of a steel frame.

Let us examine the romantic claim that causation is but an abstraction from interdependence, in the restricted and precise form set forth by Mach.[1] The founder of neopositivism was among the first to hold the functionalist view of causality; he demanded the replacement of every sort of connection, particularly causal connections, by functional relations expressing a symmetrical interdependence. Thus the relation

$$x_1 = f(x_2, x_3, \ldots)$$

can be read to mean the property designated by 'x_1' is related to the properties designated by 'x_2', 'x_3', ... in the precise way specified by the function f. As a particular case, laws of

[1] Mach (1872), *Die Geschichte und die Wurzel des Satzes von der Erhaltung der Arbeit*. His more mature thought on causality will be found in *Die Mechanik*, 4th ed., 1901, pp. 513 ff., and *Erkenntnis und Irrtum*, 1905, chap. xvi.

conservation (of momentum, energy, and so on) will be expressed in the form $f(x_2, x_3, \ldots) = \text{const.}$ (or $\partial f/\partial x_1 = 0$). Kirchhoff followed Mach shortly afterward along this path. The doctrine was adopted by most positivists, and some have even claimed that science has already achieved the complete replacement of the outmoded causal link by functional dependence; at least, this was announced in 1929 by the Wiener Kreis in its famous manifesto.[2]

We have dealt with the program of reducing causation (and, in general, determination) to functional dependence; indeed, functions establish correspondences between sets of numbers, so that they are a mathematical expression of regular conjunction, or constant association. In other words, the claim that causation is exhausted by functional dependence is more precise than, but not essentially different from, the constant-conjunction view on causation, held by traditional empiricism (see Secs. 2.3 and 2.4). However, for the sake of clarity we shall deal separately with the functional view of causation.

4.1.2. *Criticism of the Functional View of Causation*

The following objections can be raised against functionality:

(*a*) Functions express *constant relations*, fixed correspondences among two or more sets of numbers that are supposed to represent the numerical values of (metrical) properties; they may be used to state that something is invariably associated with or accompanied by something else—which may consist of another, coexistent, property of the same object in question. But functions are insufficient to state anything concerning the cause that *produces* the state or the phenomenon in question. Although functions make it possible to symbolize and to give precise quantitative descriptions and predictions of connections enormously more complex and rich than the causal bond, they fail to state the one-sided genetic connection that characterizes

[2] *Wissenschaftliche Weltauffassung der Wiener Kreis*, 1929, p. 23. See also Moritz Schlick (1932), "Causality in Everyday Life and in Recent Science", in Feigl and Sellars, ed., *Readings in Philosophical Analysis*, p. 523. The belief that causality is exhausted by functional dependence is common among nonpositivists as well; see, for example, Jeffreys (1948), *Theory of Probability*, p. 12.

causation; such a one-sided dependence of the effect upon the cause must be stated in an extra set of (semantic) propositions.

(*b*) Functional relations are *reversible* whenever the functions in question are single valued; thus, from $y = ax$ we may infer that $x = y/a$ (provided $a \neq 0$). Hence, functions serve to symbolize interdependence, whether of different features of a given object, or of properties of different events. In this special but important case of functional dependence, the variables denoting the cause and the effect are interchangeable, whereas genuine causal connections are essentially *asymmetrical* (see Sec. 2.3). In other words, if $y = f(x)$ is a single-valued function, it can be inverted unambiguously to yield $x = f^{-1}(y)$, so that the relative places of the "cause", first represented by x, and of the "effect", which in $y = f(x)$ is symbolized by y, can be exchanged.

The failure to account for genetic connections is a shortcoming of the functional relation. But not all connections in the world are genetic; in many, perhaps in most, cases we are confronted with interdependence rather than with a one-way relation of dependence, as is shown by the pervasiveness of the function concept in the sciences. A trivial illustration of noncausal interdependence of different features of a single phenomenon is afforded by the relation $y = A \sin \omega t$ between the phase ωt and the displacement y of an harmonic oscillation; it would be meaningless to point to either variable as the *cause* of the other, since no genetic connection is involved in that relation. Only a *change* in one of the variables could be read as the cause of the change in the related variable; still, this causal interpretation would not do justice to the asymmetry or unidirectionality of causation.

(*c*) A typical trait of causation is *uniqueness* (see Sec. 2.4.1); on the other hand, in general functions do not establish one-to-one (that is, biunique) correspondences. Consider the function $y^2 = x$; this is a two-valued function, as two values of y are associated with every value of x. If interpreted in causal terms, it would say that two different effects, $y = \sqrt{x}$ and $y = -\sqrt{x}$, correspond to a single cause x—and two exactly opposite effects at that (see Fig. 8).

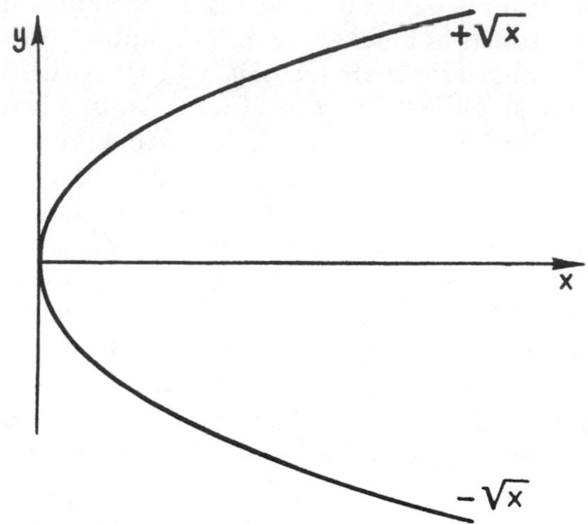

Fig. 8. The two-valued function $y^2 = x$. The same "cause" has two "effects," a positive and a negative one.

(*d*) In many cases, a variation in the "independent" variable x is accompanied by no change in the "dependent" variable $y = f(x)$; in other words, the "cause" is ineffective—hence it is not a cause at all. Within these ranges of values, variations in the "cause" have no effect; if preferred, the effect does not depend on the precise intensity of the cause, which may vary arbitrarily within the given range. Such zones of insensitivity (sometimes called plateaus) are characteristic of saturation phenomena (see Fig. 9). At least in the saturation zone the "independent" variable x and the "dependent" variable y cannot be regarded as representing the cause and the effect, respectively; in other words, plateaus correspond to the noncausal range of the law in question.

An even more drastic illustration of this point is afforded by the function $y = \sin n\pi$, which is always zero for integral values of n; n may jump from integer to integer without having any effect on y, whence it cannot be regarded as the "cause" of y.

(*e*) Functional relations may be employed to describe processes that may or may not be causal; but by itself functional

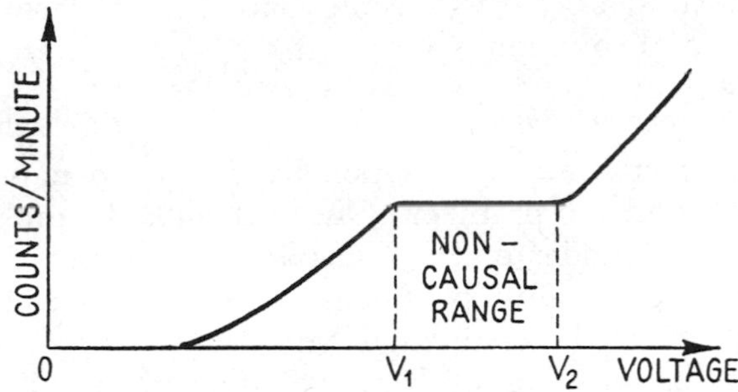

Fig. 9. Counting rate of a Geiger-Müller counter. In the range V_1–V_2, an increase in the value of the "cause" (voltage) produces no alteration in the "effect" (counting rate).

dependence does not constitute a category of determination. Functions, together with the semantic rules stating the meaning of the variables tied by them, are often useful to tell what happens and why it happens; if a causal meaning can be attached to some of the symbols intervening in a function (rarely, if ever, to all of them), such an *interpreted* function will reflect a causal connection (see Sec. 3.4.1). In other words, functions, which are syntactic forms, cannot *replace* causal propositions; at most, they may take part in the description of causal connections.

I may be allowed to recall briefly the formal nature of mathematical entities (see Sec. 3.4.1). Functions are mathematical "forms" that may be "filled" with an infinity of contents—not excluding unscientific (untestable) statements. Functions, alongside other mathematical entities, help us to express some of the features of laws of whatever type, whether causal or not; thus, distribution functions help us in reflecting random processes. Moreover, functions are by themselves neutral with regard to material (factual) truth: they may be used to state truths of fact as well as errors of fact. Scientists who are aware of this semantic neutrality of mathematics are not prepared to receive the naïve idea that, in view of the alleged scientificizing virtue of mathematical formulation, all that is needed to obtain a perfectly scientific definition of causation is to identify it with some mathematical form (function, differential equation, etc.).[3]

4.1.3. *Strange Features of the Functional View of Causation*

The functional view of causation has suggested naïve solutions to time-honored problems. One of them is the problem of teleology. The founder of the Vienna Circle stated that the replacement of causality by functional dependence leads to the breakdown of the distinction between teleology and causality, for the meaning of 'determination' is possibility of calculation or prediction; and if we can calculate E from C by means of a certain formula, we can also infer C from E—whence "it makes no difference whether we say 'the past determines the future' or 'the future determines the past'".[4] We shall not linger upon this curious result of functionalism, which depends on the idealistic identification of 'determination' with 'possibility of calculation' (see Sec. 1.2.2), as well as on the confusion between future (or final) state and 'goal for whose attainment present processes are assumed to serve'—too obvious a confusion to deserve further attention.

A second problem that can be solved in naïve terms on the ground of the functional approach is the mind-body problem. If it is stated that determination is exhausted by functional interdependence, then psychophysical parallelism ensues. In fact, the statement that the series of psychical states $\psi_1, \psi_2 \ldots$ parallels that of the physiological states $\phi_1, \phi_2, \ldots$ of the nervous system amounts to asserting the existence of a (symbolic, not necessarily metrical) functional relation $\psi = f(\phi)$, with the eventual (but unwarranted) proviso that the function $f(\phi)$ must be single-valued (to ensure unambiguous coordination). This functional view of the mind-body problem is consistent with dualism, neutral monism, and Leibniz's pre-

[3] The Pythagorean belief in the purifying virtue of mathematics is strong among contemporary positivists. See Lenzen (1938), *Procedures of Empirical Science*, in *International Encyclopedia of Unified Science*, vol. I, no. 5, p. 39: "The mathematical form of causality purges it of the last vestiges of efficiency. Causal laws are stated as functional relations between numerical measures of variable quantities".

[4] Schlick (1932), "Causality in Everyday Life and in Recent Science", in Feigl and Sellars, ed., *Readings in Philosophical Analysis*, p. 527.

established harmony—but is it consistent with the elementary remark that we can cease thinking without having to annihilate our brain, whereas the removal of the latter renders thinking impossible?

A third problem that has been given a naïve and uncompromising solution by functionalism is that of the relation between the material conditions of existence and spiritual culture. As against both the idealistic doctrine of spiritual life as the ultimate source of every cultural (social) event and the materialistic theory of the ultimate predominance of material over spiritual factors, historical functionalism states that it is not a question of genesis and predominance (whether short-run or long-run), but only of an inextricable interdependence among "factors" standing on the same footing. The undeniable fact of the reciprocal action among features actually standing on different levels is thereby emphasized—but at the price of giving up the hope of understanding the real genetic mechanism, since the "factors" involved in the functional dependence are regarded as given and, moreover, as standing on a single level, as if society were not a multilevel structure.

Functionalism is also consistent with organismic cosmologies, and is true to the extent to which organicism is true, its element of truth being that real connections are not one-sided but many-sided, and variable instead of static. The main shortcoming of functionalism is its lack of recognition of the productive, genetic character of determination. Moreover, if applied consistently, functionalism leads to the view that the whole universe is an unanalyzable muddle consisting of an infinity of "factors", all of them standing on the same footing. The isolation of factors from one another is certainly an abstraction in many cases; but such an isolation or singling out of the various features is required by a scientific treatment. What is wrong is not the tearing of partial and one-sided connections out of the general interconnection, but the remaining at such a preliminary stage of research. The truer an account of facts, the more it attains an integration of the various partial views; but such a synthesis, if scientific, will not be presupposed: it will be a result of analysis, and will include a valuation of the various factors, that is, statements about what are the relevant and decisive components or aspects, which are the ultimate determinants in a given respect.

To sum up, causality is not reducible to functionality, because functional dependence is in a sense more general than causation, since with the help of functions interdependence instead of one-sided causal connection may be adequately expressed. But on the other hand, functional dependence is poorer than causation, since it does not include the concept of genetic connection through a process, but expresses instead a mutual dependence among things or qualities that may coexist without being genetically related to one another.

4.2. Causality and Universal Interconnection: The Block Universe and Chance

Determinism, whether in its causal variety or not, asserts (at least as defined in Sec. 1.5.3) the genetic principle and the principle of legality. The former implies that there are no isolated facts in concrete reality, things being actually interconnected—and in precise and regular ways at that, adds the principle of legality. But this does not mean that determinism, and in particular causal determinism, asserts that everything in the world is connected with *everything* else in *all* respects; nor does causal determinism assert that everything is *causally* connected with everything else, a doctrine occasionally (and erroneously) ascribed to monism, which in one of its varieties asserts nomological unity, but not necessarily of the causal kind.

The doctrine of universal and necessary interdependence, stating that everything is *en rapport* with everything else, that *tout se tient*, as romantics and occultists maintain, need not be causal—although it is *consistent* with causality. The doctrine of causality is also compatible with the idea that coexistents may not be causally related to one another (parallelism). This was clearly realized by the most consistent among the early partisans of the universal, unrestricted operation of the causal bond, namely, the Stoics. In order to substantiate their claim that the world is an organic unit, a common life animating all of its parts (as Plato had suggested in his *Timaeus*), the Stoics postulated the existence of a universal *sympathy* among the different linear chains of successive causes and effects.[5] That is, the Stoa added a special, "horizontal", connection through space (but not necessarily of a physical nature) to the "vertical" causal connection; and this universal sympathy holding the various parallel causal chains together in a single bundle, this horizontal connection, was not necessarily of a causal nature. The philosophers of the Enlightenment, plainly under the spell of the Newtonian theory of universal gravitation, often stated the principle of the reciprocal action among all concrete coexistents, as a special axiom to be added to the law of the causal succession in time[6]; the romantics who came shortly afterward joined the two ideas in the single principle of the universal and dynamic *Zusammenhang* or interdependence that was supposed to hold all things together through both space and time.

Now, if the causal principle and the theory consisting of its unlimited extrapolation allow for *independent* or parallel causal lines, then they are consistent with the admission of that type of chance consisting in the crossing or encounter of mutually independent causal lines—independent, that is, till the instant of their meeting. Hence, although the causal principle excludes chance in connection with every *single* sequence, it does not exclude contingency as synonymous with independence (or mutual irrelevance) of different causal series. This sort of chance, dealt with in antiquity by Chrysippus and in modern times by Cournot,[7] is not a mere name for human ignorance—save as regards the prediction of exact place and date of the chance encounter. Even an exact foreknowledge of such a coincidence would not prevent its being a chance encounter, since up to the moment of their crossing the two lines were disconnected; they are objectively contingent relatively to each other—to the extent to which independence, or irrelevance in definite respects, has an ontological status.

Consequently, those who (wrongly) regard chance as a

[5] See Pohlenz (1948), *Die Stoa*, vol. I, pp. 101 ff.

[6] See Kant (1781, 1787), *Kritik der reinen Vernunft* (B), p. 256.

[7] Cournot (1843), *Exposition de la théorie des chances et des probabilités*, sec. 40.

wholly epistemological category, and even as nothing but a name for our ignorance, cannot reasonably claim that this doctrine on the nature of chance is based on the causal principle, or even on causalism, which—at least implicitly—makes room for chance from the moment it does not assert universal causal interdependence in all respects. (Not only does causalism make room for chance but, if consistently developed, it leads to a self-defeating recognition of contingency, as will be shown in Sec. 8.1.)

But, quite aside from the question whether causality implies rigid universal interdependence or not, a word should be said on the claims to validity of the block-universe doctrine. The perfect test of it would be the utter failure of science; for, if things were so closely tied together as the propounders of unlimited interdependence assume, if everything were relevant to everything else, then it would not be possible to know any part of the universe without knowing the totality—which obviously is not the case, for we do not even know what the whole is and whether it has a finite spatial extension or not. Conversely, no knowledge of the whole would be possible unless a complete acquaintance with every one of its parts were available. (This tragic circle was perceived by Pascal, who, in spite of this, held the doctrine of unlimited universal interdependence—a conflict that may have influenced his conversion to religious meditation.) On such a view, which organismic ontologies should propose if they were consistent, the progress of knowledge would be impossible; the very task of science would be like that of solving a logical circle.

Since we are not caught up in such a circle, since the progress of knowledge is undisputed, we may infer that the hypothesis of *unlimited* universal interconnection, whether causal or not, is false. A further test of the falsity of the doctrine of the block universe is the existence of chance (that is, statistically determined) phenomena; most of them arise from the comparative independence of different entities, that is, out of their comparative reciprocal contingency or irrelevancy. The existence of mutually independent lines of evolution is in turn ensured by the attenuation of physical interactions with distance, and by their finite speed of propagation—the most effective looseners of the tightness of the block universe.

4.3. Is Causality Fatalistic?

4.3.1. *The Other-Worldliness of Fatalism*

When Emerson wrote that "the book of Nature is the book of Fate",[8] he merely repeated another common error. It is, indeed, a widespread misconception that natural science discloses the work of Fate; and since science is currently assumed to be entirely or at least essentially causal—another wrong belief—it is concluded that causality is fatalistic. This misconception is not only disseminated among essayists, like Emerson, but even among philosophers and scientists. Phenomenalists like Renouvier,[9] accidentalists like Jordan,[10] and above all the romantics have too often asserted that causality and determinism are identical with fatalism, that is, the belief in a transcendent necessity producing inevitable results and moreover operating *ab aeterno*, in such a way that events are all predetermined, or written, as it were, in some Book (even before writing was invented).

Actually, fatalistic determinism is in a sense the very *opposite* of scientific determinism, and is in particular incompatible with causal determinism. In point of fact, fatalism is a theological or at least a supernaturalistic doctrine asserting the existence of an unknowable and inescapable Destiny; on the other hand, causal determinism claims to be a rational theory offering the means for knowing, predicting, and consequently changing the course of events. The word 'fatalism' designates the class of doctrines subordinated to some non-naturalistic belief, and asserting that a transcendent, other-wordly, uncontrollable, unpredictable, immaterial power produces all events, or most of them.[11] There is no fatalism without a *fatum* or Fate, and this is anything but the so-called

[8] Emerson (1860), *The Conduct of Life*, in *Works*, vol. VI, p. 20.

[9] Renouvier (1901), *Les dilemmes de la métaphysique pure*, sec. 56.

[10] Jordan (1944), *Physics of the 20th Century*, p. 149.

[11] See Ranzoli, *Dizionario di scienze filosofiche*, art. "fatalismo"; Lalande, ed., *Vocabulaire technique et critique de la philosophie*, art. "fatalisme".

blind rule of law, conceived as the immanent pattern of being and becoming.

4.3.2. *The Lawlessness of Fatalism*

It is important to realize the difference between fatalism and determinism (whether causal or not), because "virtually all of the objections that are ever raised against determinism turn out on examination to be objections against fatalism, which disappear as soon as fatalism is thus distinguished from non-fatalistic determinism."[12] Causality, as such, need not assume any supernatural agency; besides, according to causality, events are chained to one another, whereas for fatalism every bond is *indirect*, since it is assumed to go across a power external to the events concerned. As conceived by fatalism, necessity is not inherent in things, it is not rooted to the nature of things, but is definitely contingent upon what things are and how they change.

According to fatalism there is a primary cause, which is transcendental, and there are secondary effects, which are worldly but occur no matter what the wordly state of affairs may be or may have been. The necessity asserted by fatalism is *lawless*, and this is why its patterns, if any, are deemed to be inscrutable. To fatalists, events occur regardless of circumstances; since what is preordained must occur, nothing will prevent its happening, nothing will be able to interfere with an unbreakable external necessity—the *fatum*—that produces or directs the course of events. In other words, according to fatalism, just as according to accidentalism, events are *unconditional*, and the future is as unchangeable as the past. As Sophocles put it,

> Strange are the ways of Fate, her power
> Nor wealth, nor arms withstand, nor tower;
> Nor brass-prowed ships, that breast the sea
> From Fate can flee.[13]

[12] H. van Rensselaer Wilson, "Causal Discontinuity in Fatalism and Indeterminism", *Journal of Philosophy* 52, 70 (1955), p. 72.

[13] Sophocles, *Antigone*, IVth Song, stanza 1, vs. 950 ff., trans. F. Storr (Loeb Classical Library; Cambridge: Harvard University Press, 1912).

Fatalism and accidentalism are the two extreme forms of lawlessness. Lawfulness, and causal lawfulness as a particular case, lie midway between the unconditional necessity asserted by fatalism and the unconditional arbitrariness assumed by tychism. While fatalism holds an unconditional necessity and conceives it as transcendent, an essential mark of causality and, indeed, of every other type of lawful determinacy, is conditionality (see Sec. 2.2). Statements of causal laws and, in general, scientific laws, do not assert *that* something will inevitably happen under *all* circumstances, regardless of past or present conditions; quite on the contrary, statements of causal laws assert that *if*, and only if, certain conditions are met, certain results will follow. A change in the conditions will therefore ensure a change in the results; the rule of fate is then seen to be illusory, events not being predetermined once and for all, but improvised, so to say, along the way.

4.3.3. *The Interference of Causes Defeats Fate*

An implication of causal conditionalness is that causes may *interfere* with one another, so that the result may turn out to differ from what would follow from any of the separate causes. Since any given cause may be counteracted, or at least influenced, by another cause, *causality does not entail inevitability*; on the contrary, it leaves room for processes and deeds capable of changing the course of events, thereby affording a real—though certainly limited—ground for both chance and freedom.

Whereas fatalism rules possibility out of the world, relegating it to the epistemological sphere, causality, if rightly understood, is one of the very grounds of possibility: it renders possibility possible. Indeed, according to the doctrine of causality there are no uncaused events, but not every cause need "succeed" in producing the expected effect; a given set of causes may be hindered, for the production of their otherwise normal effects, by the interposition of other causes.[14] Leibniz had an idea of

[14] See Thomas Aquinas (1272), *Summa Theologiae*, part I, Q. cxv, a.6: "It is not true that, given any cause whatever, the effect must follow of necessity. For some causes are so ordered to their effects, as to produce them, not of necessity, but in the majority of cases, and in the minority to fail in producing them. But that such causes do fail in the minority of cases is due to some hindering cause."

this when he said that it is necessary that everything have a *cause*, but it is contingent that every cause should produce its *effects* (whence past events have been necessary while future ones are contingent). In order that a cause may produce its normal effect, other causes (background causes, or conditions) must concur; the interference of new circumstances, a change in the background, may prevent a causal connection from taking place—and this is a source of possibility.

A fortiori, general determinism does not recognize anything unconditional and, as a consequence, it does not entail any inevitability other than what results from the lawful concurrence and interplay of processes—among which the human conscious conduct may eventually intervene. Only, general determinism shows us that the set of laws enabling us to counteract or at least modify in some way any given course of events—thereby building a different future—is much richer than that imagined by causalism.

The ethical and sociological import of this question is obvious. Neither fatalism nor accidentalism allows for freedom. The former relegates freedom, if any, to the arbitrary deeds of an uninfluentiable *fatum*. And accidentalism does not make room for freedom, because it does not afford the means of achieving freedom, whether individual or social; for, if things just happen in an arbitrary fashion, if the universe is a chaotic heap of unrelated events, what can we then rely on, which traits of the world, which laws can we use for attaining freedom from outward compulsion and constraint or, rather, liberty to subject external conditions, within limits, to our will? What is there to warrant that a tendency to resist, counteract, or simply modify such a compulsion in some way may develop? Moreover, what assures us that such a compulsion exists, and consequently that there *is* an ethical problem at all?

This extreme consequence has actually been drawn by the propounder of the so-called principle of complementarity; in fact, time and again Bohr has maintained that the strife between the upholders of free will and those of ethical determinism (or, rather, *pre*-determinism) is a *inhaltlose Streitfrage*, a vacuous question, like the antinomies determinism-indeterminism and mechanism-vitalism. The gist of Bohr's argument is the following: in order to analyze experimentally the subject's behavior one would have to disturb him basically, thereby rendering his free will illusory; in fact, such an "experimental analysis" of the subject's moral conduct would require a thorough mechanical, chemical, biochemical, and physiological dissection of his body. This argument sounds like maintaining that one cannot both be a physiologist and know something about human physiology because one cannot remain alive while being taken to pieces; or like saying that it is not possible for a psychologist to know something about human behavior and consciousness because he cannot study his own behavior while watching the flux of his consciousness, or because he cannot exert his free will while being subject to a test. Such obvious difficulties do not prove that freedom is a metaphysical fiction; rather, they support the hypothesis that moral behavior cannot be "analyzed experimentally" in the manner of physical phenomena. Free will is certainly rooted to the lower levels, but it has not been shown that it can be reduced to them; this is why nobody solves ethical problems by means of physico-chemical techniques.

Tychism, no less than fatalism—and for the same reason, namely, lawlessness—involves the *servo arbitrio* of astrology and of Mohammedan and Calvinist theologies. On the other hand, the conscious and planned handling of laws, of those asserting self-determination as well as of those expressing the way in which certain agencies may be counteracted by other agencies in a regular way, is the very condition for attaining a genuine *libero arbitrio* replacing the servitude to which ignorance of such laws condemns us.[15]

4.3.4. *Are Historical Events Inevitable?*

The same is true, *mutatis mutandis*, of social events. Historical happenings are not inevitable, in the sense of occurring no matter what the conditions are; what is true is, that many historical facts occur without our *conscious* will and sometimes

[15] Both the nature of freedom and the limitations of causalism on it are treated in more detail in Sec. 7.1.6.

even despite the will of the social group concerned—but this has nothing to do with inevitability, and much to do with our ignorance of social laws or with our inability or unwillingness to apply the scanty available knowledge of them. Nothing can refute the assumption that effective and lasting influence on the course of historical events is attainable on the basis of a scientific approach to social events, an approach including the search for the laws of sociohistorical development affording the means for performing predictions of a statistical character (see Secs. 10.4 and 12.3).

The evitability of social events is noticeable even in the case of radical changes. With regard to political upheavals, a South American should be believed upon his word; as to social revolutions, let us hear the testimony of a historian. A social revolution, says Childe, "may be *necessary* in the sense of essential to further progress, but it is not *inevitable*. In Mesopotamia, Egypt, and China theocratic despotism, relations of production appropriate to the productive forces of the Bronze Age, persisted into the Iron Age. They effectively fettered the exploitation of the new forces represented by iron with the result that technology also stagnated. The whole life of those societies stagnated too; the first two eventually perished altogether. From a Marxian analysis all that one can deduce is the dilemma —revolution or paralysis. History does not disclose an unfaltering march to a predetermined goal".[16] Revolutions, however needed at critical stages for further advance, are possible only provided certain objective conditions are fulfilled; and they have a chance to succeed only if, besides such objective conditions (which ultimately, but not entirely, depend on the socioeconomic basis), there are influential groups enlightened as to ends and means, whose acts are not utopian but are rooted in the objective conditions, and who are ready to exert their will to the extreme. Inevitability, if any, is the product of carefully planned action.

To summarize, no form of scientific determinism entails fatalism. On the contrary, the assertion that things occur in a lawful way, that is, that they follow definite patterns provided

[16] Childe (1947), *History*, p. 73. Italics mine.

certain conditions are met, is the very prerequisite for avoiding the inevitability of fatalism, both in connection with the mastering of natural forces and in the remodeling of society. Freedom, far from being the negation of determination, is a form of it: it is the victory of lawful self-determination over the external compulsions and constraints fitting, in turn, other laws.

4.4. Is Causality Mechanistic—And Is Mechanics Altogether Causal?

4.4.1. *Mechanics Restricts Causes to Forces*

Further broadcast misconceptions regarding the causal problem are that causality is necessarily mechanistic and that, conversely, mechanism necessarily entails causality. The former opinion is explainable in view of the undeserved oblivion into which scholastic philosophies (which are causal but not mechanistic) have fallen; the latter can be explained in view of the current disregard for Epicurean mechanism, which included an uncaused sort of motion—the atomic swerve, or *clinamen*. The two mistakes have been indulged in by Cournot,[17] Peirce,[18] Bergson,[19] Rey,[20] and many others[21] who have been concerned with the nature and history of mechanistic philosophy.

Causality, in one form or other, can be found in most traditional ontologies, not excluding those which—like Plato's, Aristotle's, or Aquinas's—were furthest from mechanism and emphasized final causation, regarding efficient causation as ultimately dependent upon purpose. Practically all of the

[17] Cournot (1861), *Traité de l'enchaînement des idées fondamentales dans les sciences et dans l'histoire*, secs. 171 ff. According to Cournot, the concept of cause stems from that of force and is tributary to the latter.

[18] Peirce (1892), "The Doctrine of Necessity Examined", in J. Buchler, ed. *Philosophical Writings*, chap. 24.

[19] Bergson (1907), *L'évolution créatrice*, chap. i.

[20] Rey (1923), *La théorie de la physique chez les physiciens contemporains*, 2nd ed., chap. i.

[21] Burtt (1924), *The Metaphysical Foundations of Modern Physical Science*, pp. 88, 307, and *passim*. D'Abro (1939), *The Decline of Mechanism*, chap. vii. A. P. Ushenko, "The Principles of Causality," *Journal of Philosophy* 50, 85 (1953); Ushenko proposes that the term 'causal law' be used in connection with mechanical processes only, and identifies cause with initial state.

scholastics looked upon the causal principle as a *principium per se notum*, a self-evident axiom. This point will be discussed later on; what must now be stated is the peculiarity of mechanics and of mechanistic philosophy with regard to the causal problem, namely, the reduction of *cause* to *force*.

To Galileo, the true causes of physical phenomena were forces, and these should be explained, in the last analysis, by the motions of the atoms—whence the name of mechanical causes, to distinguish them from those invented by the schoolmen. To Descartes, too, force was essentially connected with change of place. It was Newton who enlarged the notion of force, the mechanical representative of cause, to include in it not only the forces *arising from* changes of place, but every agency *producing* a change in the state of motion in a body.[22] In short, what was peculiar to both mechanics and mechanistic philosophy, from Galileo to the Newtonians, was not causality but—to say it with a tautology—*mechanical causality*, as contrasted to the richer but chimerical forms of causation imagined by Aristotle and by his uncountable commentators.

4.4.2. *Self-Movement in Mechanics: Inertia*

Moreover, mechanics is not a *purely* causal discipline—nor is consequently, mechanistic philosophy; at least neither of these has been entirely causal since the 17th century. Both mechanics and mechanism constitute, in fact, a *limitation on causality*, since they restrict causes to whatever produces a change in the velocity of bodies. In another sense, modern mechanics and mechanism transcend causalism, by the fact that they employ a category of determination that makes room for change (of place) even *in the absence of causes* (forces). Indeed, modern classical mechanics, and the few consistent mechanistic philosophies built thereon, assert, in contradistinction to causal

[22] Cartesians and Leibnizians regarded Newton's gravitational attraction as an occult quality, because it was not accounted for in terms of mechanical motion; they said that attraction was not a *mechanical* cause. It was only during the 19th century that gravitation was, so to speak, materialized when (in analogy with electricity and magnetism) it was assigned to a field, that is, to a kind of extended matter. (Not to mention Lesage's 18th-century unsuccessful corpuscular theory of gravitation.)

determinism, the principle of inertia, which is a restricted version of the principle of self-movement. Since this point is capital for the understanding of both causality and mechanism, we shall consider it in some detail.

Contrary to what most immaterialist, and especially romantic, philosophers believe, modern mechanics, whether classical, relativistic, or quantal, does not regard matter as inherently passive, inert stuff; it asserts, instead, that the motion of material systems (relative to a given reference system) *need not be caused,* that is, mechanical motion need not be elicited by agents external to the system itself and summarized in the force concept. The mechanical motion (change of place) of a system endowed with mass is, moreover, never the sole result of the external forces acting upon it; only the alterations in the state of motion or of rest of the system are the effects of such forces —and this, in accordance with Newton's second law, or some of its generalizations. Indeed, according to the first axiom or law of motion of Newton's *Principia,*[23] if a body is left to itself it will not therefore cease moving, as Aristotle had taught and Aquinas had repeated, but it will continue moving until some force (originating, for instance, in collision with an obstacle) causes it to deviate or even to stop. If a new system is considered, such that it includes the sources of the forces previously regarded as external, then it will evolve of itself; from this we infer that, if the future states of this wider isolated system are not entirely determined by its present state, then it is not as isolated as we had thought, but is subject to some external perturbation.

That is, mechanics rejects the scholastic maxim "*Omne quod movetur ab alio movetur*" ("Everything that moves is moved by something else"); it acknowledges, instead, an element of spontaneity, hence of noncausality. Another basic Peripatetic maxim is thereby rejected, namely, "*Causa cessante cessat effectus*" ("The effect ceases with the cessation of the cause").

[23] Newton (1687), *Mathematical Principles of Natural Philosophy,* ed. F. Cajori, p. 13: "Every body continues in its state of rest, or of uniform motion in a right line, unless it is compelled to change that state by forces impressed upon it" (Law I). See also Descartes (1644), *Principia philosophiae,* secs. 37 and 39, where the principle of inertia is stated, and regarded as the first law of nature—but in connection with change in general.

In short, the *principle of the mechanical self-movement of matter*—that is, the principle of inertia—enunciated by Galileo, Descartes, and Newton, is openly *noncausal*, for it states that a certain type of change, the simplest of all, requires no efficient (extrinsic and motive) cause, that is, no force or external compulsion to proceed.[24]

Let us give an illustration of the foregoing discussion—a point misunderstood even by physicists with too generous notions about the scope of causality, as well as by philosophers who retain scholastic ideas about matter and motion. Consider the motion of a billiard ball (or, rather, of its center of mass) during a time interval short enough to permit us to neglect the effects of friction. At an arbitrary initial time the ball moves with velocity v_0 and is struck with a billiard cue, thereby acquiring a new momentum mv, 'm' being the mass of the ball. The total change in momentum will then be $mv - mv_0 = F\Delta t$, where 'F' denotes the intensity of the cause (that is, the force) that produces the change in velocity $v - v_0$, and 'Δt' designates the time interval during which the cause has acted.[25] In other words, the cause F, acting during the small time interval Δt, has produced a change of momentum $\Delta(mv) = mv - mv_0$. After the cause F has ceased acting, the ball continues changing its position (relative to the billiard table) in accordance with the law $d(mv)/dt = 0$, which holds to the extent to which friction forces can be neglected; from this law we deduce (by integration) that the change in position pro-

[24] The explicit recognition of the noncausal nature of inertial motion is rarely found; an exception was d'Alembert (1743), *Traité de dynamique*, part I, chap. i. The fact that Schopenhauer (1813), *Über die vierfache Wurzel des Satzes vom zureichenden Grunde*, sec. 20, regarded the principle of inertia as a corollary of the causal principle shows only that he misunderstood at least one of these principles. Aristotle, on the other hand, was perfectly aware that the ceaseless rectilinear motion of atoms in the void, as postulated by Democritus, contradicted causality; this was just one of the grounds for his rejection of ancient mechanism.

[25] Newton's second law of motion reads $F = d(mv)/dt$. In the *Principia*, ed. Cajori, p. 13, we read: "The change of motion is proportional to the motive force impressed; and is made in the direction of the right line in which that force is impressed". That is, the external forces are not the causes of the momentum mv but of the *change* of momentum $d(mv)$ during the time interval dt. This law is retained in the special theory of relativity; it also holds, though averaged over space, in quantum mechanics (Ehrenfest theorem).

ceeds in accordance with the formula $x = vt + x_0$, x_0 being the initial position and v the velocity acquired by the ball as a consequence of the collision. Thus, the force has produced only the first change; the succeeding positions, after the time Δt has elapsed, were not acquired by the ball as a consequence of new blows with the cue, nor as a consequence of air vortices behind the ball (as an Aristotelian would think); they were the result of previous states *in the absence of causes.* Moreover, the sole causes acting after the blow with the cue were retarding and not accelerating forces, namely, those of friction.

Still, it might be argued that mass, which summarizes mechanical inertia (resistance to change in the state of motion), is the *cause* of inertial motion, so that after all the latter does not violate the causal principle. In fact, the statement is often found in textbooks that "inertia *causes* matter to move along a straight line". A similar argument was employed by Carneades against the Epicurean explanation of free will as the outcome of the irregular and uncaused motion of the atoms (a mechanistic explanation that has recently been revived in connection with the uncertainty principle). In fact, according to Cicero, Carneades held that the Epicureans might have defended freedom of the will without renouncing causality, by asserting that the will acts freely, without a previous external cause, not however uncaused, "because the voluntary motion [of the soul] is by its very nature in our power and is dependent upon us, and not without a cause, for *its cause is its very nature*".[26] This reasoning amounts to regarding qualities as causes; it is, moreover, circular, for it does not explain events but merely states that it is in the nature of things that such events should happen.[27] It is nothing but a verbal attempt to save causality.

[26] Cicero, *Concerning Destiny*, 24. Italics mine.

[27] Descartes was aware that not everything can be explained in causal terms; yet, hesitating to fall into the heresy of limiting causality, which was an essential component of Peripatetism, he introduced the notion of "positive essence of a thing", in a way similar to that of Carneades. See *Réponses aux quatrièmes objections* [of Arnauld] (1641), in *Oeuvres de Descartes*, ed. V. Cousin, vol. II, pp. 65–66: "I think it is necessary to show that, between the *efficient cause* in the proper sense of the word, and *no cause*, there is something as it were in the middle, namely, the *positive essence* of a thing, to which the idea or the concept of efficient cause can be extended."

4.4.3. *Causation in the Laws of Motion of Aristotle, Newton, and Einstein*

Newton's second law of motion embodies the principle of inertia, hence it is far *more* than a strictly causal law: besides defining the cause–effect bond in the realm of mechanics, it asserts the mechanical form of the principle of self-movement. The causal component of Newton's second law is of course very important since, although the law states that changes of *place* may be causeless, it also states that every change of *velocity*, that is, every acceleration, is the effect of a force. (Note, by the way, that the law "Force causes acceleration" involves neither contiguity nor precedence of the cause over the effect.) This double meaning of Newton's second law is made possible by its mathematical form, which is a *second*-order differential equation,

$$F = m\frac{d^2x}{dt^2},$$

having nontrivial solutions ($x \neq$ const., that is, motion) even if the cause vanishes ($F = 0$). On the other hand, Aristotle's law of motion can be framed, in modern terms, as a *first*-order differential equation, namely,

$$F = R\frac{dx}{dt};$$

that is, the force is proportional to the velocity, 'R' denoting the resistance to motion. In the absence of causes ($F = 0$), this equation has only a trivial solution ($x =$ const., that is, rest). Aristotle's law of motion is then an *entirely causal* law, whereas Newton's law has only a *causal range*.

This is why Newton, in contradistinction to Aristotle (and even Kepler), did not look for the cause "pushing" the planets around the sun, but looked instead for the cause that bends their trajectories (see Figs. 10 and 11). Newton performed, in this regard, an analysis of planetary motion paralleling Galileo's celebrated decomposition of the motion of a projectile into the spontaneous (inertial) motion that it would have in the absence of gravity, and the "free" fall that would take place owing to gravity in the absence of an initial velocity.

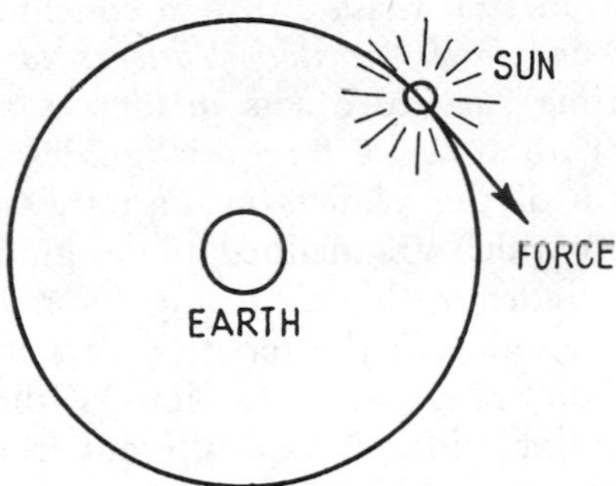

Fig. 10. Pre-Newtonian "systems of the world": a force causes the motion of celestial bodies around the earth (theory of tangential forces).

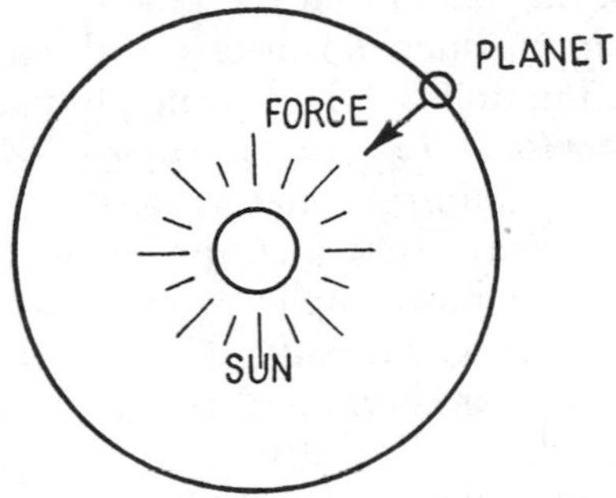

Fig. 11. Newtonian "system of the world": a radial force causes the bending of an otherwise spontaneous motion (theory of central forces).

The situation is rendered even worse for causality in the relativistic theory of gravitation. In fact, according to this theory, what produces the bending of the trajectory of a planet is not a classical force but the space-time curvature (that is, departure from Euclidean geometry) around the sun. Here again, planets move of themselves; but, unlike Newtonian mechanics (in which cause = force), in general relativity *no force* is assumed to take part in the motion of planets. As Weyl has said, "Since Galileo we conceive of the motion of material bodies as a *strife between inertia and force* . . . From the equality of gravitational and inertial mass, Einstein concluded that, *in the duality of inertia and force, gravitation belongs to the side of inertia*".[28] Space-time curvature acts in fact as a condition or constraint rather than as a force: it guides the otherwise *spontaneous* motion of the planets around the sun, in a way similar to that in which an inclined plane guides the rolling down it of a cylinder without causing its motion. General relativity therefore enlarges the territory of self-movement at the expense of the range of causation—without, however, eliminating causation, since forces different from gravity are retained in it as causes.

4.4.4. *Action–Reaction, and Inner Stress*

A second noncausal component of mechanics is the principle of the equality of the action and the reaction (Newton's third axiom), according to which no mechanical action is exerted, so to speak, with impunity.[29] This principle may be regarded as the *mechanical version of the general principle of interaction*, asserting that there are no unidirectional actions (as the causal ones are supposed to be), save to a first approximation. By stating that principle, mechanics implicitly rejects the scholastic notion, correlative of strict causality, of the passivity of material systems, that is, the notion that there are pure "patients" which do not react on the "agents".[30]

Finally, a third noncausal component of mechanics is the concept of inner stress. The mechanical theory of continuous media (fluids and elastic solids) focuses on the inner stresses that develop as a consequence of the combined action of the external forces and of the inner (intermolecular) forces keeping the molecules together. The kernel of the dynamics of continuous media is an equation relating the variations of the

[28] Herman Weyl, "Geometrie und Physik", *Naturwissenschaften 19*, 49 (1931), pp. 50–51. See also Schrödinger (1935), *Science: Theory and Man*, p. 146.

[29] Newton (1687), *Principia*, ed. Cajori, p. 13; for the various meanings of 'action' and 'reaction', see the *scholium* to the laws of motion, pp. 21–28.

[30] Newton's third law of motion has no general validity in relativistic mechanics as far as its exact quantitative formulation is concerned. But relativity does not alter the statement that there are no actions without the corresponding reactions, and this is what matters in our discussion.

components of the stress tensor. This may be regarded as the mechanical version of the basic principle of dialectics, regarding the "contradictory" nature of all concrete existents, which may be interpreted as asserting that every material object, however homogeneous it may look at first sight, is actually inhomogeneous in some respect and to some extent, and is moreover composed of mutually opposed or "conflicting" (that is, mutually disturbing) parts or features, thus being subject to an inner stress that may develop (eventually enhanced by external forces) to the point of producing a radical (qualitative) change in the object concerned. (Needless to say, *ontic* "strife" or, simply, inner opposition does not involve *logical* contradiction. Neither the assumed polar structure of concrete existents nor change is inconsistent with the "principle" of contradiction.[31])

In brief, causality need not be mechanistic, and classical mechanics is not a purely causal discipline although it contains an important causal ingredient, namely, the force concept.[32] By asserting the importance of self-movement, reciprocal action, and inner stress (or "strife", to employ the Heraclitean and Hegelian metaphor), classical mechanics transcends causality and asserts, if only in a restricted way, the germs of the dialectical theory of change—just the germs, since qualitative change, the basic notion of dialectics, is absent from mechanics.

4.5. Summary and Conclusions

Causation cannot be replaced by functional dependence, as romantics and neopositivists have demanded, because the

[31] See Kazimierz Ajdukiewicz, *Deutsche Zeitschrift für Philosophie 4*, 318 (1956), and Adam Schaff, *ibid.*, p. 338.

[32] Moreover, the force concept can be circumvented, in the realm of mechanics, in various ways, though with no obvious advantage. The first attempt to build a phenomenological formulation of mechanics is probably due to d'Alembert, whose very philosophical outlook was largely phenomenalistic. In his influential *Traité de dynamique* (1743), he did not admit a force law as a principle, and stated: "I regard Mechanics as the Science of effects rather than as the Science of causes" (p. xxxviii). A later, forceless formulation of dynamics was that of Hertz (1894). The most advanced formulations of classical analytical mechanics (those of Lagrange, of Hamilton, and of Hamilton and Jacobi) have not *eliminated* the force concept but have derived it from other, more fundamental concepts, such as momentum and energy.

category of interdependence lacks the essential component of genetic, productive connection. If a mathematical function is assigned a causal meaning, this interpretation must be stated in extra statements (semantic rules). Functionalism may do justice to all of the factors or facets concerned, but it fails to account for decisive determinants, that is, factors which in the short or the long run determine the essential characteristics of a process.

An exaggeration of functionalism leads to the organismic view of the block universe, in which there is place neither for chance nor for freedom. But causality does not imply universal causal interconnection. While the latter excludes chance, causality leaves enough holes in the universe to let chance work as an ontological category (coincidence of independent causal lines, or mutually irrelevant processes).

Two further reproaches of romantic philosophers against causality are found to be unjustified, namely, that it is fatalistic and that it is mechanistic. Fatalism is a supernaturalistic doctrine asserting the unconditional (hence lawless) operation of transcendent determiners, whereas causalism is not committed to unscientific teachings, and respects the principle of legality. Moreover, causality affords means for achieving the freedom denied by both fatalism and accidentalism: if nothing is unconditional, then in principle nothing is inevitable, but every cause can be counteracted or at least controlled by another cause.

That causality need not be mechanistic is shown by the mere existence of the Platonic and Aristotelian systems. And an examination of a few basic concepts and ideas of classical mechanics shows that this discipline, while containing an important causal ingredient (summarized in the force concept), goes beyond causality. If the romantics had been better acquainted with Newtonian mechanics, they would not have felt justified in decrying it, but might have found in that science the germs of three categories of determination of which the *Naturphilosophen* had been very fond, namely, self-movement, reciprocal action, and inner "strife" of opposites.[33]

[33] Mario Bunge, "Auge y fracaso de la filosofía de la naturaleza", *Minerva 1*, 212 (1944).

PART III

What Causal Determinism Does Assert

5

The Linearity of Causation

After having tried to defend the doctrine of causality against reproaches coming from the most distant quarters, like empiricism and romanticism, we shall now launch an open attack; the shortcomings of causalism will be critically examined in this, the third part of the book. Our first concern will be to emphasize, and then criticize, the linear character of becoming stated by the doctrine of causality.

5.1. Is Multiple Causation Strictly Causal?

5.1.1. *Simple and Multiple Causation*

The less vague formulations of the causal principle involve either a single cause C and a single effect E, or a (finite) set of causes $C_1, C_2, \ldots, C_n$ and a single effect E, or vice versa. In the former case we speak of *simple causation*, in the latter of *multiple causation*, including in it both plurality of causes and diversity of effects (see Fig. 12). Thus there are many ways of wasting a fortune (plurality of causes), the possession of which may have various alternative consequences (plurality of effects).

To the extent to which functionality parallels causality—a limited extent indeed, as was shown in Sec. 4.1—the functional correlate of simple causation is $E = f(C)$. Or if, as is usual, the intensity of the cause is denoted by the "independent" variable 'x', and the intensity of the effect by 'y', then the functional relation corresponding to simple causation is $y = f(x)$. The functional correlate of the plurality of causes will on the other hand be $y = f(x_1, x_2, \ldots, x_n)$; and the functional correlate of

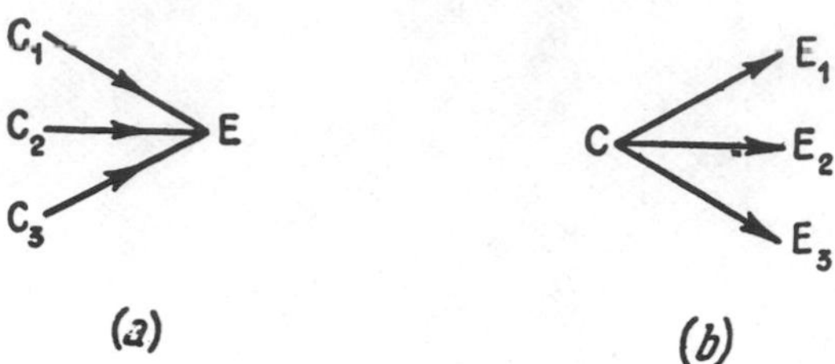

Fig. 12. Symbolic representation of multiple causation: (*a*) plurality of causes; (*b*) diversity of effects.

the diversity of effects will be the set of functions of a single variable $\{y_i = f_i(x)\}$, $i = 1, 2, \ldots, n$.

If there is both a finite set of causes $\{C_i\}$ and a finite set of effects $\{E_i\}$, the relation among the two sets will be causal provided the correspondence among them is one-to-one reciprocal ($C_i \leftrightarrow E_i$), or, if preferred, when $y_i = f_i(x_i)$. But this case falls under the heading of simple causation, since the connection is split into the set of relations $y_1 = f_1(x_1)$, $y_2 = f_2(x_2)$, . . ., $y_n = f_n(x_n)$, representing a bundle of independent, noninterfering, causal lines. If, on the other hand, no such one-to-one reciprocal correspondence exists, if there is a relation between the set of causes and the set of effects, then we may be in the presence of multiple statistical correlation, in which the characteristic linearity of causation is lost, as different variables become intermingled.

The last case to be considered is that of an infinity of factors or variables. But infinite sets of either causes or effects ($n = \infty$ in the foregoing formulas) will not do: they are not empirically verifiable. In fact, the best tests for the authenticity of a causal factor (or, rather, for the adequacy of a hypothesis stating the probable relevancy of a causal factor) are its removal and its deliberate reproduction; and a necessary condition for either test is that there be a finite and moreover a reduced number of relevant determiners alone. It could always be objected that the infinity of connections remaining beyond the reach of experimental control are not really causal, but constitute instead a noncausal remainder. Therefore, in order that a connection be entitled to be called causal, it must involve finite sets of both causes and effects.

We conclude that multiple causation is restricted to connections among a single cause and a finite set of effects, or a finite set of causes and a single effect.

5.1.2. *Conjunctive Plurality of Causes: Reducible to Simple Causation*

Needless to say, when speaking of single causes or effects in connection with either simple or multiple causation, *simple* events are not necessarily meant; the events may be infinitely complex situations, but always such that they behave as units or wholes in the respect that is being investigated. In other words, if a single effect is at stake, it may be produced—and so it is in actual cases—by the joint action of various causes; but whenever the causal connection itself is simple, that just means that the complex of conditions constitutes a single determinative unit, that is, a collection of jointly acting determiners. It may prove convenient in this case to speak of *conjunctive multiple causation*, symbolizing accordingly the total cause in the form $C = C_1 \cdot C_2 \cdots C_n$. The connection of conjunctive multiple causation may then be symbolized $C_1 \cdot C_2 \cdots C_n \rightarrow E$.

A most interesting variety of multiple conjunctive causality is that in which the causal complex can be analyzed into a hierarchical gradation. This procedure was commended by Galileo, who refused to regard all of the factors involved in an authentic causal complex as standing on the same footing: "the true and primary cause of the effects of the same kind must be a single one".[1] When such a gradation of causes is possible, we may speak of the primary cause, the first-order perturbation, the second-order perturbation, and so on. The application of the calculus of perturbations in astronomy and physics rests on the possibility of such a gradation of causal factors. In an ideal case, the numerical values C_k and C_{k+1} of the intensities of two successive terms in such a sequence would stand in a ratio like $|C_{k+1}/C_k| \cong 1/10$, a neat separation among the various

[1] Galileo (1632), *Dialogo sopra i due massimi sistemi del mondo*, 4th day, in *Opere*, vol. VII, p. 444.

factors being thus possible. In particular, the *vera e primaria causa* can then be assigned unambiguously. However, things are not always so easy.

Whether the cause can be analyzed into a neat gradation of factors or not, conjunctive plurality of causes reduces to simple causation, as the various causes $C_1, C_2, \ldots, C_n$ must be there jointly if the effect is to be produced. In other words, conjunctive plurality of causes does not belong to authentic multiple causation but is at most *a variety of simple causation.*

5.1.3. *Disjunctive Plurality of Causes: Genuine Multiple Causation*

Genuine multiple causation does not occur with the joint but with the *alternative* application of causes, that is, when *the effect is produced by each cause alone*, the joint occurrence of two or more causes not altering the effect. Let us call this the *disjunctive plurality of causes*, to distinguish it from the connection involving the conjunction of causes. An obvious illustration of disjunctive plurality of causes is the variety of processes by which heat can be produced; each of them *alone*, whether friction, combustion, or the explosion of a nuclear bomb, produces the same effect—not, of course, with the same intensity. Again, a knife, a bullet, or a poison may each *alone* suffice to cause the death of a man; and their joint application does not alter the effect—death.

The functional relation $y = f(x_1, x_2, \ldots, x_n)$ does not represent faithfully the connection between an effect and its multiple alternative causes. In fact, changes in the "dependent" variable y may be effected either by the joint variation of the "causes" $x_1, x_2, \ldots, x_n$, or by the variation of each "cause" *alone*. That is, the function $y = f(x_1, x_2, \ldots, x_n)$ is not only powerless to convey the productivity and unidirectionality of causation, but is moreover formally inadequate as a representative of multiple causation, since it can be correlated to both conjunctive and disjunctive plurality of causes. A convenient symbolization of disjunctive multiple causation is in terms of the compound 'and/or' connective, thus: $C_1 \vee C_2 \vee \ldots \vee C_n \rightarrow E$. It is just the ambiguity of '$\vee$' which renders it an adequate logical correlate of genuine multiple causation.

Strict partisans of causality, like Thomas Aquinas, as well as critics of it, like Hume, have rejected the plurality of causes. Newton's second Rule of Reasoning in Philosophy reads: "to the same natural effects we must, as far as possible, assign the same natural causes"[2]—which was a polite manner of dispensing with Providence in scientific matters. One of Hume's enunciations of the causal principle contained an explicit reference to the one-to-one reciprocal correspondence characterizing causation: "The same cause always produces the same effect, and the same effect never arises but from the same cause".[3] A follower of Hume has held that the plurality of causes "results only from conceiving the effect vaguely and the cause precisely and widely".[4] Cohen and Nagel have repeated Russell's contention, holding that "when a plurality of causes is asserted for an effect, the *effect* is not analyzed very carefully. Instances which have significant differences are taken to illustrate the *same effect*. These differences escape the untrained eye, although they are noticed by the expert".[5] Granted, this is so in many cases; analysis often succeeds in decomposing a plurality of causes (or a diversity of effects) into bundles of causal threads. But what warrants us that this will *always* occur? It may be argued that "to train the eye" in order to detect the presumed differences may take centuries, so that the assumption of multiple causation is in many cases at least a provisional working hypothesis—but so is simple causation as well.

Multiple causation has been defended, and even taken for granted, by the most diverse thinkers.[6] It hardly needs an elaborate defense—unless it is dogmatically postulated that

[2] Newton (1687), *Principia*, ed. Cajori, book III, p. 398.

[3] Hume (1739, 1740), *A Treatise of Human Nature*, book, I, part III, sec. 15.

[4] Russell (1912), "On the Notion of Cause", in *Mysticism and Logic*, p. 180.

[5] Cohen and Nagel (1934), *An Introduction to Logic and Scientific Method*, p. 270.

[6] Machiavelli (1516), *The Discourses*, book III, chap. xxi. Sánchez (1581), *Que nada se sabe*, pp. 124 ff., employed multiple causation in support of agnosticism. Mill (1843, 1872), *A System of Logic*, book III, chap. x, acknowledged the plurality of causes as possible *in re*, although his own canons of induction rested on the assumption of simple causation. Wisdom (1952), *Foundations of Inference in Natural Science*, p. 94. Ernest H. Hutten, "On Explanation in Psychology and in Physics", *British Journal for the Philosophy of Science*, 7, 73 (1956).

only one-to-one connections are possible. Rather, simple causation is suspected of artificiality on account of its very simplicity. Granted, the assignment of a single cause (or effect) to a set of effects (or causes) may be a superficial, nonilluminating hypothesis. But so is usually the hypothesis of simple causation. Why should we remain satisfied with statements of causation, instead of attempting to go beyond the first simple relation that is found?

The existence of multiple causation is one of the traits of the world that render both phenomenalism and fictionism possible. Think, in fact, of human psychology: one and the same set of circumstances may have different consequences in different persons; conversely, one and the same pattern of human behavior is usually consistent with several different intentions. Consider the following instance of diversity of effects: a fire breaks out (C); Mr. A. behaves in it as a hero (E_1) while Mr. B. runs cowardly away (E_2). The puzzle is not solved by analyzing the causes and effects more closely with the hope of explaining away the plurality of effects, but by investigating further factors, such as the conditions and predispositions of the acting individuals, that is, by disclosing *non*-causal determiners. Thus, while the mere existence of multiple causation is compatible with phenomenalism, both the acknowledgment that it happens to be multiple *causation*, and the requirement of its further *analysis*, are incompatible with phenomenalism. We may accept disjunctive plurality of causes or effects; but, whenever possible, we should attempt to explain such multiple connections in terms of further, eventually noncausal, terms.

5.1.4. *Multiple Causation Is Not Strictly Causal*

Probably the only adequate objection that can be raised against multiple causation is its name, since it is not strictly causal. In the first place, (disjunctive) multiple causation, being by definition a one-to-many connection, fits none of the more or less adequate formulations of the causal principle, all of which rightly assert a unique connection (see Chapter 2). As soon as the possibility of a plurality of causes (or of effects) is admitted, the picture of the causal chain ceases to be a possible model of becoming—or at least loses its meaning after a few branchings. In the second place, (disjunctive) multiple causation is nonadditive: the joint action of causes that might each alone produce the effect does not alter the result. Third, multiple causation goes over into statistical determination as soon as the set of possible causes attains a sufficient degree of complexity and, particularly, if the causes are all of the same kind and order of magnitude of intensity. Thus one and the same equilibrium distribution of a large assembly of molecules can be attained from an infinity of nonequilibrium situations, much in the same way as a stone can reach the floor from an infinity of positions.

To summarize, genuine multiple causation, that is, disjunctive plurality of causes (or effects), is often a more adequate hypothesis than simple causation—only, it is not strictly causal because it is not a unique and additive connection, and because it degenerates into statistical determination if the causes are very numerous and all about equally important.

5.2. Causality Involves Artificial Isolation

5.2.1. *The Universal Chaining*

The modern approach to historical problems began during the Renaissance, when the characteristic procedure of medieval historiography—the chronicling of events regarded as rather disconnected objects—was replaced by naïve causal chains that were supposed to explain historical events in a scientific fashion (see Fig. 13). The new approach was illustrated in masterly fashion by Machiavelli in his writings on the history of Florence. The following is a beautiful sample of the new method: "The city of Florence, having after the year 1494 lost a portion of her dominions, such as Pisa and other places, was obliged to make war upon him who held these places; and as he was powerful, they expended great sums of money without any advantage. These large expenditures necessitated heavy taxes, and these caused infinite complaints from the people; and as the war was conducted by a council composed of ten citizens who were called 'the Ten of the War', the mass of the people

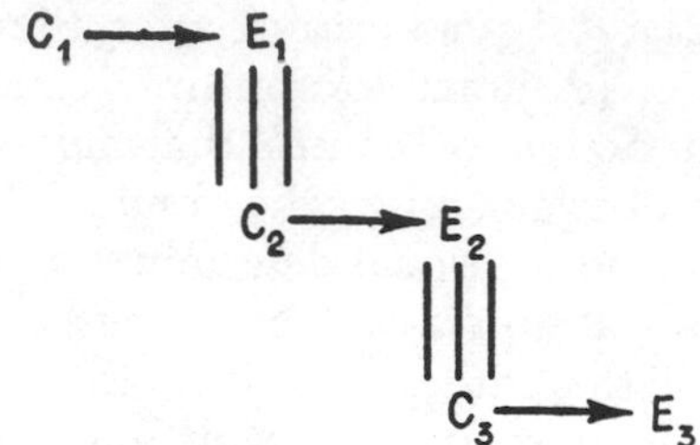

Fig. 13. Causal chain: every effect E_n is also a cause C_{n+1}.

began to hold them in aversion, as being the cause of the war and its expenses, and began to persuade themselves that, if this council were done with, the war could also be ended. Thus when the time came for reappointing the Ten, they allowed their term to expire without renewing the council, and committed their functions to the Signoria".[7]

Machiavelli's causal chain is simple enough: loss of territory—war—large expenditures—heavy taxes—popular dissatisfaction—fall of the war council. This type of description certainly meant an enormous advance over the dry medieval chronicling of separate events; it showed the links between them and was thereby both explanatory and more faithful. Still, it was an oversimplification, since actually every link in the chain is produced by several determiners, of which the *chief* one is singled out and mentioned. Besides, the specific nature of every link in the chain matters as much as the whole chain; and that specific nature, characterized as it is by a peculiar structure and by peculiar laws of self-development, remains outside the causal nexus, which is largely external. Finally, causal chains are not as self-explanatory as is usually believed: what is explanatory is not the sequence of links by itself, but rather the host of hypotheses or even theories hiding behind the words designating the various links.[8]

The extreme oversimplification entailed by the linear series

[7] Machiavelli (1516), *The Discourses*, book I, chap. xxxix, in *The Prince and the Discourses*, p. 216.

[8] See N. R. Hanson, "Causal Chains", *Mind* (N.S.) *64*, 289 (1955).

of causes and effects was dimly realized even during a century that held simplicity as an ideal—though no longer as an indubitable fact, at least in natural and social matters. Voltaire, who was quite an enthusiast of the *enchaînement universel* in the domain of metaphysics, had at the same time too deep an historical insight, and too extensive an historical information, not to perceive that the explanation of historical events in terms of such chains could lead to ridiculous extremes if applied to long stretches. He therefore made fun of the linear series of causes and effects,[9] maintained by the Jansenists while denied by the Jesuits, who were partisans of the arbitrary intervention of Providence in worldly affairs, as well as of absolute free will, which they needed in order to render punishment moral and reward meaningful. However, Voltaire did not explain why it was ridiculous to assume too long causal chains; nor did Diderot, who in *Jacques le fataliste et son maître* (c. 1774) made fun also of both predetermination and free will. Both *philosophes* were contented, in this regard, with showing that they had enough *esprit*.

The main ground why causal chains can at best work as rough approximations for short periods of time is that they assume a fictitious *isolation* of the process in question from the remaining processes. Let us deal with this point in greater detail.

5.2.2. *Isolation: Fictitious*

We saw in Sec. 5.1 that, in order that a process may be regarded as causal, either one causal factor or one of the consequences must be *singled out* of a whole constellation of determiners. Now, such a singling out—and consequently the neglect of the complement of what is singled out—is valid and even indispensable as a methodological procedure, but it is ontologically objectionable. In fact, it is empirically ascertainable that every event is actually produced by a number of factors, or is at least accompanied by numerous other events

[9] See, among others, the delightful "Dialogue entre un Brachman et un Jésuite sur la nécessité et l'enchaînement des choses", 1756, in *Oeuvres*, vol. XVIII, pp. 272 ff.

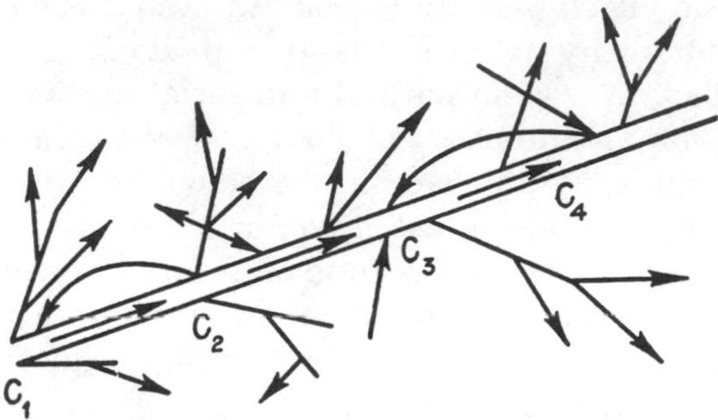

Fig. 14. The causal chain: a one-sided singling out of a rich net of interconnections.

that are somehow connected with it, so that the singling out involved in the picture of the causal chain is an extreme abstraction. Just as ideal objects cannot be isolated from their proper context, material existents exhibit multiple interconnections; therefore the universe is not a heap of things but a system of interacting systems. As a consequence, a particular effect E is not only the product of the *vera e primaria causa* C (see Sec. 5.1.2), but of many other factors linked with C as well; a whole constellation of determiners is always present in real situations. Conversely, C will produce not only E but many other consequences as well. This will be the more so if the so-called remote instead of the proximate causes of an event are taken into account. As Hegel wrote, in the "multiplication of causes which have been entered between it [the fundamental fact that is being considered] and the ultimate effect, it connects itself with other things and circumstances, so that the complete effect is contained not in that first term which is pronounced to be the cause, but only in this plurality of causes together".[10]

To put it in a different way: there are always connections among numerous sets of factors, never connections between single isolable events and qualities, as causality assumes. This being the case, the hypothesis that becoming is a causal chain appears as a one-sided selection out of a rich net of inter-

[10] Hegel (1812, 1816), *Science of Logic*, vol. II, p. 195.

connections (see Fig. 14), and also, to some extent, as a hypostasis of logical inference—hence as a partially anthropomorphic, prescientific conception. Yet, the singling out effected by causal thought, while ontologically defective, is methodologically unavoidable; here, as everywhere else, the mistake consists not in making errors but in ignoring or neglecting them.

5.2.3. *Isolation: A Methodological Requirement*

The isolation of a system from its surroundings, of a thing or process from its context, of a quality from the complex of interdependent qualities to which it belongs—such "abstractions", in short, are indispensable not only for the applicability of causal ideas but for any research, whether empirical or theoretical. Indeed, science has no use for "brute facts", for the "immediately given" imagined by Comte, Bergson, and Husserl; it is the concern of science to analyze such mazes of interconnected elements, singling out a few entities and features, and focusing on them with the hope of attaining a better understanding of the whole after the singled-out parts have finally been replaced in it. Holists complain that this procedure damages the totality concerned, and this is true; but analysis is the sole known method of attaining a rational understanding of the whole: first it is decomposed into artificially isolated elements, then an attempt is made to synthesize the components. The best grasp of reality is not obtained by respecting fact and avoiding fiction but by vexing fact and controlling fiction.

In many cases it is even possible to obtain experimentally such an isolation in an approximate manner and in certain respects, at least in a limited domain of space and for a short time interval: think of the thermos bottle. The possibility of empirically obtaining fairly complete isolations, of varying one factor without provoking substantial alterations in the remaining variables, refutes the doctrine of the universal *unrestricted* interconnection (of all existents in all respects). To realize this is both ontologically and methodologically important; but it is also important to realize that *perfect* isolation is a theoretical fiction. It is by now almost common knowledge that, though often a good approximation, isolation is never complete. Any doubt in this respect should be dispelled upon consultation with an expert in heat engines or in electricity.

Perfect isolation in some respect is not so much a fact as a *hypothesis* that the scientist is forced to make if he wishes to turn the complex interacting systems of the material world into simple, schematic—in short, ideal and therefore tractable—objects.[11] Thus it is likely that the birth of modern dynamics, dependent as it was on astronomy, would have been considerably delayed were it not for the following highly favorable and highly exceptional features of the solar system: (*a*) at the mechanical level the system is fairly isolated from the rest of the galaxy; (*b*) the interactions among the planets are also negligible to a first approximation. The importance of these exceptional traits, which make mechanical isolation nearly complete, can be fully realized if it is recalled that no means of shielding gravitational fields have been found—nor are they to be found if the relativistic theory of gravitation is right in maintaining that the gravitational field in a sense constitutes space instead of filling an otherwise empty spatial frame.

The hypothesis of isolation, or, conversely, of a noninterfering background, is, then, a methodological requirement of the sciences dealing with the material world; hence, the fiction of the isolated "causal chain" will work to the extent to which such an isolation takes place. And this is often the case in definite respects during limited intervals of time. But actually an infinity of neglected factors—Galileo's *cause accidentarie* or *cagioni secondarie*—are constantly impinging upon the main stream—the chosen "causal line"—producing in it small modifications that may accumulate, thus eventually provoking,

[11] Moreover, in some cases systems of actually interacting components can be replaced by corresponding abstract systems of independent components. An elementary illustration of this procedure is afforded by the use of normal coördinates in dynamics; for example, in order to treat a system of coupled springs, special coördinates can be introduced which represent the oscillations of the same number of independent fictitious oscillators, every one of which oscillates undisturbed by the remaining springs. Analogously, the equations of motion of coupled fields, like the electromagnetic and the electron fields, can formally be reduced to sets of equations of fictitious isolated fields (interaction representation).

in the long run, an essential modification.[12] As Bernal puts it, such "chance variations or side reactions are always taking place. These never completely cancel each other out, and there results an accumulation which sooner or later provides a trend in a different direction from that of the original system".[13] (Elementary statistical theory usually treats only the simplest case, namely, that in which chance deviations cancel out, so that no significant change in the general trend of the process is produced, the process ending up in an equilibrium state. The mass of small canceling influences coming from the rest of the universe is of this type; thus, the gravitational disturbances impinging on the earth are randomly distributed and do not produce lasting effects on the earth's orbit, which is stable. Such collections of small influences have been regarded as an "irrational remainder";[14] actually they are not irrational but unknown in detail: although the individual elements are not controllable, the whole mass of the deviations is statistically tractable.)

5.2.4. *Paradoxes of Isolation*

Isolation, an assumption underlying every causal hypothesis, entails the assertion that there are entities that remain *out* of certain causal connections; and this is enough to defeat the doctrine of the universal causal interconnection. That is, theories of universal interdependence cannot be consistently causal; or, if preferred, the more causal they are, the less consistent they are. (In Sec. 4.2 we have shown that they need not be causal.) This is the first paradox of isolation.

A second paradox concerns the usual (and wrong) identification of isolation with causality. Partisans of strict causality have argued not only that nature is to be analyzed into mutually

[12] Galileo (1632), *Dialogo sopra i due massimi sistemi del mondo*, 4th day, in *Opere*, vol. VII, pp. 484–485. Galileo drew a clear distinction between the *vera e primaria causa* (which, like the sun, he assumed to be constant) and the *cause secondarie o accidentarie*; and he pointed out that, in the long run, the latter could effect essential modifications.

[13] Bernal (1942), *The Freedom of Necessity*, p. 31.

[14] Reichenbach (1929), *Ziele und Wege der physikalischen Erkenntnis*, in Geiger and Scheel, ed., *Handbuch der Physik*, vol. IV, p. 70.

independent (isolated) causal lines but also, conversely, that every isolable process is causal, so that causal anomalies can emerge solely as a result of external perturbations, that is, as a consequence of imperfect isolation. This has also been maintained by some propounders of the empirical indeterminacy entailed by the orthodox interpretation on the quantum theory.[15] However, inertial motion goes of itself in complete isolation and in the absence of causes. On the other hand, the disintegration of naturally radioactive nuclei is both a statistical and an isolable process. Granted, the individual disintegration times are not undetermined, since every individual nucleus is supposed to "obey" some laws of physics; but the fact that the average rate of decay is insensitive to all known external circumstances shows the high degree of self-determination of this process, and consequently its noncausal character.

We conclude, then, that partial isolation is a *necessary* condition for causality to hold, yet not a *sufficient* one.

5.2.5. *Causal Chains: A First Approximation*

The picture of linear causal chains is ontologically defective because it singles out a more or less imaginary line of development in a whole concrete stream (see Fig. 15). Such an isolation of a causal series from the whole complex process, in which the singled-out part is regarded as a thread external and parallel to the remaining threads, is but a useful fiction that can be applied only for limited intervals of time, much in the same way as a small arc of a curve can be approximated by the corresponding chord. Soon the initial causal "line" will either be found to branch, as is the case with the divergent development characteristic of biological evolution, or some of the

[15] Dirac (1947), *The Principles of Quantum Mechanics*, 3rd ed., p. 4: "Causality applies only to a system which is left undisturbed. If a system is small, we cannot observe it without producing a serious disturbance and hence we cannot expect to find any causal connexion between the results of our observations. *Causality will still be assumed to apply to undisturbed systems* and the equations which will be set up to describe an undisturbed system will be differential equations expressing a causal connexion between conditions at one time and conditions at a later time." (Italics mine.) For a criticism of the doctrine according to which differential equations mirror causation, see Sec. 3.4.

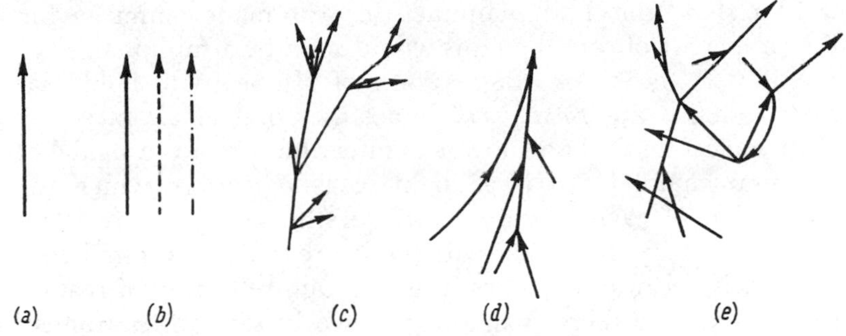

Fig. 15. Main schemes of processes. *Isolated* lines of development (causal or not): (*a*) unilinear; (*b*) multilinear (parallel). *Combined* lines of evolution: (*c*) divergent (branching with emergence of new levels); (*d*) convergent (interpenetration with emergence of new levels); (*e*) real (combination of simpler types of development).

branches will be seen to merge with other, initially independent lines, until the whole ends up—to continue with picture language—in a continuous stream (the convergent development characteristic of cultural evolution). In summary, strict causal lines or chains simply do not exist; but in particular respects, in limited domains, and for short time intervals, they often afford both a satisfactory approximate picture and an adequate explanation of the essential mechanism of becoming.[16]

At least two dangers lie in ambush for those who ignore or forget the largely fictitious character of causal chains. The first is, obviously, Leibnizian monadism; the second is the imitation of a contemporary American historian who, in his enthusiasm for causal chains, undertook to prove that all events that happened in the Roman Empire were nothing but the last links of a long one-way chain originating in the internecine wars of the Mongolian tribes. What is often a methodologically sound hypothesis may, if exaggerated, lead to ridiculous extremes.

Empiricism has rightly pointed out that rigid causal chains exist only in our imagination. A Humean philosopher has

[16] See MacIver (1942), *Social Causation*, chap. vi.

written that "the chain manufacturer who made chains having the properties of causal chains would soon be a millionaire".[17] But it is, I think, an exaggeration of Humeans to hold that causal chains are *nothing but* hypostases of logical inference. (The logical figure known as *sorites* may be taken as a model of a causal chain, for it is a chain of syllogisms in which the conclusion of each is the premise of the next.) There is no ground to deny that, if causal chains are valid in limited contexts, it is because they constitute a rough reflection of reality, that is, because there *is* in the real world something vaguely resembling the causal chain.

At any rate, the Stoic conception of the linear chains of successive causes and effects is both nearer truth and more fruitful than the conception of the world as a heap of separate, isolated events that are self-contained or at most initiate circumscribed processes detached from the rest of the universe. The former picture has a limited domain of validity, whereas the latter renders scientific research pointless. In short, causal chains are rough reconstructions of becoming; what is to be emphasized in this statement, *rough* or *reconstruction*, depends on the specific nature of the case.

5.3. Causality Requires Either a First Cause or Infinite Regress

5.3.1. *The Two Alternatives*

Causalism assumes that whatever can be explained at all is to be explained as the result of a cause or as a link in a causal chain. Now, chains may have a beginning or not. If causal chains are assumed to have an absolute beginning, then a First Cause is *eo ipso* asserted to have been operative—that is, an ultimate source of change is postulated, a First Mover is imagined, which is itself uncaused and eventually (not necessarily) unmoved by its own action—like Aristotle's Unmoved

[17] Toulmin (1953), *The Philosophy of Science*, p. 163. Another Humean has, on the contrary, exaggerated the importance of separate causal lines, to the point of regarding the assertion of their existence as the fundamental postulate of scientific method: Russell (1948), *Human Knowledge: Its Scope and Limits*, part vi. Note, by the way, that such an assumption is an ontological hypothesis.

Mover, or the One of Neoplatonists.[18] The assumption of a *causa prima* is, of course, a theological requirement.[19] It has been moreover contended that this hypothesis is not only supernaturalistic but also self-contradictory.[20] Actually, the assumption of a prime mover entails only a *limitation* on the range of validity of the causal principle, not its negation. It is not self-contradictory but it is certainly inconsistent with the principle of sufficient reason—which is not a law of change but a principle of knowledge (see Sec. 9.3); in fact, this principle demands, in connection with our problem, that a *reason* be afforded for the existence of a First Cause.

In order to refute causality in the name of the hypothesis of the *causa prima*, the so-called secondary (that is, natural) causes have to be deprived of every efficiency. This is actually what Malebranche did; according to his doctrine of occasional causes, God alone has causal efficiency, while natural (secondary) causes are not really causes (producing agents) but rather occasions for the intervention of the First Cause. However, neither science nor religions are very interested in this doctrine, which renders science useless and punishment immoral.

If, on the other hand, no first beginning is admitted, then causality demands the postulate of the infinite regress of causes and effects. In this way the unrestricted validity of the causal principle is saved, and the extrascientific assumption of an uncaused First Mover is avoided. Unfortunately, the *regressus ad infinitum* is ontologically fictitious and gnosiologically inoperative. That the infinite regress is ontologically fictitious follows from the fact that the linear series of causes and effects is an oversimplification, valid as a first approximation, but quite useless after a few branchings and crossings (see Sec. 5.2).

[18] For a description of the Neoplatonic theory of the hierarchical degradation of causes starting from a First Cause, see [Pseudo-] Dionysius the Areopagite, *The Mystical Theology* and *The Celestial Hierarchies*. A description of the influential *Liber de causis*, belonging to the same trend, will be found in Gilson (1952), *La philosophie au moyen âge*, pp. 378 ff.

[19] See Aquinas, *Summa theologiae*, I, Q. 46, art. 2, reply to obj. 7.

[20] Schopenhauer (1813), *Über die vierfache Wurzel des Satzes vom zureichenden Grunde*, sec. 20. As usual, Schopenhauer here offers the painful spectacle of the irrationalist resorting to argumentation—without, of course, succeeding in his attempt.

And the infinite regress is gnosiologically sterile, for, instead of explaining the unknown in terms of the known, it does just the opposite; the device is, in fact, supposed to explain the present in terms of a mostly unknown past.[21] But the main shortcoming of the infinite regress is that it does not enable us to begin at any definite stage in the development, requiring instead a continuous unending recession; it does not make room for definite *stages* in processes, for fresh starts associated with the emergence of qualitatively new modes of being, hence of qualitatively new modes of becoming. It is precisely the existence of levels characterized by laws of their own, and emerging discontinuously in the course of time, that enables us to dispense with the causal *regressus ad infinitum*—though not with every finite regress.

5.3.2. *Evaluation of Infinite Causal Regress*

It should, however, be realized what a tremendous conquest of thought the concept of regressive chain was, what an enormous progress it meant to acquire the habit of receding back from consequents to antecedents, viewing the latter in turn as results of an unending process. This way of thinking had been practiced by a few Greeks and was thereafter practically lost; during the Christian Middle Ages, infinity was regarded as the monopoly of God. Schoolmen usually dealt with problems separately, each "thesis" being discussed as a self-contained unit, by means of syllogisms that were eventually connected with one another but not necessarily chainwise as in *sorites*, and which always ended up in axioms that were regarded as self-evident, hence as indisputable. (The syllogism, rather than the whole logical discourse, was the focus of medieval logic.) What the Middle Ages knew was not so much the chainwise process as the static Neoplatonic Chain of Being, arranged once and for all in agreement with divine law and according to the ethical dignity and power of the links concerned.

It is during the Renaissance that the tendency emerges to

[21] Aristotle, *Metaphysics*, book II, chap. ii, 994a, argued that causes cannot form an infinite chain but must end in a First term, just because infinity is *unthinkable*. This argument is, of course, indefensible, save on a strictly empirical theory of knowledge (such as Locke's or Berkeley's), or on the equally empiricist mathematical finitism of Borel and the intuitionists.

treat every fact in its actual cosmic context. Theoretical interest shifts from isolated facts to series of facts standing in a genetic relation with one another. In correspondence with this transformation in the scientific and ontological approach, the center of the Renaissance logic becomes what Descartes calls the *longues chaînes de raisons*—though often under the disguise of attacks on formal logic. As was recalled above (Sec. 5.2.1), Machiavelli inaugurated the explanation of historical facts, which in the West had been merely chronicled, and explained them as links in long causal chains. (The restriction to the Christian Middle Ages is meant to make room for the astonishing scientific approach of Ibn Khaldun.) The seeming chaos of historical events is then transformed into an orderly whole—perhaps into an oversimplified contrivance in which no room is left for either chance or design. The Stoic picture of the cosmos, in which everything is tightly connected to everything else, is rediscovered. The *regressus ad infinitum* became a valuable ontological instrument for such a conversion of the chaotic maze of appearances into a lawful, hence intelligible and controllable, cosmos; the chain of reasons was its logical instrument.

To sum up: like so many other concepts, the infinite causal regress played a progressive role, until the very wealth of data that it contributed to accumulating showed its inadequacy. There is infinite regress, but not of the causal type: the first member of this compound sentence is a philosophical hypothesis (whereas the assumption of a First Cause is a theological one); the qualification follows from Secs. 5.1 and 5.2—and, indeed, from the whole present book.

5.4. Causality Involves Continuity of Action

5.4.1. *Ground and Consequences of Continuity of Action*

The causal principle involves continuity of action between the cause and the effect—that is, the absence of gaps in causal lines.[22] This requirement is almost obvious, as every dis-

[22] The compatibility of causality and continuity has been challenged by Alfred Landé, "Continuity, a Key to Quantum Mechanics", *Philosophy of Science* 20, 101 (1953).

continuity in the causal chain would have to be assigned to the action of an extra cause. (Continuity of action does not, however, involve spatial contiguity or action by contact; this was shown in Sec. 3.1.)

Another feature of the continuity characteristic of causality is expressed by the maxim *Small causes have small effects*, meaning that a gradual change in the cause has a gradually changed effect. (The contrary proposition, namely, "Small causes have great effects", is, on the other hand, often regarded as describing chance phenomena—but it also applies to instability.) In the beginnings of modern science this condition was found not only ontologically plausible but even methodologically necessary for the application of the differential calculus to the description of natural phenomena—which is so characteristic of modern science. In fact, the simplest differential equation employed to reflect natural processes is $dy/dx = f(x)$; if the function $f(x)$ is not continuous enough, it will not be possible to ensure that the equation has a single continuous solution. In other words, if $f(x)$ has not some property of continuity (for instance, piecewise continuity), the process that the solution $y(x)$ of the equation is supposed to reflect may not go *uniquely* from one state to another, whence it will not satisfy an essential requirement of causality, namely, uniqueness. Continuity and uniqueness are, in short, closely related.

A consequence of the hypothesis of the gradual character of all change—a hypothesis clearly formulated by Leibniz[23]—is that the products of change, too, form a continuous series of intermediate beings. This consequence, of the existence of a *plenum formarum*, was of course noticed by Leibniz, who adopted

[23] Leibniz (1703), *Nouveaux essais*, Introduction: "Nothing takes place all at once, and it is one of my most important and best verified maxims that *nature makes no leaps*. This I called the *law of continuity* when I spoke of it in the first *News from the Republic of Letters* [1687]; and the use of this law in physics is considerable: it means that the passage from the small to the great and back again takes place always through that which is intermediate, both in degrees and in parts, and that a motion never arises immediately from rest, nor is reduced to it except through a smaller motion, just as we never manage to traverse a given line or length without first traversing a shorter line . . . To think otherwise is to have but little knowledge of the immensely subtle composition of things, which always and everywhere include an actual infinity."

the old idea of the continuity of living forms, a hypothesis certainly not warranted by facts.

5.4.2. *An Argument Against the Continuity of Causation*

Let us consider the following argument against the compatibility of causation and continuity. A causal bond will be *inferred* solely provided in a sequence of events there appears in some part of it a neat *cut*, such that a part of the sequence can be called the cause, and the remainder the effect. This is, no doubt, a condition for acquiring a "feeling" of causation, or for inferring that we are concerned with a causal process, not with a mere sequence of associated or concomitant events. The argument looks powerful if causation is regarded as only a category of experience; in this case it may be viewed as destructive either of the causal principle (which, as we saw, requires continuity) or of the empiricist reduction of causation to constant association or to uniform succession. It has in fact been employed in the latter sense, and in the following way: "The dynamical ground of causality, whether subjective or objective, is perturbation, interruption. As long as a sequence of changes is regular or harmonious, the feeling of causation does not appear. But if an accustomed sequence is interrupted, the feeling of causation awakens instantaneously. Hence, there is an *associative* thinking, a thought of mere sequence, and a *causal* thinking, which appears when the sequence is interrupted."[24]

This argument is valid if causation is assigned no place at all outside experience, that is, if it is regarded as but a connection among sensations, thoughts, and so on. On the other hand, if the causal bond is regarded as obtaining whether we experience it or not, and if experience at a given moment is seen as illuminating solely a section of objective reality, then the foregoing argument does not show that causality as such involves discontinuity. Indeed, it might be assumed (as Leibniz did with his theory of imperceptible sensations) that the empirical

[24] Ruckhaber (1910), *Des Daseins und Denkens: Mechanik und Metamechanik*, p. 120.

cause-effect bond (that is, the link of which we may be aware) is resolvable into a factually continuous causal line.

It is, however, true that causalism tends to focus on separate (though connected) events rather than on processes; moreover, it tends to focus on instantaneous events—which are, of course, only a convenient fiction that can be used as a first, rough, approximation to real events. This contradiction between the requirement of continuity, on the one hand, and the habit of causal thinking, of concentrating on separate events, on the other hand, must be faced as a characteristic antinomy of the doctrine of causality, as an unavoidable shortcoming of it.

5.4.3. *Criticism of Hypothesis of Universal Validity of Law of Continuity*

I will not argue against the law of continuity. But against the belief in its *universal* validity—that is, against the doctrine which Peirce[25] called synechism—various objections can be raised, among them, the following:

(*a*) *Instability.* The simplest, most tractable, facts of the real world are those that take place in conditions of static or dynamic equilibrium or stability. This simplicity is the main reason why such a large part of our sciences of nature and of society are concerned with them in spite of the fact that stability has always a limited range, and that the most interesting events are departures from stability or transitions from unstable to stable states. Now in the case of stable states it is true that, whenever causes are really at stake, "small causes have small effects"; this rule may even be used to *define* stability (see Fig. 16 (*a*)). But in the case of unstable states, exactly the opposite is true: a small perturbation may lead the system to a very different state—more exactly, to a whole range of states. Moreover, the states following a state of unstable equilibrium are not entirely determined by the cause of the departure from the first state; the cause only *triggers* a process that it does not entirely control. A typical phenomenon of this kind is the one represented in Fig. 16 (*b*). A more complex illustration of instability is the discontinuous change (known as "jump

[25] Peirce (1892), *Philosophical Writings*, chaps. 25 and 26.

Fig. 16. (*a*) *Stability:* after small departures from the equilibrium position (bottom) the marble returns to it: "Small causes have small effects." (*b*) *Instability:* small departures originate large effects: "Small causes have large effects."

effect") in the amplitude of the oscillations of a nonlinear oscillator when the frequency of the driving force attains a certain value (Fig. 17). Instability is, in short, a source of causal discontinuity in classical physics. As to quantum theory, according to the usual interpretation, one kind of discontinuity at the atomic level arises just from the existence of stationary states; the transitions between stationary (stable) states are discontinuous. Another kind of discontinuity is the sudden reduction of wave packets, an event which, according to the orthodox interpretation, takes place "when the observer takes cognizance of a result of measurement". Nothing warrants, however, that these discontinuities may not be reduced, in the future, to swift continuous processes.[26]

(*b*) *Quantitative discontinuities.* In some cases, if the intensity of the cause is not large enough its effect, far from being proportionally small, is exactly null. This is characteristic of situations in which energy thresholds or minima are involved. If E, C, and T denote the intensities of the effect, the cause, and the threshold, respectively, the cause-effect relation in such cases can be represented in the form $E = f(C - T)$, with the proviso that $E = 0$ for $C < T$. An illustration of this type of law is the all-or-none law of the "firing" of neurons: in this case E is the intensity of the response, and $C - T$ the excess of

[26] For a criticism of the sudden collapse of wave functions upon measurement, see Henry Margenau, *Physical Review* 49, 240 (1936).

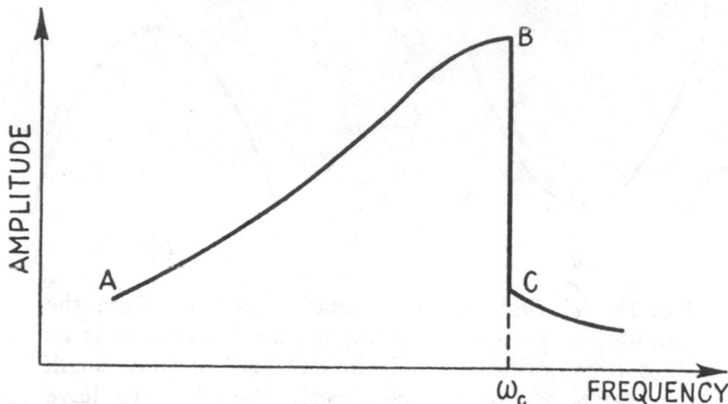

Fig. 17. "Jump effect" in a nonlinear oscillator. When the frequency of the impressed force attains the critical value ω_c, the amplitude of the oscillation jumps from *B* to *C*. [After McLachlan, *Theory of Vibrations* (New York: Dover, 1951), p. 56].

stimulus *C* over threshold *T*. Similar situations are met with in everyday life: think of the minimum force that is required to start the motion of a car, or to open a door. Such a "quantization" is even more conspicuous at the atomic level: it suffices to recall the photoelectric threshold, below which photocells do not respond (see Fig. 18). Again, visual perception, which at first sight looks like a continuous process, has not only a definite threshold but is moreover composed of a discrete string of events; every visual impression lasts for a nonvanishing time interval, so that facts occurring in between cannot be recorded by the eye, still busy as it is with the former impression—which accounts for stroboscopic phenomena. A simpler instance of quantitative discontinuity appears in processes describable by functions of the type $\tan x$, which at certain critical points change sign abruptly on passing through an infinite value (see Fig. 19). The effect not being in this case a continuous function of the cause over the whole range of the variable, the principle of continuity would require us to refrain from using the (infinite) class of functions of this kind in the factual cases, relegating them to the realm of intermediate mathematical

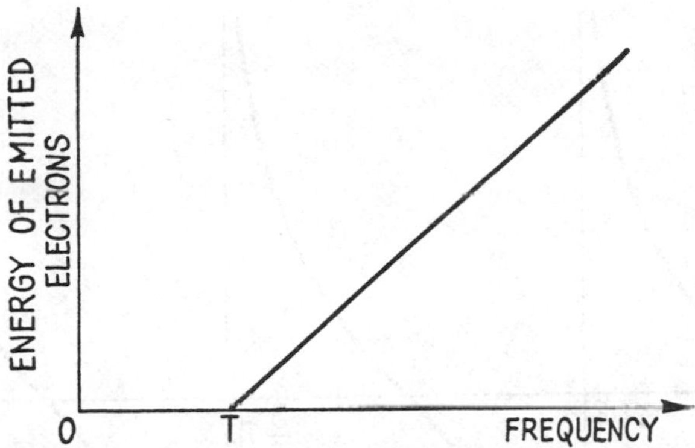

Fig. 18. The discontinuity in the photoelectric effect: below the threshold frequency *T*, the incident light, however intense, is ineffective.

steps. I wonder whether all continuists would be prepared to accept such a mutilation.

(*c*) *Qualitative discontinuities.* There are qualitative gaps among classes of coexistents, and qualitative jumps occur at critical points in processes (this being why they are called 'critical'). An elementary illustration of the latter is afforded by light absorption: a light quantum, in colliding with an atom, may be absorbed by it, hence annihilated as such (first qualitative change); and, if its energy is larger than the atom's ionization potential, an electron will be emitted by the atom (second qualitative change).

If there are qualitative discontinuities, qualitative jumps in processes, there can be, strictly speaking, no continuity of forms, no *plenum formarum.* This is confirmed by the bare fact that classifications of material existents are possible; if there were no abrupt qualitative differences at least in *some* respects, most classification would be impossible, since classifications are based on qualitative differences. A "principle of discontinuity" is inherent in all of our classifications, whether of mathematical, physical, or biological objects.[27] There are types (though not necessarily fixed) in all domains; and such types, by their very

[27] D'Arcy Thompson (1942), *On Growth and Form*, 2nd ed., p. 1094.

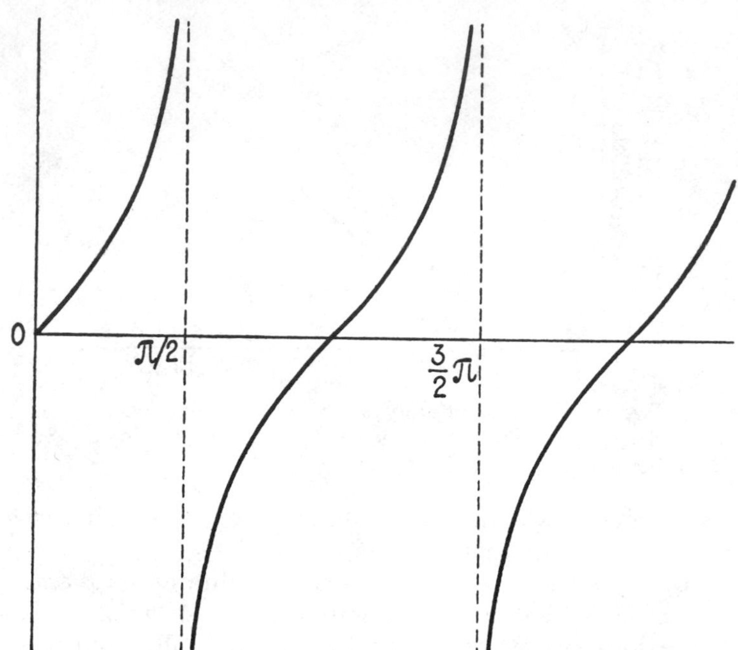

Fig. 19. The function tan *x*. A small change in the cause near any of the critical points would have catastrophic effects.

existence, refute the hypothesis of the strict continuity of forms —which does not invalidate the Darwinian hypothesis that species arise out of varieties. (The existence of ambiguous or borderline cases has nothing to do with it; such exceptions do not show that classification is impossible, as is often argued, but rather that it must be improved.) The theory of the endless small continuous variations has then a limited range of validity. And it cannot be argued that the gaps in the series of forms might well be the holes produced by selection in a primeval continuum; this would require an actual infinity of beings in all realms—for which the least of necessary conditions, namely, room, would be wanting. Paradoxical as it may seem, real evolution is inconsistent with *strict* continuity in all respects, just as it is inconsistent with the discontinuous emergence of newness out of nothing.

5.4.4. *Continuity: A Hypothesis With a Wide but Limited Range of Validity*

The hypothesis of continuity has, in short, a very wide but—like every other known law—limited domain of validity. In both nature and society there are processes involving definite quantitative and qualitative changes that violate the continuity required by causality; there is, besides, and as a result of such jumps, the discontinuity of existents.

However, it does not follow from this that discontinuities are in *all* respects ultimate or irreducible; jumps are to be regarded as real but not necessarily as absolute, that is, independent of contexts and levels. Thus, for instance, we now know that quantum jumps need no longer be viewed as instantaneous transitions, but can be regarded instead (at least according to some interpretations of quantum mechanics) as continuous motions in space and time. Again, the process of evaporation of a liquid is actually discontinuous from a molecular point of view, as molecules and even whole droplets evaporate one at a time; but this atomicity is entirely lost at the molar level, at which the process is continuous.

Similarly, the emergence of new qualities or of whole bundles of new qualities (as in the appearance of new stages of development) certainly constitute leaps in some regards, but not necessarily in all. Continuity in some respects is consistent with discontinuity in others, just as change in some respects is consistent with permanence in others. Continuous variables do not replace but supplement integers. Besides, new qualities do not arise out of nothing but always as a result of a process (often one of sheer quantitative variation), and moreover against a continuous background. That which constitutes a leap at a given level may turn out to be the result of a continuous process at some other level, and vice versa. Both continuity and discontinuity are conspicuous traits of the world, and they are not external to each other but arise one from the other.

In short, the "law" of continuity, while it does not enjoy an unlimited validity, is a most valuable tool in the analysis of becoming—not excluding the explanation of discontinuities at other levels. Faith in continuity helped Faraday in founding the field theory of electromagnetism—but it also prevented him from inferring the atomic nature of electric charge, which was almost displayed in his laws of electrolysis. Continuity and discontinuity are complementary categories; the problem is not to *reduce* one to the other, but to investigate whether continuous processes at one level are made up of jumps at another level and, vice versa, whether jumps are resolved into continuous processes at a different level. The consequence of all this for causality is obvious: the mere fact that the hypothesis of continuity is not universally valid entails a limitation upon the range of validity of the causal principle.

5.5. Summary and Conclusions

Only simple causation (to which multiple conjunctive causation can be reduced) complies with the usual formulations of the causal principle, all of which entail the uniqueness of the causal bond. Multiple disjunctive causation is often a more adequate picture of change, but owing to its ambiguity it is not strictly causal; moreover, when the complex of determinants is complex enough, and when they are all about equally important, multiple causation goes over into statistical determination.

Simple causation involves an artificial isolation or singling out of both factors and trends of evolution; it may reflect the central streamline but not the whole process. Isolation is a simplifying hypothesis rather than a fact. It is indispensable and even approximately valid in many cases; nevertheless, it is never rigorously true.

A consequence of causalism is the need of making a choice among an uncaused First Cause or infinite regress. The former is a theological, the latter a philosophical, fiction. Infinite causal regress has no cognitive value, since the knowledge of the present is thereby made to hang from the whole infinite ignored past. There is regression, but it is neither linear nor, in particular, causal. The emergence of new levels, that is, of definite qualitative discontinuities, dispenses us from tracing back the entire past history in every single case. Causal chains are valid along limited stretches; their validity is sooner or later ruined by branching, convergence, or discontinuity. Continuity is essential to causality, but no more essential to the world than discontinuity, with which it is intimately connected.

In summary, the linearity of causal lines or chains is one of the characteristics of causality that restricts its validity, while on the other hand it tempts us with the paradise of simplicity. Yet, such a linear character of causation is not altogether fictitious; it does work in definite respects and in limited domains. Causal chains arc, in short, a rough model of real becoming.

6

The Unidirectionality of Causation

We have seen (in Sec. 4.1) that the romantics criticized the unidirectionality of the causal bond, as contrasted with the allegedly universal mutual dependence. This was a positive contribution of the romantic critique of causality; unfortunately, the critique was spoiled by its exaggeration and by the unscientific, foggy, and metaphorical way in which it was stated and argued. We shall try to pick out the intelligible and testable kernel of the romantic criticism of the unidirectionality or asymmetry of the causal link. But we shall also show the limitations of the interaction category, which can by no means be regarded as a substitute of the causation category in every instance. That is, we shall point out the limitations of causality with regard to reciprocal action, but we shall likewise show that functionalism, or interactionism, has in turn severe limitations.

6.1. Causality Neglects the Response

6.1.1. *Asymmetry of* Actio *and* Passio: *Essential for Causality*

The causal principle asserts not only a linear but also a one-sided dependence of the effect upon the cause: that is, it reflects only the Cause → Effect way, it accounts for activity but neglects reactivity. This is clearly shown in the following precise definition of physical causation: "From a physical point of view, causality is introduced in the following way: Let S be a system capable of acting upon a second system, S'; and let us consider a given modification of the system S', supposing that a certain modification of the system S corresponds to it; if the modification of S' is entirely defined, in intensity as well as in direction, by the modification of S, then the latter is called the 'cause' of the modification of S'".[1]

The conception of causes alone as active and productive, and of effects as their passive consequences, has its correlate in the peripatetic dichotomy of substances into agents and patients; its logical correlate is the belief that the work of reason is nothing but an extraction of consequences out of premises—as if a critical examination of the theorems were unable to suggest the need for a renewal of the very starting points (axioms). The unidirectionality attributed to the causal nexus is perfectly consistent with the doctrine that identifies causation with unique and uniform succession of states, since obviously a state cannot react back upon a previous, already nonexistent, state. But we have already seen (see Secs. 3.3 and 3.4) that uniquely determined and regular time series may but need not be causal; we pointed out that causes cannot be defined as states, both because the claim of causalism is just that every state is produced by a cause, and because there are processes in which the successive states are not brought about by any causes.

As commonly conceived, and particularly as conceived by the doctrine of efficient causation, natural causes are always changes (events or processes) originating other changes. Now, for this type of object, the unidirectionality of causation is often, yet not always, inadequate. Let us begin by giving a few illustrations of this inadequacy.

6.1.2. *Reciprocal Action in Physics*

The category of reciprocal action, now universally accepted in science, was quite a novelty toward the end of the 18th century; some philosophers of the next and even of our own century went so far as to regard mutual action as ridiculous; this was, for instance, the case with Schopenhauer. Yet the simplest of the fundamental physical theories, namely,

[1] Max Morand, "Sur les fondements logiques de l'axiomatique physique", *Comptes rendus du 2ème Congrès National des Sciences* (Brussels, Académie Royale de Belgique, 1935), p. 177.

dynamics, has at its basis a principle asserting not only the existence of a reaction accompanying every action, but even the quantitative equality of the two (see Sec. 4.4.4). In this, as in other cases, science preceded philosophy.

The Newtonian law of gravity is usually regarded as an illustration of causality—and even as the paradigm of causality. However, the connection between two gravitating masses is typically *non*-causal, for it consists in an interaction, not in a one-sided action. In other words, Newton's law of universal attraction,

$$F = G\frac{m_1 m_2}{r^2},$$

is not a causal law, for it is in general meaningless to assert either that the mass m_1 is the cause of the acceleration of m_2, or vice versa. Every change produced by m_1 on m_2 reacts back on m_1, the consecutive action of which will consequently differ from what it was previously; gravitational attraction is a mutual change, not a unidirectional process. Only if one of the masses is much smaller than the other (for example, a stone as compared with the whole earth), can the greater mass be regarded as the *cause* of the acceleration of the smaller one, and the reaction of the latter's motion on the motion of the larger mass be *quantitatively* neglected. Thus, we explain the fall of human-sized bodies to the floor as due to the attraction of our planet, that is, as a one-sided action, although the falling body in turn modifies the earth's gravitational field. Even though in such cases the quantitative differences (in the accelerations involved) are negligible, as a matter of principle the situation is not altered; in the classical theory of gravitation we are not confronted with causation but with reciprocal action. (And, in general, forces denote one side of interactions rather than actually unidirectional actions; see Sec. 6.1.3).

A similar illustration is provided by the coupling of any charged body with an external electric field; the charge is surrounded by its own field, so that it reacts through its proper field upon any externally applied field and modifies it. If the body happens to be an electron moving in an external macro-

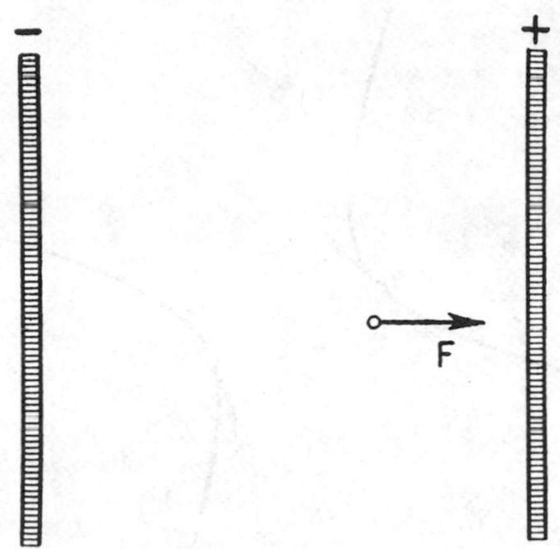

Fig. 20. Causal approximation. An electron is accelerated by the electric field between the plates of a condenser; the reaction of the electron's field upon the cause (external field) is quantitatively negligible.

scopic field, such as the one between the plates of a charged condenser, then the reaction of the electron's field on the external field is quantitatively negligible, and we can describe the latter as the *cause* that modifies the state of motion of the electron (Fig. 20). We apply, in short, what can be termed the causal *approximation*.

Generally, in classical as well as in quantum physics, we are in principle able to solve the equations of motion of a particle in an *external* field of force, in a field that is supposed to be known and that is not appreciably modified by the particle's motion; in other words, we are able to solve problems of interaction in the causal approximation. In some cases this involves no error at all; thus, the dynamical problem of two interacting bodies can, by a suitable choice of the reference system (change from laboratory system to center-of-mass system) be transformed into the fictitious causal problem of a single (fictitious) body moving under the action of an external force; a physical interaction problem is thereby transformed into an ideal causal problem. Moreover, the causal approximation can be applied

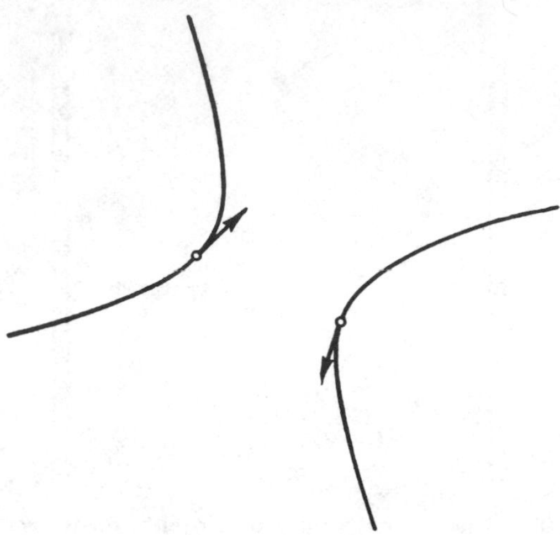

Fig. 21. Irreducible interaction. In repulsion of two neighboring electrons, no unidirectional cause–effect bond can be inferred.

provided the interaction is *pairwise*, that is, if the sum total of the interactions among the various parts of the system concerned can be described with the help of an array of functions H_{ij} of the relevant characteristics of every pair of components (such as the mutual distance r_{ij} between the ith and the jth molecules of a body). On the other hand, if the system consists, for instance, of polarizable charged particles, the principle of pairwise decomposition of the interaction will not apply; in such cases, one-sided cause–effect bonds will be inadequate idealizations of the actual net of mutual actions. The general problem of finding the motion of two or more particles interacting through their fields with noncentral forces cannot be solved exactly; irreducible reciprocal action is a harder bone than causation. Thus, in the case of two neighboring electrons, each of which moves in the field of the other (see Fig. 21), there is no external field, no field common to both particles whose variations are independent of the electron's vicissitudes; the mechanical and the electromagnetic equations are so intimately coupled, in correspondence with the symmetrical character of the interaction, that the causal approximation is not valid in this case.

6.1.3. *Force as One of the Poles of Interaction*

This is perhaps the place to point out that particle dynamics (not, however, the whole of mechanics) treats forces as if they were *external* to the bodies on which they act; more exactly, particle dynamics deals only with external causes of departure from inertial motion. This has not, however, the *ontological* meaning that is usually imagined; it does not mean that mechanics denies self-movement (see Sec. 4.4.2). It has merely a *methodological* value; for, as soon as we enlarge the system in question so as to include in it the bodies producing the forces in question, thereby obtaining a free system, we realize that, *in re*, it is a question of reciprocal actions among different portions of matter rather than of forces acting *ab extrinseco* on a given portion. This point was brilliantly elucidated by Maxwell: "The mutual action between two portions of matter receives different names according to the aspect under which it is studied, and this aspect depends on the extent of the material system which forms the subject of our attention. If we take into account the whole phenomenon of the action between the two portions of matter, we call it Stress . . . But if . . . we confine our attention to one of the portions of matter, we see, as it were, only one side of the transaction—namely, that which affects the portion of matter under our consideration—and we call this aspect of the phenomenon, with reference to its effect, an External Force acting on that portion of matter. The other aspect of the stress is called the Reaction on the other portion of matter."[2]

Physical action and reaction are, then, two aspects of a single phenomenon of reciprocal action. But, in order to treat phenomena with the available conceptual apparatus, it is often convenient to focus on one aspect at a time; then, causation is artificially enhanced at the expense of reciprocal causation—as has so often been remarked by dialecticians.[3] The fact that

[2] Maxwell (1877), *Matter and Motion*, pp. 26–27. See also Hertz (1894), *The Principles of Mechanics*, pp. 184–185.

[3] See Engels (1872, 1882), *Dialectics of Nature*, p. 174.

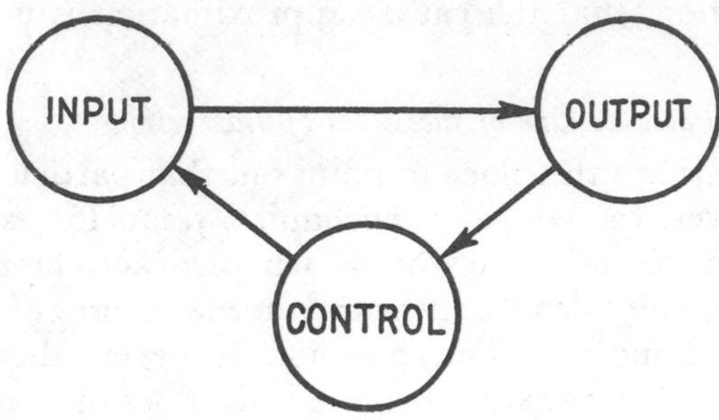

Fig. 22. Feedback loop.

most tractable cases are precisely those in which the causal approximation is valid makes us forget that such cases are quite exceptional, reality being on the whole feebly causal—or at least a lot less unidirectional and linear than most philosophers of causation have imagined. If preferred, reality is a richer causal structure than the one assumed by the theory of causality.

6.1.4. *Causality and Feedback*

The inadequacy of asymmetrical causation, as opposed to reciprocal causation, is being acutely felt in contemporary technology, where the concept of feedback—or rather its conscious use—has become indispensable, at least in the domain of control. As is well known, in devices endowed with negative feedback, such as automatic heating systems, a part of the output (effect) of the system is sent to a control instrument and is then fed back, as a correction "signal", to the motive force or energy source (cause), as a consequence of which the output itself is changed (see Fig. 22).

One of the oldest devices for automatic control is the governor (see Fig. 23) invented by Watt (1788). When the engine runs too fast, the balls B move outward, and by so doing they tend to close the throttle, thus slowing down the speed of the machine. And when the engine runs too slowly, the balls tend to open the throttle. An engine with a governor, or with an equivalent automatic control device, is a self-regulating

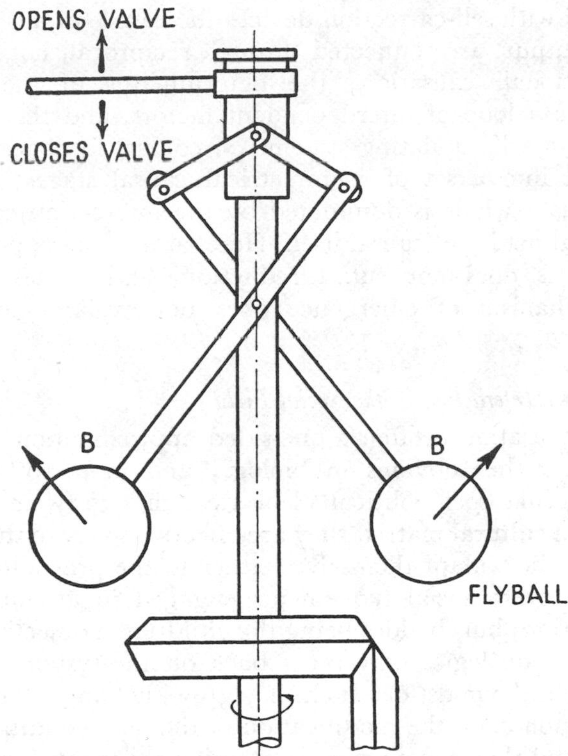

Fig. 23. Watt's flyball "governor": illustration of causal cycle.

system, in which a small part of the input energy is used to compensate for undesirable changes in its functioning, for example, those produced by random variations in the environment.

The process that goes on in automatic machines has been called a *causal cycle*, in contradistinction to linear and one-sided causal series.[4] This designation is correct, for in mechanisms

[4] Mechanisms with feedback have also been called, strangely enough, "teleological" and even "purposeful"—as if the correct proposition "Purposeful behavior involves negative feedback" could be inverted to read "Negative feedback involves purposefulness". See Arturo Rosenblueth, Norbert Wiener, and Julian Bigelow, "Behavior, Purpose and Teleology", *Philosophy of Science*, *10*, 18 (1943).

endowed with self-correction devices the cause (input) and the effect (output) are connected through reciprocal, not through unidirectional, causation; the determinants of the system constitute a loop of interdependent factors, and this is what makes it a self-regulating system. Of course, feedback can be analyzed into a set of asymmetrical causal stages; but the process as a whole is dominated by the interaction category: the causal net is not causal itself. Here, as elsewhere, possibility of analysis does not entail reduction, and explanation of the mechanism of emergence does not explain emergence away.

6.1.5. *Interaction in the Social Field*

That causation is often a one-sided approximation to interaction is rather obvious in biology, and more so in social matters. Take sociopolitical ideologies: they grow in a given social and cultural matrix, they are effective solely to the extent to which they adapt themselves either to the prevailing social structure or to social movements engaged in attempting its modification; but, besides being the children of objective social situations, ideologies may react back on the system of social relations in the midst of which they grow, helping either in the conservation or in the modification of the very conditions that had elicited their emergence. (Orthodox cyberneticians would attempt to frame all this in technological terms, saying perhaps that conservative ideologies are nothing but instances of positive feedback, since they tend to enhance the cause, whereas revolutionary ideologies illustrate negative feedback, as here the effect tends to oppose the cause.)[5]

Despite the almost obvious fact that the various aspects of social life act upon one another, most sociological theories and metahistories assert a one-way action of one of the factors over the remaining ones. Suffice it to recall, in this connection, the following doctrines: geographic determinism, racial determinism, historical idealism, and economic determinism. However, most contemporary social scientists would be prepared to

[5] For a criticism of the metaphors of cybernetics, see Mario Bunge, "Do Computers Think?", *British Journal for the Philosophy of Science*, 7, 139, 212 (1956).

admit that interaction rather than causation is the prevailing category of determination in social matters. Yet nobody, except functionalists, regards society as a muddle of reciprocal actions standing all *on the same footing*. The various social functions are usually acknowledged to rest *ultimately* upon work, upon material production, upon economics, just as the highest functions of the organism are *ultimately* dependent on the intake of food, oxygen, and heat, however much this feeding may in turn be affected (in particular, controlled) by the higher functions. In either case, the dependence upon the material basis is ultimately one-sided in the *majority* of *important* cases; but at crucial points the arrow may be reversed, and the dependence is rarely direct unless the levels are contiguous (as happens, for instance, with the economic and the social levels). Moreover, the dependence is not a strictly causal one, since a host of categories of determination intervene in it; it takes a causal aspect if the inner mechanism is overlooked. Once again, pure causation is a feature of superficial analysis rather than the typical pattern of change.

6.1.6. *Interaction in the Theory of Knowledge*

Typically causal theories are the naturalistic and the sociologistic views of knowledge, according to which *all* human knowledge is merely a *reflection* of the human natural or social environment in the human mind, so that no department of organized knowledge can evolve, at least partly, according to its peculiar, inner "logic"; according to these theories, nature or society are the cause of knowledge, and knowledge is a passive effect. In particular, many sociologists of knowledge, whether they call themselves Marxists or not, tend to regard all sorts of products of spiritual activity, not excluding formal logic, as if they were mere reflections of the relations of material production, and as if the work with feelings and ideas had not laws of its own.

To the materialist trend of empiricism, the inner world of sensations merely reflects the external world, without adding anything of its own, save the eventual distortion of the image. (To subjectivistic empiricism, on the other hand, things *are* experiences, so that there is no division between the inner and the outer world—or, if preferred, everything is in the inner world—and consequently no problem concerning the adaptation of the one to the other arises. At most, the question of the internal coherence between the perceptual and the conceptual strata can arise in this theory; thus, Mach wrote about the progressive "adaptation of thought to fact", and even of the copy [*Nachbildung*] of fact in thought—but his "facts" were just sensations and, in general, experiences, so that the adaptation was an affair of *internal* coherence between two layers of knowledge, the sensible and the conceptual.)[6] The variety of empiricism that grants the autonomous existence of the external world usually holds that man is solely able to *reflect*, in a passive manner, a reality that is given to him and that he "apprehends"—as when we pick up a coin lying on the floor; the human mind can only combine ready-made elements supplied to it by the senses, but is incapable of creating new, ideal (say mathematical or philosophical) objects lacking an empirical counterpart. Ideas are, then, either products or recollections of previous impressions (including those furnished by that vague "internal sense" which Locke and Hume employed as an emergency entrance whenever the empirical counterpart of an idea was not apparent). All ideas stem, in short, from the senses; and only their combinations or, rather, juxtapositions can produce anything different from what is experienced. The subject–object relation is thus reduced by this branch of empiricism to an instance of the Aristotelian dichotomy of patient and agent.

This example of a causal theory of knowledge shows, I think, how intimately connected epistemology and ontology are, and how general ontological hypotheses, like the law of causation, may favor or hinder the clarification of epistemological issues. The doctrine of causation has, in fact, contributed to framing both objective idealism and the reflection theory of knowledge. Conversely, the realization that in the vast majority of cases

[6] Mach (1883), *The Science of Mechanics*, p. 580: "Sensations are not signs of things; but, on the contrary, a thing is a thought-symbol [*Gedankensymbol*] for a compound sensation [*Empfindungskomplex*] of relative fixedness."

there are no one-way roads, owing chiefly to the larger or smaller extent of autonomous activity in every material existent, suggests that there are no one-way roads to knowledge or from knowledge.

In recent times, the question of the relation of ideas to the specific structure of the society in which they arise and spread (or succumb) has been discussed from many angles. This question has at least two sides, a psychological and a sociological one. The psychological (or rather psychosociological) problem concerns the way in which certain ideas are suggested by the prevailing social organization and by the characteristics of material production. The other facet of the problem—sometimes referred to as pragmatics—is the responsiveness or the indifference of a given social organization to a certain idea; this is perhaps the central concern of the sociology of knowledge. Although the systematic approach to the former problem is quite recent, one result of it is certain, namely, that *some* general ideas on the constitution of the world and on the nature of divinities are suggested by the prevailing social organization; for example, the early Mesopotamian and Mexican divinities were organized in the image of the primitive democracy peculiar to the first city-states. But it is not possible to expect a *precise* reflection or an *exact* adequacy of ideas to social facts. Granted, ideas are the product of an active process that gathers much *raw* material from the external world; but ideation has laws of its own, which at the logical level are neither physical nor physiological nor psychological nor social—and, moreover, ideas, when grasped by people, are apt to react back on the soil that nourishes them. Of all the ideas that are produced during a given period by a given social group, the group adopts only a few or even none; this should suffice to show that the alleged "reflection" is at least very distorted.

In the domain of knowledge, it is not a question of unidirectional influence of the environment on the cognitive subject, but rather of a process of interaction and integration. Knowledge of fact, as contrasted with formal knowledge, which is not directly concerned with material objects—except in a psychological sense—is true to the extent to which it is coordinated with the external world.[7] But adequacy (or, better, faithful *reproduction*, which is metaphorically called "reflection") is anything but an engraving on a *tabula rasa*; it is, on the other hand, an outcome of a peculiar activity of the knower, of a process of creation in which the subject places himself in mutual action with his natural and artificial surroundings.

6.1.7. *The Relation of the Category of Causation to That of Interaction*

Essentially three views have been propounded with regard to the relation of reciprocal action to causation: one is the placing of the two categories side by side, the second is the subsumption of interaction under causation, the third is the converse subsumption of causation under reciprocal action. The first view was held by Kant, the second by Russell$_{1948}$, the third by Hegel. We shall not deal at this stage with the view according to which interaction is just entanglement of causal lines,[8] whence causation is the primary category, because this claim of causalism has been examined earlier in this chapter.

Having before him the successful all-explaining theory of universal gravitation, which furnished at last a scientific basis for the old doctrine of universal interdependence, Kant realized that causation does not exhaust determination; he therefore included reciprocal action (under the names of *Gemeinschaft*, community, and *Wechselwirkung*, reciprocal action) in his table of categories,[9] though he retained the scholastic notions of agent and patient. Reciprocal action was also one of

[7] According to the materialist theory of knowledge, factual knowledge is *true* to the extent to which it *reflects* the piece of the external (natural or social) world to which it refers. The same criterion of truth obviously could not apply to abstract (ideal) objects, such as those of logic, mathematics, and philosophy, despite the claims of vulgar materialists. The word 'reflection', as employed in the materialist theory of knowledge, should solely mean that the existence of the world is prior in time to the existence of ideas on the world. But it is definitely a misnomer as long as it suggests the Lockean notion that knowledge is a passive reflection, an engraving of the environment on an initially blank tablet. 'Reproduction' should be preferred to 'reflection', since it is a creative process with laws of its own.

[8] See Russell (1948), *Human Knowledge: Its Scope and Limits*, p. 500.

[9] Kant (1781, 1787), *Kritik der reinen Vernunft* (B), p. 106.

the Kantian "analogies from experience", side by side with causation and restricted to the instantaneous mutual action among existents[10]—much in the same spirit as the Stoa had postulated the universal sympathy side by side with the causal chains. Moreover, reciprocal action was to Kant the fundamental law of coexistence, providing a connected picture of the world at a given instant, just as causality provided—to indulge in an anachronism—a motion picture of the universe. As is well known, instantaneous interconnections are now rather outmoded, though not entirely excluded as a possibility; what matters for our present concern is not, however, the unavoidable aging of Kant's ontology, but the fact that he placed the principle of causation and the principle of reciprocal action *side by side* as mutually independent axioms—which, incidentally, illustrates Kant's characteristic petrification of differences.

Hegel, on the other hand, maintained that causation, far from being external to interaction, is nothing but a particular case of it: he described mutual action as *gegenseitige Kausalität* (reciprocal causation); moreover, interaction was regarded by Hegel as a stage in a process.[11] According to Hegel, then, cause and effect are but the two poles of the interaction category, which "realizes the causal relation in its complete development".[12] Besides, in Hegel's system of objective idealism, the category of interaction enjoyed an ontological status, whereas Kant had treated it, alongside the remaining categories, as a purely epistemological element, and even as prior to experience.

The materialist disciples of Hegel retained and developed his teaching on the interaction category, holding that the cause–effect connection should not be conceived as an irreconcilable antithesis but rather as a stage (or *Moment*, in the Hegelian jargon) in a many-sided and changing connection; cause and effect "become confounded when we contemplate that universal action and reaction in which causes and effects are eternally changing places, so that what is effect here and now will be

[10] Kant (1781, 1787), *Kritik der reinen Vernunft* (B), p. 256.
[11] Hegel (1812, 1816), *Science of Logic*, vol. II, pp. 203 ff.
[12] Hegel (1817), *Encyklopädie der philosophischen Wissenschaften*, "Logik", sec. 156.

cause there and then, and vice versa".[13] Although Marx and Engels regarded causation as "a hollow abstraction", holding instead that interaction is the dominant category, they did not reject the notion of causal nexus altogether but recognized its working at exceptional points, "during crises".[14]

Among dialecticians there is a definite tendency to subsume causation under reciprocal action; however, neither Hegel nor his materialist disciples indulged in functionalism in the working out of special theories; in practice, if not always in philosophic theory, they did not *replace* causation by reciprocal action but recognized and even emphasized the existence of decisive factors or primary agents, such as mind in Hegel's system and matter in the theories of his materialist followers, from Feuerbach to dialectical materialists. Let us take a closer look at functionalism, a preliminary criticism of which was advanced in Sec. 4.1.

6.1.8. *Exaggerations of Interactionism*

Let us agree to call interactionism, or functionalism, the view according to which causes and effects must be treated on the same footing, in a symmetrical way excluding both predominant aspects and definitely genetic, hence irreversible, connections. While this view is legitimate in many cases (notably in dynamics), it has no general validity, and its universalization may be regarded as a hasty extrapolation of the mechanical principle of the equality of the action and the reaction. Interactionism certainly supplies a faithful account of facts in simple cases like static gravitational interaction, where the one-sided picture of causation utterly fails (see Sec. 6.1.2). But interactionism fails as a universal theory, if only for the simple reason that material objects are in a state of flux, so that generally the action has over the reaction the definite "advantage"—to use an anthropomorphic expression—of priority in time.

[13] Engels (1878), *Anti-Dühring*, p. 36. See also Letter to Mehring, July 14, 1893, in Marx and Engels, *Correspondence*, pp. 512 and 517.

[14] Engels, Letter to C. Schmidt, Oct. 27, 1890, in Marx and Engels, *Correspondence*, pp. 477 ff.

The systematic replacement of genetic connections by functional interdependence (in which the interacting factors are regarded as equivalent aspects of one and the same thing) may lead to a muddle, as illustrated by most meson theories of nuclear forces. These theories aim, among other things, at accounting for the transformation of some particles into particles of a different kind, such as the conversion of pions π into muons μ with the emission of neutrinos ν according to the scheme $\pi \rightarrow \mu + \nu$. In this case, the parent particle (π) is unstable; it decays spontaneously (that is, without any known extrinsic cause, though presumably as a result of an inner process) with a lifetime of about two-hundredths of a microsecond. This is an irreversible, typically genetic process, the thing furthest from an interaction—despite which meson theories usually treat this process as if it were a mutual action among coexistents. More exactly, the parent–child connection existing between the pion and its descendents is described as an interaction eliciting that very transition—despite the fact that the products are not yet born.[15]

If functionalism is pushed far enough it can even lead to a teleological view of physical reality. Indeed, if, as usual, only ordinary (retarded) forces are employed in electrodynamics, then the mechanical law of action and reaction is violated in a certain sense; in fact, the action of charge 1 on charge 2 is followed by the reaction of charge 2 on a charge 1 that has already passed into a different state, since charge 1 will have changed during the time interval between the action and the reaction; hence, the reaction will not equal the action. But the chain of action and reaction can be completed so as to satisfy functionalism, if *advanced* forces are introduced alongside retarded ones.[16] Such advanced forces behave in a "premonitory" way; moreover, they represent the action of a still unborn future on the present. To avoid relapsing into medieval,

[15] To make things worse, this alleged "interaction" is handled as a small perturbation of the disintegration products on the parent particle, which no longer exists. Faith in calculation for calculation's sake is so deep that there are people who wonder why these theories do not work.

[16] J. A. Wheeler and R. P. Feynman, "Classical Electrodynamics in Terms of Direct Interparticle Action", *Reviews of Modern Physics 21*, 425 (1949).

teleological science—a return that, needless to say, no experiment forces upon us—we have to relinquish the functional view of the complete exchangeability of cause and effect; we must reject, in short, the typically romantic notion that the distinction between them is superfluous[17] or pointless[18]—unless we are willing to fall into the state of the man who, upon remembering his burial, hoped to be born.

But where interactionism can be and has been more misleading is in the field of the social sciences, and particularly in the sociology of culture. Here again we find the soothing functionalist theory according to which it is pointless to seek for ultimate or long-run determiners, since everything is resolved into a net of interactions. It suffices to recall, in this connection, the fluctuations of the French materialists of the 18th century, who could not decide whether ideas mean nothing as compared with the milieu that gives them life, or whether the social environment is produced by opinion.

6.1.9. *Does Dialectics Require the Subsumption of Causation Under Interaction?*

The comfortable and uncompromising doctrine of interactionism, often hailed as dialectical, can be no less confusing and sterile than the view that regards every determination as the result of the one-sided action of an agent on a patient. It is certainly important to discover the mutual actions among the components of a system, to account for the reactivity as well as for the activity; but the disclosure of interactions need not always *exhaust* the problem of determination, unless an extreme symmetry is at stake. It is also important to realize that:

(i) *Not every interaction is dialectical*, unless it is agreed to attach the label 'dialectical' to *every* sort of change, whether qualitative or not—which would render the word 'dialectics' inoperative. Not every interaction produces a qualitative change; only the "interpenetration" type of interaction, characterized

[17] Mach (1905), *Erkenntnis und Irrtum*, chap. xvi.

[18] Wheeler and Feynman, reference 16, p. 428. For a criticism of teleological ideas in quantum electrodynamics, see Mario Bunge, "The Philosophy of the Space-Time Approach to the Quantum Theory", *Methodos* 7, 295 (1955).

by the mutually determined *qualitative* modifications in *heterogeneous* interacting systems, should properly be described as dialectical[19]—whence dialectics cannot be regarded as an exhaustive theory of change.

(ii) The interplay of a *large* number of entities, whether similar or not, may result in a system having qualities of its own, which do not characterize the individual components. This may happen whether there is a direct interaction among the individual components or not. The growth of heavenly bodies through accretion of small particles, determined by gravitational interaction, illustrates the first type of emergence of wholes; an example of the formation of totalities in the *absence* of significant interaction among the parts is afforded by the statistical aggregate of a large number of free, or nearly free, molecules constituting a mass of gas; in this case interplay within a common environment, rather than reciprocal action, is what determines the new (molar) level of qualities that emerge out of the microscopic level.

(iii) Decisive changes are produced not only as a result of the "interpenetration" of heterogenous entities and as a consequence of the interaction or the interplay of numerous homogeneous entities; qualitative changes may also be brought forth by the definite *predominance* of one of the concerned determiners, even if such a predominance is not constant but obtains in the long run, despite relatively infrequent reversals of the dependence bond.

In short, interaction is not a universal substitute for causation —nor is causation to take the place of reciprocal action; the relation between them should be investigated in every particular case, instead of deciding once and for all that one of these levels of determination is the fundamental one.

6.2. Causality Involves the Superposition of Causes

6.2.1. *Summative Character of Causes: Necessary for Causalism*

The neglect of interaction, as well as the isolation or singling out that characterizes causal chains, results in the hypothesis

[19] The contrary thesis has been defended by Edward G. Ballard, "On the Nature and Use of Dialectic", *Philosophy of Science* 22, 205 (1955).

of the "summative" character of causes and effects. The hypothesis of the summation or superposition of causes holds that the factors that make up an effect act *independently* of one another; that is, even if they act jointly they behave as an aggregate and not as a combination, synthesis, or whole having qualitities of its own that had been absent from the separate addends. According to causalism, whether mechanistic or not, causes may certainly add or subtract, as when two oppositely directed forces are applied to a body; they may interfere constructively or destructively, as in the case of waves. But the association of causes will not constitute a novelty in itself. In general, such a conjunction will produce a quantitative change only, that is, an increase or decrease of certain effects, not however the emergence of new traits; this is the hypothesis of the summative character of causes.

The additive character of causation is involved in the very notion of the *causal analysis* of a given situation. In fact, such an analysis consists in the decomposition of a complex determinant into a sum of causes, or in the separation of a given result into the separate contributions of the individual components or aspects of a system or process. (The 17th century called this the analysis of the given "idea" into "simple ideas".)[20] Causal analysis is analogous to mechanical analysis but not identical with it; the former decomposes determiners into causes, the latter decomposes things into mechanical units such as simple machines.

That the hypothesis of the independence or superposition of causes and effects is vital for causality can be seen not only from the fact that it follows logically from the isolability of causal lines (see Sec. 5.2) and from the neglect of reciprocal action (see Sec. 6.1), but also from the fact that the nonadditive joint action of different causes—that is, the production of "causal" syntheses or wholes—might well end up in noncausal entities. The synthesis or integration of causal factors is increasingly being recognized. A fashionable illustration is the perception or

[20] Enriques (1941), *Causalité et déterminisme dans la philosophie et l'histoire des sciences*, pp. 17–18.

recognition of perceptual universals, such as squareness or sweetness; the external configuration or Gestalt of perceptible entities seems to be perceived as an integrated whole rather than as an aggregate or sum of elementary perceptions. However, the nonadditive composition of causes is not an exclusive property of the higher levels, although it seems more conspicuous in them—or at least more clearly discernible—than in the inorganic levels. As we shall see in the following, the so-called "organic" type of connection occurs in physics, and the nonintegrative and yet noncausal formation of wholes out of large numbers of nearly independent entities is typical of statistical determinacy. In either case the result, or effect, is qualitatively different from the components.

6.2.2. *Nonlinearity as Illustration of Nonadditive Connection*

"Organic", or mutually integrative, connections can be found in all sectors of material reality. This fact has been used in support of organismic ontologies and intuitionistic epistemologies; it is, however, the common property of all ontologies that do not restrict themselves to causal determinism, mechanistic reductionism, or both.

In physics, nonlinearity illustrates nonadditive connectivity. Nonlinear systems do not "obey" the "principle" (theorem) of superposition (of forces, displacements, and so forth), a law that plays a central role in most physical theories, such as mechanics, optics, electromagnetic theory, and quantum mechanics.[21] The "principle" of superposition may be regarded as the specific form taken in physics by the hypothesis of the independence of causes. If applied to forces (mechanical causes) it states that the joint effect of a number of forces applied together equals the (vector) sum of their effects regarded as acting separately from one another; in connection

[21] The usual criterion of nonlinearity is the nonlinear character of the differential equations that express the behavior of the system. A property of linear differential equations is that, if y_1 and y_2 are solutions, then $y_1 + y_2$ (and, in general, $c_1y_1 + c_2y_2$) is also a solution of the same equation; this superposition does not hold for nonlinear equations, in which the "dependent" variable is either multiplied by itself a number of times, or by some of its derivatives.

with vibrating systems, the hypothesis states that any given oscillation can be decomposed into a sum (or an integral) of separate elementary oscillations.

Whenever the "principle" of superposition is abandoned, the components concerned (oscillations, forces, and so on) are no longer regarded as acting independently of one another; they are instead allowed to *interact*, producing phenomena that differ widely from those happening in linear systems, such as instability of certain states, discontinuity of certain changes, "occupation" of certain modes of vibration, and so forth. Therefore, only nonlinear field theories account for the reciprocal actions of particles when the latter are regarded as condensations or singularities of fields; this is why the relativistic theory of gravitation, a nonlinear field theory, does not need to postulate any extra equation of motion for the gravitating masses besides the field equations, which contain the description of the motion of the particles.

The fact that nonlinear theories are rare is not so much a peculiarity of nature as a sign of the infancy of our science. Nonlinearity involves large mathematical difficulties; besides being mathematically clumsy, it affects the very symbolic representation of physical entities. Thus forces that add nonlinearly (as gravitational forces do) cannot be exactly represented by vectors, since the addition of the latter conforms to the superposition "principle". From the moment it was discovered that the laws of ferromagnetism are nonlinear, it has been more and more clearly suspected that *all* physical phenomena may turn out to be at least weakly nonlinear, linearity being only an approximation which is excellent in some cases but only rough in others.

Since nonlinearity entails noncausality, we see once again that causality is a first approximation, that it is, so to speak, a linear approximation to determinism.[22]

[22] Bridgman (1927), *The Logic of Modern Physics*, pp. 88 and 174, saw clearly that nonlinearity entails a failure of causality, since the effects can no longer be decomposed into a sum of partial effects every one of which can be traced to individual events; but Bridgman did not conclude that causality is an approximation of the same kind as linearity.

6.2.3. *Randomness as Further Illustration of Nonadditivity of Causal Factors*

Equilibrium states, such as are studied by statics, are the result of the canceling out of opposite forces; likewise, dynamical equilibrium, as studied by dynamics, is a state of motion resulting from the "harmonious" (nondestructive) interconnection of opposite trends, in which none of them prevails overwhelmingly over the others. In either case, nothing radically new arises from the joint action of several homogeneous entities; mechanical behavior results here from equally mechanical conditions. Statistical equilibrium, on the other hand, is characterized by radically new qualities, although it may emerge from the joint action of independent homogeneous factors. This is why, from a molar point of view, it is uninteresting to trace the macroscopic states of a mass of gas to the individual trajectories of the molecules. Not only would it be practically impossible to do so (at least with the available means) owing to the enormous complexity of the microscopic situations and to the experimental impossibility of distinguishing the separate molecules without in addition introducing significant disturbances in their states of motion; it would, moreover, be quite *useless* to perform such a causal analysis, since the thermodynamic properties of such a molar body—its temperature, specific heat, entropy, and so forth—arise out of its atomic properties but stand on a different level of their own, being meaningless in connection with the individual components of the large system.

An even more striking result of the nonadditive interplay of nearly independent processes is the distribution of the results of observation or measurement of a given property. A series of independent measurements of the intensity of a given quality constitutes a whole rather than a mere heap of altogether disconnected elements; indeed, some of the properties of every one of them are determined, at least partly, by its membership in the statistical ensemble. Thus, the probability of every individual error of measurement in a given series will depend on statistical, collective properties (such as the mean value, the scatter around it, and so on). The over-all and long-run behavior of large collections of whatever kind is studied by statistics; in this science, which is a real propaedeutic to the so-called empirical sciences, the large number of intervening entities is not an obstacle to successful analysis but is, on the contrary, a condition of success, since averages, scatters, correlation coefficients, and other collective properties are the more stable the larger the number of constituents, and because the usual concept of probability can be applied provided a certain stability (of the frequency ratios) exists, a stability requiring, among other things, a large number of entities, so that a canceling out of small individual deviations can occur.

The case of statistics shows, I believe, that, contrary to what organismic philosophers claim, totalities or wholes can arise in the *absence* of systematic integration of the individual components: that is, wholes can be formed not only on the basis of a strong coupling among the parts of a system, but also by the mere existence of an environment common to mechanical or nonmechanical elements behaving almost independently of one another, that is, in such a way that there are no constant relations between any two elements in the collection. In other words, a definite regular collective behavior may arise from an aggregate of individual random (that is, mutually irrelevant) motions.

6.3. Summary and Conclusions

A severe shortcoming of the strict doctrine of causality is that it disregards the fact that all known actions are accompanied or followed by reactions, that is, that the effect always reacts back on the input unless the latter has ceased to exist. However, an examination of real processes suggests that there are often *predominantly* (though not exclusively) unidirectional $C \rightarrow E$ actions. Causality may be a good approximation in cases of extreme asymmetry of the cause and the effect, that is, when there is a strong dependence of the effect upon the cause, with a negligible reaction of the output upon the input—and, of course, whenever the cause has ceased existing.

In other words, the polarization of interaction into cause and effect, and the correlative polarization of interacting objects into agents and patients, is ontologically inadequate; but it is often a hypothesis leading to adequate approximations and, more often than not, it is the sole practical course that can be taken in many cases, owing to paucity of information and theoretical instruments; hence, it is methodologically justified in many cases. But such a success is not justified in itself; if it is successful, it must be because it is rooted in the nature of things, because in reality most reciprocal actions are not symmetrical.

The frequent asymmetry of interactions, as well as the fact that processes in which the antecedent disappears altogether cannot be described as interactions (although they involve reactions upon different objects), renders interactionism inadequate as a universal doctrine. Causation cannot be regarded as a particular case of interaction, because the latter lacks the essential component of *irreversible* productivity.

Real determination is probably neither wholly causal nor strictly functional. However, in some cases determination can be approximately described as causation, and in other cases as interaction—which suggests that sometimes we are in the presence of *predominantly* (but not exclusively) causal processes, whereas at other times we are confronted with *predominantly* (but not exclusively) functional dependencies. It is likely that in most events both causation and interaction take part, in combination with other determination categories.

In general, the interaction among individual entities, and sometimes even the mere interplay in the presence of a common environment, may produce supraindividual entities having qualities of their own and behaving, at least in some respects, as self-contained units. Hence the independence of causal factors, assumed by causality, does not entail the independence of the effects. More often than not, independence is a hypothesis necessitated by the paucity of knowledge or by technical difficulties encountered in the application of available methods. In the long run, the hypothesis of the superposition of causes is found insufficient in this or that field of research, as initially independent lines of development may fuse together, producing a trend with important noncausal aspects. This is illustrated by nonlinear systems and by statistical populations; in both cases, what may be termed causal syntheses—in contradistinction to causal aggregates—are at stake.

Yet, the tremendous historical and methodological importance of the hypothesis of the (approximate) independence and hence superposability of causes should be realized. An entirely organismic approach to reality, as the one preached by contemporary holistic philosophies, would be powerless to perform that *dissectio naturae* demanded by Bacon, which gave us and is still giving us modern science and its applications. The more or less explicit recognition of the principle of superposition of determiners, on the other hand, makes an analysis of real situations possible; most of our science involves it. The exteriority of causes, like the remaining defects of the doctrine of causality, should then be criticized from a progressive standpoint, that is, from a point of view which, instead of proclaiming the utter impotence of the analytic method, acknowledges instead that causal analysis is not the sole kind of analysis needed in the scientific treatment of problems of determination.

A constructive critical attitude toward the problem of the superposition of causes should rely on the recognition that the neat separation and isolation of determiners, while not the *last* stage of research, is a very important preliminary stage, whereas the tenet of the unanalyzability of wholes blocks *ab initio* every advancement of knowledge. The hypothesis of superposition is, then, neither an absolute truth nor utter nonsense; like so many simplifying hypotheses of science and philosophy, it is true to a first approximation.

Once more, we conclude that causation does not exhaust determination, but the latter necessarily entails the former as one of its varieties.

7

The Externality of Causation

Efficient causes are, by definition, extrinsic determinants. An important contribution of the criticism of causality performed by the Renaissance philosophy of nature and by the Romantic *Naturphilosophie* was to point out the insufficiency of the externality of causation. The emphasis on the inner springs of change, suggested to both Renaissance and Romantic philosophers by the processes of life and mind, as well as by the hermetic tradition, was, however, both exaggerated and largely speculative; these philosophers overestimated the relative weight of intrinsic determiners at the expense of environmental conditions, and they were not eager to offer some sort of test of their statements.

As in the criticism of the unidirectionality of causation, I shall here attempt to show the profile of the rational kernel of the Renaissance and the Romantic criticisms of the externality of causation. As elsewhere in this book, most illustrations will be drawn from contemporary science, which effects the long-sought synthesis of external and internal determiners, thereby retaining and limiting the rival doctrines of the omnipotence of external factors and of the sufficiency of self-determination.

7.1. Causality: Restricted to Extrinsic Determination

7.1.1. *Efficient Causes: External by Definition*

As understood in modern times, causal determinism asserts the universal operation of efficient causation. Now, by definition, of all kinds of cause, the efficient cause is the motive or active one; it is, moreover, an agent acting on things *ab extrinseco* and one that cannot act on itself. The efficient cause is, in short, an *external* compulsion; hence, an essential mark of (efficient) causation is externality (see Sec. 2.1).

On the other hand, although the term 'internal cause' has occasionally been employed,[1] inward "principles", conditions, drives, and so forth are in modern times usually described as noncausal. Thus Bruno[2] distinguished clearly two contributions to the constitution of things: the *principle* or inner component, and the *cause* or external component, the former being "that which intrinsically concurs to the constitution of the thing and remains in the effect", while the cause is "that which extrinsically concurs to the production of things". Intrinsic determinants—such as inner stresses—might of course be counted among causes; but such a linguistic convention would run counter to the traditional nomenclature and would suggest an extrapolation of causality to cover the whole range of types of determination.

However, it should be warned that, although causal explanations are framed in terms of environmental agents, not every account of being or becoming in terms of the action of the environment need be causal. In order to be causal, an explanation must assign the whole power of origination or production to what is outside the object concerned. The recognition that extension, duration, and mass are not intrinsic but relational properties has therefore nothing to do with causality as long as the very *existence* of the properties is not attributed to the action of the environment. On the other hand, Mach's attempt to regard the mass of every body as the result of the sum total of its dynamical connections with the rest of the universe matches

[1] Thus Duns Scotus granted internal causes but regarded them as imperfect on account of their lack of self-sufficiency. On the other hand, Thomas Aquinas, who ruled internal causes out of inanimate matter, emphasized their importance in the ethical domain, in order to be able to justify sin; see *Summa Theologiae* (1272), part II, 1, q. lxxv: sin is the effect of internal causes (ignorance, sensitive appetite, malice) and external causes (God, the devil, man); "something external can be a cause moving to sin, but not so as to be a sufficient cause thereof: and the will alone is the sufficient completive cause of sin being accomplished".

[2] Bruno (1584), *De la causa, principio e uno,* in *Opere italiane*, ed. Gentile, vol. I, p. 178.

with causalism: indeed, despite Mach's abhorrence of causality, his account of mass is causal, because he assigns the very origin of that property to the environment—and, moreover, in a way reminiscent of Aristotelian dynamics, since the whole weight is put on the environment.[3]

7.1.2. *The Peripatetic Principle* "Omne Quod Movetur ab Alio Movetur"

The slogan of causal determinism in connection with the problem of change is the Peripatetic dictum, "*Omne quod movetur ab alio movetur*"—"Whatever moves is moved by something else".[4] To causalism nothing can move by itself, nothing can change on its own account, but every change evidences the presence of an efficient cause, of an agent acting extrinsically upon the patient. A contemporary Thomist schoolman has put it clearly enough: "The subject of change is never chang*ing*, it is *being* changed. The thing which is undergoing change is a *patient*, a 'possible', and it needs an agent distinct from itself to effect the change."[5] This axiom, vital for theology, is also a capital principle of classically oriented systems of metaphysics (such as N. Hartmann's), as well as of the ontology of common sense. Before atoms, fields, and radioactivity became pieces of common knowledge, even scientists could be found to share the belief that "brute matter" is a homogeneous, unorganized, and quiescent stuff entirely lacking spontaneity[6]—the matter, in short, dreamt by immaterialist philosophers. From the fact that every experiment is an encroachment on matter, they jumped to the Aristotelian conclusion that matter is nothing but the barren receptacle of forms—a belief still held in esteem

[3] The theory of relativity has retained the thesis of the relational nature of mass, as propounded by Mach against Newton, but it has not espoused Mach's *causal* explanation of mass.

[4] See Aristotle, *Physics*, book VII, chap. 1, 214b, 242a; book VIII, chap. 4, 254b; chap. 5, 257. Thomas Aquinas, *Physics*, book VIII, lectures 7 and 10; *Summa contra gentiles*, part III, chap. vii, sec. 1.

[5] McWilliams (1945), *Physics and Philosophy*, p. 27.

[6] Bernard (1865), *Introduction à l'étude de la médecine expérimentale*, p. 123: as contrasted to the spontaneity of living beings, in the domain of inorganic matter "the phenomenon is always the result of the influence that an external physicochemical exciter exerts on the body."

by those quantum theorists who hold that it is the experimenter who produces all atomic-scale phenomena.

The inner processes of the changing object do not count and may not even exist for the doctrine of efficient causality; every change is conceived by this theory as the inevitable result of a cause external to the changing thing. Nothing but God can be self-caused or *causa sui*; everything else is the effect of a *causa transiens*, never the result of a *causa immanens* or of the joint action of external and internal determiners. The whole of nature is therefore a *principium passivum*, not a *causa se movens*; if nature were self-moving, God would be supernumerary.

The consistent elaboration of this doctrine leads to the denial of every sort of natural determination since, applied to the sum total of existents, the thesis of the externality of determination means that only a power external to the whole universe may elicit changes in it. This step was actually taken by Malebranche, who, from the assumed passivity of bodies (Platonic prejudice) and souls (empiricist tenet), conjoined with the Cartesian assumption of their radical heterogeneity, concluded that "there is no causal relation whatever between a body and a spirit. What am I saying? There is none between a spirit and a body. Moreover, there is none between one body and another, none between one spirit and another. No creature, in short, can act upon any other creature with an efficiency of its own."[7]

7.1.3. *Causal Determinism Opposes Self-Movement*

Let us begin by agreeing to call 'spontaneous' *whatever develops from within* a given object, and not that which is lawless, since according to an initial assumption of ours (see Sec. 1.5.1) everything is determined in accordance with law. Now, causal determinism neglects and even denies the inner, spontaneous activity of things, which the German romantics had termed *Selbsttätigkeit* (self-activity). Modern causal determinism is, in a way, more restrictive—but at the same time more realistic—than the Aristotelian doctrine of causes, which,

[7] Malebranche (1688), *Entretiens sur la métaphysique*, IVème Entretien, XI, vol. I, p. 138. See also the 7th dialogue, "On the inefficiency of natural causes, or the impotence of creatures. That we are immediately and directly united but to God."

besides the external causes (efficient and final), recognized two intrinsic causes in every existent and in every change, namely, the material and the formal causes (see Sec. 2.1).

It is, of course, possible to employ the term 'internal cause' in explaining how internal conditions combine with external ones in a given process.[8] But such a discourse will transcend the scope of the modern doctrine of efficient causation, which is usually meant to deal with efficient, motive causes alone—at least since the breakdown of scholasticism.[9] Chance phenomena might also be described as the result of fortuitous or accidental causes, as was common in antiquity; but, again, such a way of speaking is misleading, since it overflows the doctrine of causality, which does not admit fortuitous causes although it may grant the existence of *accidental effects* (brought forth by the conjunction of initially independent causes). When self-movement and the consequent self-determination are taken into account, when events are regarded not solely as results of external circumstances and forces stamping their seal on passive stuff but also as the outcome of inner conditions, then it is not usual and, moreover, it is misleading to speak of causation *tout court*. In such cases the scholastic terms *causa immanens* and *causa sui* are sometimes employed in contrast to what the schoolmen call *causa transiens* (transitive cause). The corresponding determination will be called self-determination (the word *Selbstbestimmung* having actually been used by Hegel).

7.1.4. *The Doctrine of Self-Movement*

The schoolmen held that *nihil est causa sui*—nothing is the cause of itself—with the sole exception of God, who exists *per se*.

[8] For example, S. Lilley, "Cause and Effect in the History of Science", *Centaurus* (Copenhagen) *3*, 58 (1953), a brilliant study on the interplay of internal and external determiners in the history of science. See also R. Bastide, " La causalité externe et la causalité interne dans l'explication sociologique ", *Cahiers Internationaux de Sociologie 21*, 76 (1956).

[9] Hegel (1812, 1816), *Science of Logic*, vol. II, p. 196, protested against the confusion of the "external stimulus", or efficient cause (which cannot produce the "growth of great out of small"), with the "inner spirit" of processes—which may "use" an indefinite number of different external stimuli "to unfold and to manifest itself."

The Renaissance rediscovered the Greek materialistic ideas on the self-movement of matter and the self-sufficiency of nature, which had just vegetated during the Middle Ages. Bruno stressed the ceaseless lifelike inner activity of all material objects; Descartes, more cautious, held that, to exist, substance needs only God's concurrence, being otherwise self-subsistent. Spinoza restated the Brunian thesis with vigor and contributed powerfully to spreading it; to him, substance is not only self-subsistent but also self-existent, self-moving, and self-caused: *substantia est causa sui*; besides extrinsic efficient causation, Spinoza introduced an efficient intrinsic causation,[10] which was to become Leibniz's indwelling force or activity inherent in every monad.

Though seldom explicitly recognized by postromantic philosophers—with the exception of dialectical materialists, Whitehead, and Collingwood—self-movement is by now a solid philosophical acquisition of the sciences. In no department of science are scholastic patients recognized any longer. On the contrary, material objects at all levels of organization are more and more being regarded as entities having an activity of their own, conditioned but not entirely determined by their surroundings. The ancient dialectical thesis is being increasingly—though unwittingly—recognized, that nothing changes exclusively under the pressure of external coaction, but every concrete object participates with its own inner flux in the ceaseless changing of the material universe; or, to put it otherwise, the sole static objects are ideal ones.

Our central concern is not, however, the theory of the inner sources of change; here we have to deal with the limitations that this theory imposes on causal determinism. The main consequence that the theory of self-movement has for causality is that *extrinsic causes are efficient solely to the extent to which they take a grip on the proper nature and inner processes of things.*

Let us consider a few cases in support of this hypothesis.

[10] See Spinoza (c. 1661), *Improvement of the Understanding*, sec. 92. The first definition set forth in the *Ethics* (c. 1666) is that of cause of itself. Schopenhauer held that the notion of *causa sui* is a *contradictio in adjecto*; he would have been right had he maintained that the concept concerned is contradictory to the notion of efficient causation—but this would not have sounded novel.

7.1.5. *External Causes Combine with Inner Conditions*

The process of fire extinction is an almost trivial illustration of our thesis. In order to slow down the combustion rate we must act with the right means; these may consist either in shielding the available combustible, or in adding some substance that decreases the rate of the chain reaction. To employ semi-Aristotelian language, the new "forms" we wish to inhere in the given body (the burning thing) cannot be contingent upon it, but must match with it, otherwise our effort will be fruitless, that is, the causal agent will not have the desired effect.[11]

As an illustration taken from the level of life, consider the invasion of an organism by noxious germs. According to the classical germ theory of disease, every sickness is but the modification of the organism, produced by invading germs, without any active participation of the "patient". On this, typically causal, theory there does not arise in the organism a conflict of opposing entities, the one coming from the outside and the other developing within; nothing new, save a quantitatively abnormal behavior (rise of temperature, loss of weight, and so on) appears as a result of the invasion. This causal theory—which of course meant an enormous advance over the maleficent-spirits theory of disease (which was also causal, but supernaturalistic, whereas the germ theory was materialistic)—the germ theory of disease, I say, had to be corrected when it was discovered that, when the germ invasion first occurs, something *new* emerges in the organism to defend it, something that had been neither in the "agent" nor in the "patient" before the action of the former, namely, antibodies. Thus, according to modern pathology, sick organisms are not patients but active fighters. Moreover, whether sick or healthy, organisms tend to maintain a constant inner condition—Bernard's *fixité du milieu intérieur*, the *homeostasis* of recent physiology and cybernetics—despite the variations of the environment. Organisms are,

[11] See F. Bacon (1620), *Novum Organum*, book II, 7: the transformation of things requires not only a mastering of the efficient and the material causes, but also a knowledge of the Latent Process, of the Latent Configuration, and of the Form (law). In II, 3, as well as in *The Advancement of Learning* (1605), book II, the efficient cause is called *vehiculum formae*, the vehicle of forms (qualities).

to a large extent, self-contained units that, so far as they survive, succeed in escaping, counteracting or transmuting extrinsic, causal determiners. As Hegel said, "whatever has life does not allow the cause to reach its effect, that is, cancels it as cause"[12]; what Hegel could not suspect in his time is that the same transmutation of external causes by inner conditions, the same specificity of the response, is to be found everywhere in reality, not only in the domains of life and spirit.

Mutatis mutandis, much of what is true of infections is also true of affections, whether at the level of sensation or at the level of feeling. With regard to sensations, it is a well-known fact that the responses of the sense organs are specific even if the stimuli are not, that is, there may be a common stimulus, but the effect will depend on the organ that responds. Thus mere mechanical pressure on the eye produces the sensation of light, and an electric current in the tongue produces the sensation of taste. In these cases of disjunctive multiple causation (see Sec. 5.1.3) the type of cause is rather immaterial as compared with the specific nature of the reacting organ. And as regards feelings, love, for instance, is in most cases unexplainable in terms of external determiners alone. The standard (and envious) question, What could she find in him? can only exceptionally be credibly answered in terms of the male's charm or appeal. From this, the conclusion is usually drawn that love is not only an irrational but also an unexplainable affair—a conclusion apparently grounded on the mistaken Aristotelian identification of explanation with *causal* explanation. If love does not stand causal analysis, it may on the other hand be explained, at least in principle, in terms of *her* own predisposition to fall in love with whosoever meets a certain set of requirements—usually a meager one, but at any rate determined by her own state. In this, like many other human affairs, particular circumstances and individuals are essentially vehicles or occasions for the fulfilment of inner processes. The

[12] Hegel (1812, 1816), *Science of Logic*, vol. II, p. 195: "that which acts upon living matter is independently determined, changed, and transmuted by such matter . . . Thus it is inadmissible to say that food is the cause of blood, or certain dishes or cold or damp the cause of fever, and so on".

environment does not act *on* a person, but rather *through* a person.[12a]

7.1.6. *Freedom: Is It Restricted to the Ethical Domain?*

The problem of self-determination, spontaneity, or freedom has traditionally been regarded as an exclusively ethical one. In a few, but important, cases, however, self-determination has been recognized as a property of *all* sorts of existents, not only as a quality of conscious (and preferably wealthy) humans. Thus Chrysippus[13] held that he who pushes a cylinder or a cone imparts to them an initial impulsion but does not confer upon them their own way of rolling, which is not the same for cylinders as it is for cones. Chrysippus went so far as to maintain that the initial impulsion was solely required to *start* the motion of bodies, which moved henceforth in virtue of their proper nature. Unfortunately, it seems that this outstanding Stoic did not develop his valuable remark into a physical theory, but employed it solely as an illustration of his doctrine on the operations of the soul. These operations he regarded, so to speak, as *triggered* but not as *produced* by external stimuli; he believed, in fact, like Plato before him, that the main source of change in the soul was not an external cause but the soul's own inner spontaneity.

Paradoxical as it may seem at first sight, determinism cannot be regarded as complete unless what may be called the *freedom* or spontaneity of *every* concrete (material or cultural) object is taken into account. The refusal to acknowledge it may be traced to the Platonic tenet that only the soul is self-moving, whereas bodies are entirely passive. This belief is vital for idealism, but is inconsistent with modern science, which shows that the activity and spontaneity of the "soul" (that is, of the psychic functions), far from being its privilege, are rooted in the fact that every bit of matter is buzzing with activity. Self-determination, spontaneity, or freedom is what makes things what they essentially are, but is of course never complete. No

[12a] Cf. H. Cantril, A. Ames, Jr., A. H. Hastorf, and W. H. Ittelson, "Psychology and Scientific Research", *Science*, 110 (1949), 517.

[13] See Cicero, *De fato*, 42–43.

concrete object can be *entirely* free,[14] but can only have a certain amount of spontaneity, both because of its intrinsic limitations and because it is actually connected with an infinity of other existents. On the other hand, no ideal object can be free according to any scientific theory of knowledge, since ideal objects are not self-existent but are products of the human brain's functioning; contrary to Hegel's teaching, not spirit but matter is self-contained existence (*Bei-sich-selbst-seyn*).

Freedom, in the general sense in which it is here understood, need not be conscious; and it is not an undetermined remainder, an arbitrary, lawless residue, but consists in the *lawful self-determination* of existents on whatever level of reality.[15]

Various levels of freedom should, however, be distinguished: as many as integrative levels. The highest, but by no means the sole, degrees of freedom are found in humans; here we have, in the first place, freedom of *choice* among externally given alternatives. In the second place, as a higher kind of freedom we find the power of *creating new conditions*, that freedom described by Rinuccini (1479) as *potestas agendi* (power to act), which Vico (1725) referred to as *libertà di fare* (liberty to do). In the third place we have *conscious* freedom of creation, described by Engels (1878) as the recognition of lawfulness and the consequent application of this knowledge of laws to the attainment of control over the environment and over ourselves. In this case, not a lack of dependence but an active and conscious self-assertion is at stake, the control of bondage rather than its impossible absence, the conscious mastering of determination rather than the unawareness of it or the illusory escape from laws.[16]

[14] Complete (hence ideal) freedom can be defined in the way Spinoza did it in the *Ethics*, I, def. VII: "That thing is called free which exists from the necessity of its own nature alone, and is determined to action by itself alone." To Spinoza, freedom is not a *negative* concept; it means freedom *to* rather than freedom *from*, active self-assertion rather than mere lack of bondage. This is the core meaning of 'freedom' adopted in the text.

[15] Needless to say, if this concept of general freedom is adopted, the doctrine according to which moral freedom consists of or can be grounded solely on physical *indeterminacy* deserves no more than a footnote.

[16] However, consciousness of necessity is a *necessary*, but not a sufficient, condition for the attainment of freedom, just as knowledge of physiology is necessary but not sufficient for the relief of pain. See Ayer (1954), *Philosophical Essays*, pp. 277 ff. Engels attributed to Hegel the idea of freedom as the appreciation of

By negating the element of freedom in every existent, causal determinism remains defenseless in the face of those who grant the individual "soul" a certain amount of lawless spontaneity (the free will to sin); it further consecrates the various philosophies of society and of history based on the belief that external circumstances are almighty—a belief to which a special section will be devoted.

7.2. Does Man Make Himself?

7.2.1. *Anthropological Environmentalism*

A characteristic feature of the various applications of causal determinism is the exaggeration of the role of the environment. A celebrated surviving species of the environmentalist genus is the biological doctrine according to which organisms are capable of responding only to external stimuli, and lack the capacity of a spontaneous activity flowing—to employ picture language—from an inner source, that inner "drive" which vitalism has exaggerated to the point of charlatanry. A subspecies of environmentalistic biology is the wrongly called Neo-Lamarckian theory of heredity, which regards the environment as the sole source of genetic change—a theory philosophically as suspect as the orthodox theory of the genetic self-sufficiency of organisms. Just as nativism leads to absolutism, environmentalism leads to biological (and cultural) relativism. In fact, environmentalism tends to regard adjustment as the highest biological (and cultural) value; now, if successful molding by the environment, or exact adaptation to it, is the highest value, then obviously ants are as advanced as men, and modern society is not better than the primitive horde.

Some of the causal anthropological theories, in a legitimate attempt to account for cultural differences, stress the importance of the physical milieu (physiographic determinism); others emphasize both the geographic and the political circumstances (geopolitical determinism); a further group of theories overestimates the action of the neighbors of every social group while neglecting the peculiarities of the group itself (diffusionism). Probably all of them would subscribe to the famous

necessity; however, at least in his *Vorlesungen über die Philosophie der Geschichte* (1837), Hegel retained the popular notion of freedom as lack of bondage.

maxim—typical of the Enlightenment—according to which "man is the product of circumstances". This belief has had and still has an enormous importance in anthropology, psychology, education, history, and sociology. Let us mention one of the most influential applications of anthropological environmentalism to psychology and epistemology, namely, the *tabula rasa* doctrine.

After the Renaissance emphasis on the inner sources of change in all existents, after Bruno had sung the excellencies of the individual and of the *artefice interno*, which "forms matter and figure from within"[17]—after that revival of the doctrine of self-movement, empiricism revived the discredited Peripatetic doctrine of the passivity of the soul. (Empiricists would not have had such a success had it not been for the dogmatic Cartesian insistence on the Platonic notion of innate ideas, with which Descartes meant to struggle against Aristotelian empiricism.) As a consequence of the *tabula rasa* doctrine, which was adopted by nearly all 18th-century *philosophes*—and which survives in classical behaviorism—the inequality of minds had to be attributed exclusively to differences in individual experience, hence partly to differences in nurture or education. Thus Helvetius, in his urgency to refute the inept tenet of the biopsychical origin of social inequality, went so far as to maintain that men are born equal in all respects, being entirely the product of their environment and, particularly, the children of education; even geniuses were to him nothing but the product of the circumstances in which they had been placed, and the whole art of education was reduced to placing young people in a favorable *concours de circonstances*.[18] Along with other progressive thinkers of his age, Helvetius did not fail to draw the social moral from the belief that education makes us what we are, namely, that the happiness of mankind is essentially a question of improving education. Needless to say, historical events have not confirmed this optimistic doctrine

[17] Bruno (1584), *De la causa, principio e uno*, in *Opere italiane*, ed. Gentile, vol. I, p. 180.

[18] Helvetius (1758), *De l'esprit*, in *Oeuvres complètes*, disc. III, chap. xxx, pp. 379–380 and *passim*.

based on the empiricist theory of knowledge; rather, reforms in education came as results of social revolutions.

It is quite a recent discovery that—to employ the beautiful title of one of Gordon Childe's books—man makes himself. This discovery had of course precursors in antiquity; it was rediscovered during the Renaissance and was again clearly formulated by Giambattista Vico (1725), who added that this is why we can know the "civil world" (society)—because *we* make it. But this doctrine has definitely spread only with historical materialism and the various metahistories and metasociologies influenced by it. According to these trends, there is a reciprocal action between man and his environment, a mutual modification and not merely a one-sided action of one over the other. Both providentialism (man is God's creature) and the causal varieties of anthropology (external circumstances are almighty) are thereby overcome in a synthesis of nature and nurture.

7.2.2. *Externalism in Sociopolitical History*

The limitations of causal determinism are particularly manifest in connection with the interpretation of concrete historical events. Consider the much-debated question of the downfall of the Old Kingdom of Egypt, which is usually explained as the effect of the penetration of some unknown Asiatics in the Nile Delta; or the explanation of the collapse of the Western Roman Empire, as brought about by Barbarian invasions only; or, finally, the explanation of the collapse of Byzantium as effected by the sole military efficiency of the Turkish troops. Is it not more likely that extrinsic determiners were successful only at certain critical points (after having failed during centuries), and that they gained in efficiency as the inner breakdown became more and more intense? Is it not more reasonable to suppose that, in the absence of an inner breakdown, the attacks of foreign enemies would have been either ineffective or far less efficient? Historical explanations in terms of external causes alone look as inept as explanations in terms of subsidiary inner determiners, such as moral corruption—which, after all, is but an aspect of the over-all inner breakdown of a human social group. External determiners succeed in producing deep changes in a community if they enhance existing inner stresses to the point of collapse—as Hèrnán Cortés must have known when he conquered the Aztec empire with a handful of soldiers aided by an army of rebel warriors.

Cheap politicians are fond of an easy sort of causal explanation of the social transformations that they are either unwilling to bring forth or unable to check. Thus the French revolution of 1789 was attributed, by the propaganda of the *émigrés*, to Russian and American agents; the Russian October revolution was occasionally attributed to agents of the German kaiser, and time and again the distressing events of our time are attributed to the almightiness of this or that kind of currency—as if corruption were politically efficient in the absence of social maladjustment. In general, police-minded people are fond of the notion that social conflicts are caused by foreign agents; it is obviously easier to catch a few foreigners and to discredit the ideas assigned to them than to relieve deeply rooted social stresses. This sort of causal explanation is easier and has less dangerous consequences than the scientific description of inner conflicts, such as those between the economic interests of different social groups—which description, by the way, might be able to account for the eventual success of foreign agents when they happened to exist outside the imagination of the police.

On causal philosophies of history every independent, original attainment of new levels of cultural growth is unexplainable; that is why the achievements of apparently isolated cultures, such as those of Mycenae and Easter Island, are still regarded as miraculous. And this is also why the naïve belief in a "Greek miracle" has been replaced by a clever explanation in terms of borrowings from Egypt and Mesopotamia—which indisputable borrowings do not explain, of course, why the Greeks and not other peoples made the most out of the ancient Near East heritage. Diffusion does not account for the original mechanism of the Hellenic processing of those imported cultural goods, just as food intake does not explain assimilation. Only the study of inner processes in their actual contexts explains satisfactorily the emergence of novelty.

7.2.3. *The Doctrine of Borrowing in the History of Ideas*

The doctrine according to which external influences are the sole motive force of change is perhaps most inadequate in the case of intellectual history, which certainly is affected by factors belonging to other levels (biological, economical, social), but is obviously not directly *produced* by them. One such theory is the doctrine according to which knowledge is but the reflection of nature or society in the human mind (see Sec. 6.1.5). A less obvious manifestation of the causalist exaggeration of external relations is the habit of seeking the origin of the ideas set forth by a given thinker in the ideas propounded by previous thinkers, without caring for either the inner motivations or the conditioning of the cultural atmosphere and *Zeitgeist* in which the author in question lived.

Among historians of ideas who neglect or entirely ignore the sociocultural background and the eventual long-run social efficiency of the ideas they investigate, it is customary to trace intellectual influences "directly" from author to author. At first sight this procedure is able to tell the inner story, as it dispenses with external, nontheoretical (so-called anecdotic) circumstances; it is, however, none the less external, for it also dispenses with the inner world of the thinkers in question. In fact, it consists in manufacturing pretty causal chains of the type "Author N read Author $N - 1$, from whom he took the idea under consideration; but $N - 1$ had in turn derived that idea from writer $N - 2$, who read it in a book by $N - 3$, who borrowed it from a manuscript by $N - 4$, who copied $N - 5$ . . ." To this sort of intellectual diffusionism belong outstanding historians of ideas such as Duhem[19] and Lovejoy.[20]

This historical mismethod, of tracing influences without caring to enquire into the personal contributions of the thinkers concerned and into the social circumstances and cultural atmosphere in which they lived, is, of course, able to explain

[19] Duhem (1913, 1917), *Le système du monde*, 5 vols. Duhem's documents are most valuable, but his most important conclusions are marred by his faulty historical method, by his religious bias, and by his many petty antipathies.

[20] Lovejoy (1936), *The Great Chain of Being*. In this fascinating work, European thought is treated almost as a string of comments on Plato, who is made to play the role of Unmoved Mover of Ideas.

many things in a simple way. It explains everything except (*a*) how author No. 1 managed to get his ideas—unless it be assumed that they were revealed to him; and (*b*) why writer number N was influenced just by author number $N - 1$, refusing to let himself be influenced by the thinkers belonging to other chains. That is, the theory according to which scholars are what they have read is as inadequate as the doctrine according to which thinkers are what they eat, as it fails to explain originality as well as why, out of an enormous number of acting influences, only a smaller number are effective, while the largest mass of them do not succeed in starting the processes of intellectual creation.

7.2.4. *Man, the Self-Domesticated Animal*

It is one of the ironies of history, one of the inconsistencies showing how far history is from logic, that the ideologists who prepared the French revolution in the sphere of ideas held—with the almost single exception of Diderot—the belief in the preëminence of external conditions. This is funny, for such a belief is not only incapable of explaining social change in terms of human actions, but fits moreover admirably with social regimes in which human life is subjected to coaction, whereas spontaneity is the privilege of a chosen few—and even this in a restricted domain.

No less amusing is the spectacle, which can still be witnessed in our days, of the thinker who displays an enormous scholarly activity with the aim of demonstrating that intellectual work is socially ineffective, on the ground that ideas are nothing but the passive reflection of society and cannot in turn affect anything—another typically causal conception. Such is not, of course, the view of revolutionaries, to the extent to which they are consistent; social transformers cannot consistently accept those varieties of determinism which, like Buckle's or Taine's, lay the main emphasis on extrinsic determination and consequently deny that man is his own creature, that—as contrasted to lower animals—human beings are primarily the children of their own material and spiritual work. This does not, however, amount to underrating the value of education in the narrow, technical sense of the word. The doctrine that in the changing of his environment man changes himself and thereby builds himself up implies that the best-contrived educational method will be ineffective unless it matches the prevailing social conditions, which have to be changed first because they act as dominant determiners; it is further implied that adequate methods of education cannot be framed in detail before it is known what the changed social conditions and the changed men will be.

The continual self-change of man as a result of his self-domestication does not preclude, however, the existence of certain human constants, of certain characteristics that have not changed appreciably in historical periods—a constancy without which no definition of humanness would be possible. One such constant is precisely the property—common to all concrete existents, but qualitatively richer at the higher levels of reality—of possessing a varying amount of spontaneity or self-movement, and consequently self-determination. This essential, though not exclusive, trait of man is implied in most progressive trends in contemporary education. The teachers of old, saturated with empiricism (whether Aristotelian or Lockean), conceived of education as the engraving of a fixed, perfect curriculum on a passive *tabula rasa*. Progressive contemporary methods, on the other hand, stress activity; they emphasize (sometimes to the point of exaggeration) the active search after interesting information, and put aside the passive and resigned reception of uninteresting information; they regard a pupil as a being endowed with a certain amount of autonomy, whose free activity must be stimulated and channeled rather than suppressed. Following the Socratic model,[21] the best of these modern trends conceive of upbringing as the task of helping to give birth to the pupil's own personality. To the extent to which it succeeds, the maieutic method refutes causalism.

[21] Plato, *Theaetetus*, 150 and 157c. The correct Socratic thesis that the most effective education consists in guiding the student's own active search was based on two erroneous beliefs: first, that the soul is entirely and unrestrictedly self-moving (*autokinētos*); second, that knowing is nothing but remembering, lifting innate ideas up to the level of awareness.

7.3. Causality Requires the Persistent Maintenance of the Cause to Secure the Continuance of Process

7.3.1. *The Peripatetic Maxim* "Causa cessante cessat effectus"

If every change, however small, is due to an efficient cause, then whatever is in process of change must have a cause *sustaining* that change; whence the Peripatetic maxim "*Causa cessante cessat effectus*"—"The effect ceases after the cause has ceased."[22] This rule is not an isolable piece of scholastic cosmology but a necessary constituent of the doctrine of efficient causation. It is, in fact, a corollary of the disregard for the inner activity of things, a neglect that is characteristic of causalism (see Sec. 7.1); the negation of that maxim would amount to the acknowledgment that there may be uncaused events.

The rule is moreover indispensable for theology, since it entails that some Creator is necessary (even in the atomic age) in order to keep the world running. But, like so many rules enjoying a persistent popularity, this one *has* a ground in the common-sense account of daily experience, which teaches us that, in human affairs, things do not get done by magic but by hard work. False as a universal rule, our Peripatetic maxim has a wide field of practical applicability.

Let us review a few examples of self-sustaining processes that falsify the foregoing rule.

7.3.2. *Instances of Self-Sustained Processes*

The Greek atomists in antiquity, and Galileo and Descartes at the beginning of the modern era, found the first example of a noncausal process, of a movement that goes on by itself without any sustaining cause. It was mechanical motion (change of place), which, once started, proceeds of itself without needing any causal agent pushing or pulling the moving body along its path; the effect (motion) does not cease after the cause (force) has been removed. As Newton put it, the impressed force "consists in the action only and remains no longer in the body

[22] Thomas Aquinas, *Summa theologiae*, I, 96, 3, ob. 3.

when the action is over. For a body maintains every new state it acquires, by its inertia only".[23] Efficient causes produce alterations in this, the simplest sort of change, whose successive states are uncaused. An electrical analogue of this sort of change is the current in a circuit: "if the current be left to itself, it will continue to circulate till it is stopped by the resistance of the circuit"[24]—its energy being converted into heat or mechanical energy or both. A magnetic analogue is the permanent magnetization induced in a piece of steel, as contrasted to the temporary magnetization induced in soft iron. And an optical analogue of inertial motion is the propagation of a light ray in vacuum after it has been emitted, that is, once the "cause" of the light (usually accelerated electron motion) has ceased. The electromagnetic theory of light shows that the radiated field becomes independent of its cause or source—this decoupling being why the velocity of light does not depend on the state of motion of its source. Moreover, the propagation of light, which geometric optics takes for granted and does not regard as a process but as a stationary state, is explained by the theory of Maxwell and Hertz as the result of the inner "opposition" between the electric and the magnetic components of a light wave, which generate each other chainwise in accordance with Maxwell's field equations. The emission of a train of light waves is caused by a certain motion of electric charges, and every link in the electromagnetic chain can be regarded as the cause of the next. But, once started, the process as a whole is not causal but rather dialectical, since—to employ the metaphorical language of the romantic period—it proceeds through a "strife of opposites".

Other well-known chainwise and self-supporting processes are chemical and nuclear chain reactions; the extrinsic cause does nothing but unchain an internal process that not only goes on thereafter of itself, but does it in an increasingly amplified way, so that the final effect is enormously larger than the

[23] Newton (1687), *Principia*, ed. Cajori, def. IV, p. 2.

[24] Maxwell (1873), *A Treatise on Electricity and Magnetism*, vol. II, p. 197. The inertia of electricity had been pointed out by Faraday, and it is especially apparent at very low temperatures (superconductors).

triggering cause—thereby falsifying a further scholastic dictum, namely, *causa aequat effectum.* An equally familiar aftereffect is phosphorescence; the cause producing the radiative instability of the material—namely, the light rays impinging upon the phosphor—may disappear, despite which the body will continue emitting light for some time. There is, of course, a motive for the reëmission of radiant energy, but it lies in the structure of the phosphor itself, which is disturbed by the extrinsic cause; the efficient cause has not been sufficient to *produce* the effect, but has merely initiated a process leading to it. In general, modifications produced in complex physical systems by external agencies last for a time after such agencies have ceased to act. In ideal cases (inertial motion, light wave in perfect vacuum, isolated magnet) the effect may last forever; in actuality that time is finite, and particularly in everyday life—where sources of dissipation of energy, such as friction, are intense—the time of recovery is often very short. When this interval (the relaxation time) is negligible for all practical purposes, the maxim *causa cessante cessat effectus* seems to apply. But rigorously speaking it never applies, for it is based on our ignorance of the processes that render the cause effective and the recovery rapid.

It is hardly necessary to show that self-supporting processes are conspicuous at higher integrative levels, characterized as they are by a richer variety of forms of self-activity and hence of self-determination. True, a great deal of physiology and psychology still employs the fiction of the direct, immediate, stimulus-response connection, and behaviorists still reject as meaningless the investigation of the inner mechanism connecting such extreme links; yet, the poverty of both the causalist and the phenomenalist approach is increasingly being realized. Recent work on the physical basis of nerve signals has shown that the nerve pulse is not a mere propagation of an externally impressed excitation, but consists instead of a self-sustained chain reaction. An electric stimulus beneath a certain critical strength produces no nerve response at all; but having reached a certain threshold, it produces a large "action potential" which is insensitive to the precise value of the stimulus, provided

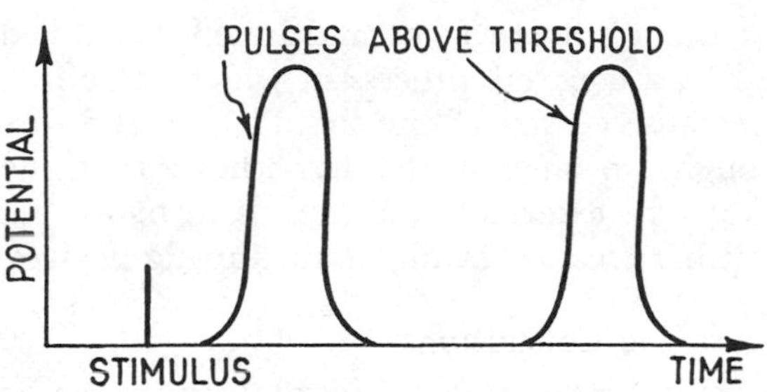

Fig. 24. A chain of electric pulses initiated by an external stimulus in an isolated nerve fiber.

the latter is larger than the threshold value. Moreover, the nerve fiber is not a mere *conductor* of electric impulses, but a *source* of new pulses that are much stronger than the stimulus, and that are propagated along the axon by means of a regenerative process that compensates for losses (see Fig. 24). And when nervous tissue builds a differentiated nervous system, the transmitting and self-exciting functions combine to constitute a shock-absorbing mechanism that protects the whole organism from its environment (Pavlov's inhibitions and Freud's defense mechanisms).

That sensations, images, and ideas may "persist" even after the removal of the agents triggering the processes ultimately producing them is a fact of immediate daily experience. It was not, of course, denied by the empiricist school of the *tabula rasa*; but memory was regarded by this school as an engraving, as a more or less permanent modification in the percipient subject, not as a process lasting for a time. The contemporary tendency, on the contrary, is to regard memory not as a static storing place but as a permanent dynamic process, not unlike Hartley's vibrations. (According to the theory propounded by David Hartley two centuries ago, sensations do not die as soon as their causes or stimuli cease, but are stored in the nervous system in the form of miniature vibrations, which may last for years.) Modern theories of memory reject, then, our Peripatetic maxim, and to that extent they are noncausal.

Examples of self-sustained processes that last after the cause unchaining them is over can be multiplied. What is difficult to find is, rather, a concrete process in which the effect ceases as soon as the cause ceases. If one should be found, previous experience ought to suggest the hypothesis that the time of recovery from the externally induced disturbances is simply so short that it has escaped available measuring devices.

7.4. Summary and Conclusions

Those who accept the principle concerning the activity inherent in every concrete (material or cultural) object, and the thesis that the world as a whole has no cause but is self-existent should not fail to realize that the consideration of efficient causes as sole determiners leads to an *incomplete determinacy*, that is, to a partial indeterminacy. Indeed, if we know that the concrete object contributes its own specific mode of being and of ceasing to be—a change that may even oppose and consequently modify the external agencies acting on it—then it is obvious that efficient causes do not determine effects completely, that is, that causation only contributes to the determination of the effects but does not produce them entirely. Lack of freedom is, then, not the rule of natural law but the serious hampering or distortion of self-determination by external constraints and disturbances; freedom is not a violation of lawlike behavior but a type of it, namely, self-determination.

In the field of the sciences of man, causal determinism leads to exaggerating the importance of the environment and of the *concours de circonstances*; it suggests the distorted image of man as a passive toy at the mercy of unsurmountable powers, whether natural (such as geography) or artificial (such as society). On this essentially conformist view, liberty is largely illusory, since it is conceived solely as a negative value, that is, as the (impossible) absence of bonds; it is freedom *from*, not freedom *to*, the active and eventually consciously planned self-determination. As soon as self-determination is taken into account, as soon as it is realized that nothing is the exclusive consequence of external conditions, however important they may be, an element of voluntarism illuminates the historical scene: not, of course, a voluntarism asserting the arbitrary, irresponsible, and lawless will of enlightened *Führers*, but one consistent with scientific determinism. Liberty is then seen as a positive value, namely, the active endeavor to attain the optimum self-determination. And fatalism is then seen to be a bad dream, for nothing can be regarded as the inevitable consequence of either past occurrences pushing events along predetermined paths or present causes operating *ab extrinseco* beyond the reach of man's material and spiritual work.

The act of releasing a bow is usually regarded as the *cause* of the arrow's motion, or, better, of its acceleration; but the arrow will not start moving unless a certain amount of (potential elastic) energy has previously been stored in the bow by bending it; the cause (releasing of the bow) triggers the process, but does not determine it entirely. In general, efficient causes are effective solely to the extent to which they trigger, enhance, or dampen inner processes; in short, extrinsic (efficient) causes act, so to say, by riding on inner processes. Or, as Leibniz put it, "every passion of a body is spontaneous or arises from an internal force, though upon an external occasion".[25] The principle that the efficiency of extrinsic determiners depends on the intrinsic ones may be regarded as an induction from the most varied fields; moreover, it is implicit in Francis Bacon's golden rule, "*Natura non vincitur nisi parendo*"—"Nature can be mastered only by obeying her."

As a consequence, causal determinism is incomplete. If causation is regarded as the sole possible form of determination, *in*determinism is unavoidable, as illustrated by the story of natural radioactivity which, being—so far as we know—a spontaneous process, was wrongly regarded as an instance refuting determinism in general, while it actually just contradicts *causal* determinism. However, the opposite exaggeration, which consists in regarding inner processes as if they could actually arise and subsist in *complete* isolation from external circumstances, may lead to equally ridiculous extremes, such as solipsism. Every concrete object is self-acting, but nothing in the world, save the world as a whole, is self-sufficient. Concrete

[25] Leibniz (1695), *Specimen dynamicum*, in *Philosophical Papers and Letters*, ed. L. E. Loemker, vol. II, p. 733.

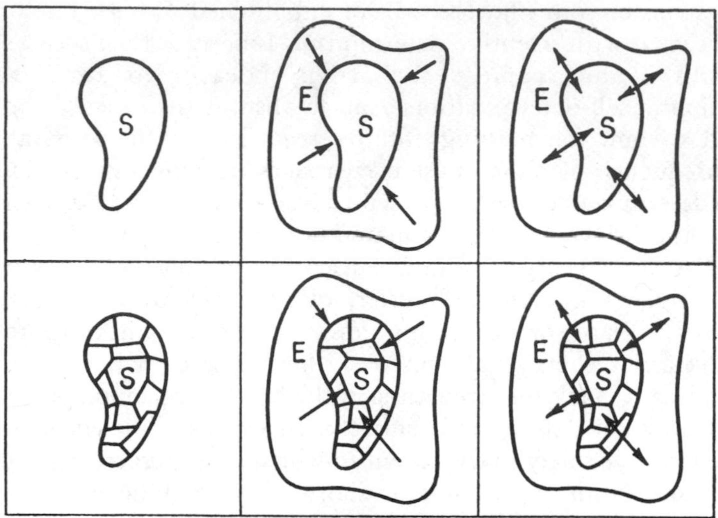

Fig. 25. Main types of system–environment relations. *Left:* self-movement—isolated system. *Center:* causality—system under action of environment. *Right:* interaction—*S* and *E* act upon each other. *Upper:* first-order approximation—both *S* and *E* are treated as wholes. *Lower:* second-order approximation—the internal structure of *S* or *E* is accounted for.

things are not self-sufficient because they actually stand always in interaction with an unlimited number of other existents; nor are ideal objects, which lack self-movement altogether, self-sufficient, since for their very existence they depend on some mind.

It is a common event in the history of knowledge that what at first is regarded as altogether spontaneous is shown on closer investigation to be caused or at least influenced by external determiners, and vice versa; and what at first appears as structureless ends by displaying its internal complexity. The explanation of the object's properties in terms of external causes alone is but a stage of a process in which various further categories appear in succession, notably those of internal structure, interaction, and self-movement (see Fig. 25).

Complete self-determination, that is, total freedom, is as illusory as efficient causation (understood as external compulsion) is insufficient. An adequate picture is provided by a *synthesis of self-determination and extrinsic determination*, in which external causes are conceived as unchainers of inner processes rather than as agents molding a passive lump of clay. The two exaggerations of environmentalism and innatism, with their sequels of relativism and absolutism, are thereby avoided. Now, if extrinsic efficient causes are regarded as efficient solely to the extent to which they inhere in inner processes, room is made for the *may* at the expense of the *must*; novelty is seen to be possible, not only as a result of the external chance encounter of initially independent lines of development, but also as emerging from the nonadditive grafting of one process on another process. But the problem of newness deserves a new chapter.

8

Causality and Novelty

Causality is undoubtedly a theory of change. But does it succeed in explaining radical change, that is, the transformation that consists in the emergence of things endowed with new properties? This is the problem that will concern us in the present chapter.

Two extreme answers to this question have been set forth within the frame of the theory of causality. According to one of them, efficient causes alone can bring forth novelty; on the other hand, what can be regarded as the strict doctrine of causality has often been invoked just against the possibility of the emergence of newness. If the synthesis of self-determination and extrinsic determination, sketched in the previous chapter, is accepted, the two solutions will be found inadequate, and a synthesis will again be sought. The one here suggested is simply this: causation is insufficient to produce qualitative changes, but it usually does take part in them.

Yet before approaching the change of properties we should consider the impact of causality on the very concept of property.

8.1. Causalism Entails the Scholastic Dichotomy Substance–Attribute

8.1.1. *The Impact of Causalism on the Theory of the Substance–Attribute Relation*

If, as the doctrine of causality holds, a thing can acquire its determinations solely by the action of efficient causes external to it; if properties are assumed to cling *ab extrinseco* to a passive substance whose sole peculiar property is the capacity of receiving properties, being otherwise immutable, independent, and simple; if, in short, substance is incapable of changing and serves only as the receptacle, carrier, or support of attributes (*substans accidentibus*), then any of the following consequences is possible:

the substance and the attributes can exist separately from one another, the former as the unqualified *ens* of the schoolmen, and the properties as Platonic ideas;

the substance can exist by itself as an unqualified entity, but the attributes cannot exist without a substratum to dwell in—like Aristotle's forms, which inhere in a passive "matter" without having themselves an autonomous existence, or like Aquinas' "accidents", which are added to the immutable essence and are dependent on the substance;

the properties can exist by themselves, the existence of the substratum being contingent.

In all of these cases the substance and the attributes are regarded as unchanging, as if there were in the world a limited stock of different qualities or properties that can add externally and stick on the *re substante* as ready-made labels. These views on the nature of properties and change exclude the emergence of authentic novelty in the sum total; they just allow for the circulation, not for the appearance of novelty. Notice that they are not merely worn-out pieces of scholastic metaphysics, but forcible consequences of the externality of efficient causation.

Needless to say, every one of the foregoing consequences of the doctrine of causality is refuted in the theoretical processing of experience,[1] that is, in science, since, save by abstraction, we never meet anything devoid of qualities and standing apart from change; nor do we find, except by abstraction, qualities outside objects endowed with them—whence the expression

[1] In everyday language we often say that *experience* refutes this or that idea. However, by itself experience can neither confirm nor disconfirm a hypothesis; only one idea can verify or refute another idea within a given theoretical context or body of ideas; in particular, ideas *suggested by experience* (never unambiguously!) can confirm or disconfirm other ideas having a factual content. Contrary to traditional empiricism, experience is not self-sufficient but is always instrumental, because we never compare ideas with experiences, but always among one another.

'emergence of qualities' should be understood as just a convenient metaphor for 'appearance of new qualities in an object'. The above-mentioned conceptual possibilities following from causalism amount to regarding the substance-attribute connection as a *contingent relation*, that is, as one that does not stem from the nature of either. That is, causalism leads to the conception of the attributes of things as accidental determinations (the scholastic *accidentia*), as contingent either relatively to the very essence of things, or relatively to each other, or both. This phenomenalist view of qualities and of qualitative change is of course possible solely on the assumption that things lack a self-movement enabling them to "assert" their peculiar being—their "personality", to indulge in an anthropomorphic term.

8.1.2. *Contingency of Attributes in Hegelianism and Positivism*

The thesis of the contingency of attributes is not, however, exclusive of causal ontologies. It can also be found, in a more or less explicit form, in two influential systems of metaphysics which, at first sight, share nothing but their dislike for causalism, namely, Hegelianism and empiricism.

Like Leibniz before him, Hegel held that conceptual entities are necessary while natural ones are contingent, that is, they might have been different; and they are contingent because they are "external to their own concept", constituting accidental, unbound aggregates of externally related properties.[2] The properties of the concrete object (*das unmittelbar Konkrete*) are logically external to one another, are not deducible from one another, whence their existence in a given object is not necessary but "more or less indifferent".[3] That is, according to Hegel, properties neither follow analytically from one another (which is right) nor constitute systems (which is wrong); they are mere heaps or aggregates of accidental marks. As a consequence, the play of forms has in nature the character of the "unbound, unrestrained chance" (*ungebundene zügellose Zufällig-*

[2] Hegel (1817), *Encyklopädie der Philosophischen Wissenschaften*, "Naturphilosophie," secs. 248–250.

[3] *Ibid.*, sec. 250: "Das unmittelbar Konkrete ist eine Menge von Eigenschaften, die aussereinander und mehr oder weniger gleichgültig gegeneinander sind".

keit).[4] Nature is, in short, the kingdom of contingency and randomness. This belief, revived later on by Boutroux (the philosopher),[5] was one of the reasons why Hegel held nature in contempt—the other reason being presumably nature's obstinate refusal to conform herself to his schemes.

The problem of the substance-attribute relation hardly arises in empiricism, which regards substance as a figment of the imagination—not, however, on the ontological ground that there is no matter without motion of some sort, but on the epistemological ground that such a constant and simple support of qualities is transperceptual. On the other hand, empiricism has usually held that things are *unorganized aggregates of sense data*, lacking both an independent carrier and an inner necessary bond; that events just are (in experience), without there being any genetic connection among them, or among the qualities of every complex or aggregate of sensations —unless such a bond is superimposed on them by God, as Berkeley (the most consistent empiricist) thought. From this doctrine of qualities and change it follows, of course, that "Everything we see could also be otherwise",[6] and that the laws of nature and of society are nothing but summaries of experience, empirical rules or directions (*Vorschriften*)[7] that might have been different. According to empiricism, laws are, in short, *contingent* relatively to the nature of things, and this for the simple reason that there are not things placed in an independently existing external world, but only lumps of sensations; and that which experience establishes further experience may demolish.

[4] Hegel (1817), *Encyklopädie der Philosophischen Wissenschaften*, "Naturphilosophie,' sec. 248.

[5] E. Boutroux (1874), *De la contingence des lois de la nature*. The whole of this book is based on the fallacious identification of 'law of nature' (or law_1) with 'law statement' (or law_2), that is, the human, approximate knowledge of the former.

[6] Wittgenstein (1922), *Tractatus Logico-Philosophicus*, 5. 63.4.

[7] Mach (1905), *Erkenntnis und Irrtum*, chap. xxiii, defined laws of nature as "restrictions on expectation". Schlick, "Die Kausalität in der gegenwärtigen Physik", *Die Naturwissenschaften 19*, 145 (1931), p. 156, called laws rules of procedure. Who doubts that they are? The question is whether they are, in addition, conceptual reconstructions of objective patterns—which alone seems to explain why, being directions, they are not arbitrary.

Thus two varieties of immaterialism, namely, objective idealism and radical empiricism, attain, with regard to the problem of qualities and their changes, conclusions similar to the scholastic theses; in point of fact, they all ultimately end up in some sort of fortuitism in the domain of nature, while retaining strict necessity in the domain of discourse. In all three cases—scholasticism, Hegelianism, and empiricism—the ontological thesis of the contingency of nature is a consequence of idealism; in the case of most scholastic philosophies of nature a further source is—paradoxical as it may seem at first sight—causal determinism.

8.1.3. *Beyond Causalism and Accidentalism*

When carried consistently to its extreme consequences, the doctrine of efficient causality may lead to indeterminism; it could not be otherwise, since the scholastic concept of quality and alteration as *accidentia* that are alien to the essence of things, and that are stamped on them from the outside like removable tags, is complementary to the belief that nothing (save God) can be self-caused, but everything must be determined *ab extrinseco* by something else.

A way of avoiding indeterminism while sticking to the belief in the externality of causation is to adopt Malebranche's doctrine of occasional causes—which, in the domain of ethics, leads of course to a serfdom no less complete than the hopeless helplessness to which indeterminism leads. If matter were nothing but extension capable of being pushed and pulled hither and thither, and if mind were nothing but a *tabula rasa* capable of receiving external stimuli, as Father Malebranche thought, then he would have been right in regarding change as brought about by the sole efficiency of God. But Malebranche's occasionalism, grounded on the identification of matter with passive extension, and on the tenet of the essential passivity of "finite spirits", is altogether inconsistent with modern science, which exhibits the dynamic character of the real world and the large extent of spontaneity in the behavior of every bit of both sensing and inanimate matter. Moreover, this sort of causal determinism is self-defeating, since it withdraws causation from the world, turning it into the exclusive privilege of God, the sole active and efficient "substance".

The extension of causality to include self-determination led Spinoza and Leibniz to modify the traditional definition of substance, which etymologically means that which does not change.[8] They asserted that the very substance of things consists in their activity, and they more or less clearly recognized that efficient causes are not sufficient producers of change, as their efficiency depends on how deeply they disturb inner processes. Leibniz went so far as to declare that "the natural changes of monads come from an *internal principle*, since an external cause could not influence their interior".[9] The scientific discoveries of Leibniz's own time (modern dynamics, "corpuscular philosophy", embryology, and so forth) certainly supported the doctrine of self-movement; still, they did not warrant his exaggeration of it, an exaggeration to which he was led partly by the hermetic writings, which had always emphasized the "inner principles" of things and of man.

The adequate course is, I believe, the synthesis of extrinsic and intrinsic determination, suggested in the previous chapter; and such a synthesis stops the whole babble about the *accidentia*, rendering both quality and qualitative change intelligible.

8.2. Causality Renders Genuine Novelty Impossible

8.2.1. *The Principle* "Causa aequat effectum"

If the joint action of several causes is always an external juxtaposition, a superposition, and in no case a synthesis having traits of its own (see Sec. 6.2) and if the hypothetical patients on which the causal agents act are passive things incapable of spontaneity or self-activity—incapable, in short, of adding something of their own to the causal bond (see Sec. 7.1)—then it follows that, in a sense, *effects preëxist in their causes*.

According to this extreme but consistent doctrine on the

[8] For a clear definition of the "obscure and relative idea of substance in general", see Locke (1690), *An Essay Concerning Human Understanding*, book II, chap. xxiii, 2.

[9] Leibniz (1714), *Monadology*, sec. 11, in *Philosophical Papers and Letters*, ed. L. E. Loemker, p. 1045.

nature of causation, *only old things can come out of change*; processes can give rise to objects new in number or new in some quantitative respects, not however new in *kind*; or, again, no new qualities can emerge. A world running on a strictly causal pattern is such as yogis, Thomists, and 18th-century Newtonians imagined it, namely, a universe without a history, making no step forward and running (as Hegel said of nature) around a perpetual circle after the model of the apparent motion of the heavens (before the "corruptible", that is, changeable nature of celestial bodies was discovered by Tycho Brahe and Galileo). The formula summing up this extreme version of causalism is, "*There is nothing in the effect that has not been in the cause*".

Now this thesis admits two variants, both of which exclude novelty (at least as a natural event). According to one of them the effect includes *the same* as the cause (symbolically, $C = E$), whereas in the second version the effect contains *less* than the cause (symbolically, $C > E$). In the former case we have the old formula, *Causa aequat effectum*—which by no means should be admitted by consistent believers, since it entails the divine nature of the Creator's works (pantheistic heresy). This is why Thomism asserts that "like begets like", that cause and effect resemble each other or at least cannot be too dissimilar, as "every agent intends to induce its own likeness in the effect, so far as the effect can receive it"[10]—with the important proviso that the effect never receives the *whole* content of the cause.

Before analyzing this important and widespread theory of change, it will be convenient to place it in its historical context.

8.2.2. *Archaic Origins of Belief in Immutability*

One of the oldest written statements of the above-mentioned view of change is found in the *Upaniṣads* (1,000–500 B.C.), in which it is metaphorically said that "whatever there is belonging to the son belongs to the father; whatever there is belonging to the father belongs to the son".[11] This assertion of the identity of antecedent and consequent was also the

[10] Thomas Aquinas (c. 1260), *Summa contra gentiles*, book II, chap. 45.
[11] See Radhakrishnan (1931), *Indian Philosophy*, 2nd ed., vol. I, p. 181.

kernel of Solomon's celebrated wisdom: "The thing that hath been, it is that which shall be; and that which is done is that which shall be done: and there is no new thing under the sun. Is there any thing whereof it may be said, See, this is new? It hath been already of old time, which was before us".[12] A sociologist of knowledge would perhaps say that these pieces of scriptural wisdom reflect the boredom of pastoral and agricultural life, with its recurrent cycles, and that they were probably supported by the wish to preserve the corresponding social structure.

The oldest theoretical elaboration of the view that the effect is similar to the cause and is even contained in it, so that no newness can ensue from causation, is perhaps the *Sāṅkhya* religious-philosophical doctrine,[13] which emerged around the 7th century B.C. and was one of the most influential of all (it is the philosophical partner of the yoga system). The tenet of the qualitative immutability held such an important place in the *Sāṅkhya* system that its followers called themselves *Satkāryavādinah*, which derives from *sat-kārya-vāda*, preëxistence of the effect in the cause. In contrast to the *Nyāya-Vaiśeṣika* school, which held an emergentist view, the *Sāṅkhya* asserted that it is not licit to speak of the *birth* of something but only of its *manifestation*, since nothing is qualitatively created or lost[14]; whatever seems to arise had been latent and is freed from the state of cause by an agent acting from the outside. This view, which is sometimes regarded as a typical illustration of Eastern conservatism, has been the most conspicuous theory of change in the West since the end of the Greek classical period.

8.2.3. *Conservative Evolution: From Thomism to Mechanism*

A similar theory of change as the unfolding, growth, and manifestation of preëxisting potentialities under the influence

[12] *Ecclesiastes* 1: 9 and 10.

[13] See Radhakrishnan, *Indian Philosophy*, vol. II, pp. 256 ff.; Hiriyanna (1949), *The Essentials of Indian Philosophy*, chap. v, esp. p. 109.

[14] Notice the similarity of this archaic view with the widespread misinterpretation of the principle of conservation of energy, according to which the latter would state that "nothing is gained and nothing is lost". This popular interpretation is wrong; what the principle asserts is a *quantitative* invariability (of the quantity of energy) through *qualitative changes* (of forms or types of energy). See Sec. 8.2.6.

of external causes, was held by Aristotle and worked out in detail by Moslem and Christian schoolmen. According to the Peripatetic schools, change is nothing but the actualization of that which had always been there, though only in potency, that is, in a hidden, nonmanifest way.[15] Hence whatever is in the effect must previously have been in some way in the cause: "*Quidquid est in effectu debet esse prius aliquo modo in causa*", the maxim so brilliantly ridiculed by James.[16]

This scholastic aphorism, entailing as it does the assertion that change is nothing but the unrolling or unfolding of preexisting potentialities, excludes flatly the emergence of authentic novelty; while asserting conservative evolution, it denies epigenesis, what Bergson called creative evolution. In particular, that theory of change excludes higher novelty, and consequently progress. Progress is indeed explicitly denied by another scholastic maxim, namely, "The cause is higher than its effect".[17]

Contemporary irrationalists, in their defense of the mysteriousness of emergence, often claim that the negation of the

[15] The theory that every change is nothing but the transition from potency to act under the influence of an efficient cause is related to, but by no means identical with, the *germ theory* of novelty, or preformation theory, proposed by the Stoics, adopted by Augustine, and independently revived by Leibniz. The essential difference between the actualization theory and the germ theory of change lies in the fact that the former assigns the passage from potency to act to external causes, whereas the germ theory emphasizes inner development and finality.

[16] James (1911), *Some Problems of Philosophy*, chap. xii. James's criticism of Peripatetic causality (which he called the "conceptual view of causation") is one of his finest pieces. However, it had been anticipated by his much-hated predecessor Hegel, and his own view (the "perceptual view of causation") was basically irrationalistic, since it regarded causation as "just what we feel it to be, just that kind of conjunction which our own activity-series reveals".

[17] Albert the Great, *On the Nature of the Intellect*, chap. iii, in McKeon (ed.), *Selections from Mediaeval Philosophers*, vol. I, p. 332: "It is true in all things caused univocally with respect to nature, that whatsoever is present essentially in the thing caused, that is present more powerfully and more nobly and more clearly and prior and more perfectly in the cause of that which is caused". Thomas Aquinas (1272), *Summa theologiae*, q. 105, 6: "The higher cause is not contained in the lower one, but conversely". In view of the success of the theory of evolution, Neo-Scholasticism now grants that biological progress, though impossible with the sole help of "secondary" (natural) causes, becomes possible with the assistance of the First Cause. See George P. Klubertanz, S.J., "Chance and Equivocal Causality", *Proceedings of the XIth International Congress of Philosophy*, 1953, vol. VI, p. 203.

possibility of radical change is typical of modern science. Thus Bergson[18] held that science and, in general, intelligence (in contrast to intuition), does not grasp newness but regards everything as given, being solely interested in repetition. As we have seen, however, the doctrine that change is the mere unfolding of preëxisting capacities has remote origins, being in point of fact at least as old as the Bergsonian (and holistic) thesis that the emergence of novelty is basically unintelligible.

A similar inability to account for novelty has been attributed —this time rightly—to mechanism. True, mechanistic philosophy does not grant the emergence of novelty but aims at leveling everything to the plane of mechanical actions and reactions. Remember Diderot's definition of that most intriguing type of matter: "Life is a succession of actions and reactions . . . To be born, to live, and to pass away, is to change forms".[19] But mechanist philosophers did not *invent* the theory of change without novelty; mechanism just adopted, reinforced, and rationalized the picture of change as a circulation of a limited stock of forms—which is quite understandable, since the doctrine of conservative evolution satisfies both the scientific demand that providential arbitrary interventions be dispensed with from the moment the world machine has been created, and the dogma of creation without antecedents. Unlike schoolmen, mechanist thinkers during the 17th and 18th centuries usually did not grant the Deity the ability to create novelties after the sum total of qualities (which were but a few primary ones) had been built into matter. God, when acknowledged at all, was either an absentee landlord (as with most deists), or a mechanic repairing the cosmic machine from time to time, though only in a quantitative sense—for instance, by restoring the quantity of motion lost during nonelastic collisions, as Newton thought.

Just as most schoolmen had asserted that the cause is *higher* than the effect, Descartes[20] maintained that the cause is *higher*

[18] Bergson (1907), *L'évolution créatrice*, pp. 29 ff.

[19] Diderot (1769), *Rêve de d'Alembert* (publ. 1830), in *Oeuvres*, p. 930.

[20] Descartes (1641), *Méditations*, 3rd med.: "It is obvious by the light of nature that there must be at least as much reality [or perfection] in the total efficient cause as in its effect".

than or *equal* to the effect (symbolically, $C \geqslant E$); and all the remaining mechanistic philosophers accepted—at least concerning natural affairs—the thesis of the *exact equivalence* between the cause and the effect. In fact, explicitly or not, mechanists rejected the progressive degradation required by most systems of theodicy and held the principle *Causa aequat effectum*. This conservative principle once played a progressive role (now almost forgotten) because, among other reasons, it entailed the rejection of miracles. (Its further positive features will be mentioned in Sec. 8.2.6.) But the price paid for such a rationalization of the world was, again, other-worldliness—though certainly a very intellectual one. In fact, if the material universe was a self-running mechanical contrivance in which nothing new could emerge, the "first origins" of things necessitated a first (and almost always last) miracle. This was clearly realized by Newton: "Blind metaphysical necessity, which is certainly the same always and everywhere, could produce no variety of things. All that diversity of natural things which we find suited to different times and places could arise from nothing but the ideas and will of a Being necessarily existing".[21]

Since the strict doctrine of causality rendered radical novelty impossible, the emergence of newness had to be either denied or assigned to that which, by definition, was absolutely autonomous, free, spontaneous, namely, spirit, with or without a capital. (Some contemporary occultists, like Ouspensky, hold that the sole possible source of novelty is psychical free will, the foundation of magic; orthodox holism, too, regards emergence as a sort of magical mystery beyond the scope of science.) As to progress—which by definition involves the emergence of higher novelty out of previously existing levels of being — it was of course unaccountable in mechanical terms, since mechanism is essentially reductionistic; for mechanism, 'higher' can only mean more complex, never qualitatively richer.

8.2.4. *Qualitative Immutability and Causation in Kantianism*

Essentially the same qualitatively unchangeable picture of the universe was adopted by Kant—who, however, managed

[21] Newton (1687), *Principia*, ed. Cajori, gen. schol., p. 546.

to impoverish it by depriving mechanics of (inertial) self-movement[22] and by imagining the existence of an insurmountable barrier between immutable essences and changeable appearances, a barrier which, by the way, was in the purest Platonic tradition. According to Kant's ontology, causation is a form of thought (see Sec. 1.6), a category belonging to the conceptual canvas of categories on which human experience is painted—whence no experience might conceivably falsify the causal principle. Now, experience reaches phenomena, which flow over the immutable kernel of things-in-themselves, to which causality docs not apply. Conscqucntly, as a Nco-Kantian has explained, according to the critical philosophy, "despite all phenomenal changes the world remains essentially the same; the cause is just the form taken previously by the effect, and the effect is the form that the cause takes. The true meaning of the causal law is, then, that there is nothing new in the world; and so far, essential acquisitions of modern physics, in the first place the principle of conservation of energy, and the law of conservation of electric charge, must be regarded as expressions of the causal law".[23]

The Kantian view that causation brings forth no novelty seems to have been introduced in Britain by William Hamilton, according to whom the whole meaning of "the intellectual phenomenon of causality" is that "all that now is seen to arise under a new appearance, had previously an existence under a prior form . . . There is thus conceived an absolute tautology between the effect and its cause. We think the causes to contain

[22] Even during his youthful period of enthusiasm over Newton's theories, Kant (like Voltaire before him) only grasped what had been popularized of them, namely, the theory of universal gravitation (to which Kant added a universal repulsion). Motion, as conceived by Kant, was the result of two opposite forces, one of attraction, the other of repulsion; motion was therefore always caused. See his *Allgemeine Naturgeschichte und Theorie des Himmels* (1755), esp. chap. i. The current opinion that Kant's philosophy was but the metaphysical translation of Newtonianism cannot be held with regard to this work.

[23] Lothar von Strauss und Torney, "Das Kausalprinzip in der neuen Physik", *Annalen der Philosophie* 7, 49 (1928), p. 72. But conservation of charge has nothing to do with the alleged equality of cause and effect, nor has any other invariant quantity; and conservation of energy, far from meaning immutability, means the quantitative equivalence among qualitatively different forms of energy.

all that is contained in the effect, the effect to contain nothing which was not contained in the causes".[24] According to this conception, then, change is ultimately reduced to permanence, and variety to identity; a causal picture of the universe is asserted, which need not be wholly static but which is at most periodic, since it allows only for repetitions. Change is granted only to show how similar the present is to the past. As the French saying goes, *Plus ça change plus c'est la même chose*—The more it changes, the more it is the same thing.[25]

Such is the poor result of the elaborate Kantian play with categories, antinomies, and analogies from experience—a wrong archaic solution of the central problem of ontology, namely, the rational and verifiable explanation of the possibility of newness.

8.2.5. *General Lawfulness Accounts for the Novelty Excluded by Causalism*

The poor picture of becoming held in various ways by the *Sāṅkhya*, by scholasticism, and by mechanism, and adopted by Kant, began to change towards the end of the 18th century under the pressure of two movements that were otherwise largely antagonistic, namely, romantic philosophy and science. Natural science discovered nonmechanical patterns of change, notably biological evolution; and some of the romantic philosophers, particularly Hegel, elaborated—certainly in a fantastic and mostly unintelligible way—the dialectical theory of change, the nucleus of which is the hypothesis according to which radical change is brought about by the tension and final synthesis of opposite trends. From that time on, it is less and less

[24] W. Hamilton (1858–1860), *Lectures on Metaphysics and Logic*, lect. xxxix, II, pp. 377–378, quoted in Mill (1865), *An Examination of Sir William Hamilton's Philosophy*, vol. II, pp. 25–26.

[25] Meyerson, who was in many ways close to Kantianism, stretched this view on change and causation to its extreme limit by asserting that the law of causation is but a form of the logical principle of identity; see *Identité et réalité* (1908), p. 33 and *passim*. In "Hegel, Hamilton, Hamelin et le concept de cause", reprinted in *Essais* (1936), pp. 28 ff., Meyerson saw Hegel's influence in Hamilton's teaching on causality. However, Hegel had *decried* causality just for the "tautology" it entails, and had asserted that causality is not valid in the realms of life and of spirit; see Hegel (1812, 1816), *Science of Logic*, vol. II, pp. 92 ff.

disputed among scientists that there is no such thing as an equality sign between *every* cause and its corresponding effect, that sometimes there may be "less" (as Aquinas thought), but at crucial points there may be "more" in the effect than in the cause. Genuine novelty, unexplainable on strict causalism, is understandable—at least in principle—with the help of the totality of categories of determination, not excluding causation.

The scientific description, explanation, and prediction of change, whether quantitative or qualitative, is performed with the help of law statements (see Chapters 10 to 12); among laws there are those with a strong causal component, like Newton's law of motion, and laws of emergence, like those of the formation of molecules out of atoms. Now, if the emergence of an isolated new property (new, of course, in a given context) need not entail the correlative emergence of new *laws*, what about the emergence of entire *systems* of new qualities? It seems that such a basic change does bring with it a change in the laws—or, if preferred, the addition of new laws, as is the case with the laws of social behavior with regard to those of life. Such a newness in the laws (of nature or of society), which seems to accompany the emergence of a new integrative level is, of course, *relative* to the given context. The emergent laws need not be entirely and absolutely new, that is, new regardless of contexts; they merely have to be new with regard to the laws followed by the object concerned, which evolves to the point where the new modes of behavior appear. A few such emergences of new systems of qualities, presumably accompanied by the emergence of new laws, may have occurred and may occur in the future for the *first* time in the history of the universe —as presumably the levels of life, thought, and social organization occurred.

In any case, whether original or just relative novelty is at stake, the appearance of entities characterized by entire sets of new properties (that is, the emergence of a new level of being) seems to involve changes in the laws of the mode of being and becoming of such entities—since, after all, laws are patterns of being and becoming, and such as relate qualities. Consider the laws of chemical binding: we know that they do not find any application outside a certain temperature range. Thus, before any experiment performed on the spot, we are pretty sure that no chemical reaction takes place in the inside of the known stars, on account of their high temperature. But as soon as a bit of matter in the stellar atmosphere cools down to the point where chemical reactions become possible, at such a critical point a radical change in the system of qualities characterizing the bit of matter in question may occur. The very laws of nature may change in it, at least in the sense that new laws (new relative to the ones that had been working in that bit of matter till the change concerned) may emerge from the old laws, or—if a mechanistic way of speaking is preferred—new laws will be *added* to the former laws. To be more specific, in a mass of matter nothing but mechanical, thermodynamic, and electrodynamic changes will occur within a certain temperature range. But beyond a critical temperature the laws of thermonuclear phenomena may begin to operate; and below another critical temperature chemical laws may begin to work.

This has an important consequence for the theory of causation, namely, that causal connections should not be isolated from the set of associated laws; or, if preferred, a causal law will obtain provided the remaining relevant laws do not change. For, if the background laws do change, then the same cause C may no longer produce the same effect E; that is, under changed conditions the connection between C and E may be altered. Thus, an electric discharge will produce a set of *physical* phenomena E within a certain temperature range, but if the temperature is conveniently raised or lowered, the same discharge may elicit a *chemical* result E' (emergence of molecules out of atoms).

Even mere changes in the reference system may produce qualitative changes. Think of the (classical) magnetic field of a charged body in motion: it simply does not exist in the charge's own reference system, although it does exist in systems in motion relative to the charge concerned. Such a qualitative change does not require any change in the *laws* of the variable phenomena. Moreover, the *laws* are assumed to be independent of the reference system (principle of covariance of the laws of nature); but the *solutions* of the corresponding equations do depend on the reference system, and it is precisely the solutions that are correlated to the phenomena. In other words, the uniformity of nature does not apply to the phenomena (which are relative) but to the laws of phenomena (which are absolute); or, if preferred, while it is usually *assumed* that place and time are irrelevant to the laws of nature, it is *known* that events are relative to their space-time context.

There might be a further way in which phenomena could vary while keeping the laws unchanged, namely, if some of the so-called fundamental constants of nature (the gravitational constant, the velocity of light in vacuum, Planck's constant, and others) were to change in time. If this were the case—and there is no a priori ground for rejecting this possibility—physical laws would remain constant, but phenomena would change in time. Consequently we would no longer be entitled to assert that "If *C* happens then *E* will *always* be produced by it", but we would have to add a qualification concerning the time range within which such a regularity is approximately found to obtain.[26]

We may, in brief, account for novelty without adopting James's "perceptual view of causation". The rejection of causality, or, rather, the realization of its limited scope, need not lead us to admit the emergence of novelty out of nothing and to bow before it "with natural piety", as the upholders of emergent evolution advise us to do.[27] Science is engaged in disclosing the specific ways of emergence of newness in the various sections of reality, as well as in investigating the

[26] However, if fundamental physical constants do change in time, they must have varied at an extremely slow pace since life has been on the earth, that is, in the last billion years; in fact, life is possible solely within a narrow fringe of values of certain physical variables, such as temperature, the values of which depend essentially on those constants. See Edward Teller, "On the Change of Physical Constants", *Physical Review 73*, 801 (1948).

[27] The recognition of the emergent or "creative" character of evolution does not entail a commitment to the irrationalistic doctrine of emergent evolution, which negates the possibility of understanding the phenomenon of emergence of a new quality. Determinacy (that is, lawfulness and productivity) accounts for emergence, at least in principle—and provided scholastic notions of change are not retained.

appearance of whole new sections of reality (levels). And the central problem of a scientifically grounded ontology may be regarded as that of inferring from science the most general ways of qualitative change—a task that may eventually lead to suggesting the search for further forms of emergence.

8.2.6. *Positive Features of the Invariance Asserted by Causality*

The positive features of the principle *Causa aequat effectum* should, however, not be missed. One of them concerns quantitative invariance, another immutability, and a third identity of structure. The former restriction on change is true, hence positive, insofar as it states that there are constant quantities and relations amidst the ceaseless flux of phenomena. The insistence on qualitative invariance is positive to the extent to which it warns us not to expect novelties of a radical kind at every step, newness being the exception rather than the rule. Finally, the contention that the cause and the effect are structurally identical may be regarded as one of the principles of scientific inference, or, rather, as one of the ontological principles underlying scientific inference. Let us deal separately with the three interpretations of *Causa aequat effectum.*

The equality of causes and effects has been interpreted, from Aristotle down to our own time, as synonymous with lawfulness. Thus, when the Philosopher defined the force as the cause of motion, he established a constant relation between a cause (force) and its effect (velocity); in modern terms, Aristotle's law of motion reads $F = Rv$, where 'F' symbolizes the force, 'R' the resistance to motion, and 'v' the velocity (see Sec. 4.4.3). Two millennia afterwards Leibniz[28] made an explicit use of the principle of the equality of the cause and the effect, in order to demonstrate what we now call the theorem of conservation of mechanical energy. By adding to that philosophical principle the definitions Cause = Force, and Force = Living force (twice the kinetic energy), he is able to demonstrate his

[28] Leibniz (1695), *Specimen Dynamicum*, in *Philosophical Papers and Letters*, ed. L. E. Loemker, p. 726: "I assume it to be certain, however, that nature never substitutes for forces something unequal to them but that the whole effect is always equal to the full cause". See also *Theodicy* (1710), sec. 346.

laws of motion and, what is more important, feels himself entitled to set forth his hypothesis of conservation of "force" (energy).

One and a half centuries later, Julius Robert Mayer reasoned in a similar way when trying to demonstrate a priori the principle of conservation of energy (which he called force, as was then usual). The kernel of his (partially wrong) reasoning —or, rather, of his own ulterior reconstruction—was the following: "Forces are causes: accordingly, we may in relation to them make full application of the principle—*causa aequat effectum*. If the cause c has the effect e, then $c = e$; if, in its turn, e is the cause of a second effect f, we have $e = f$, and so on: $c = e = f = \ldots = c$. In a chain of causes and effects, a term or a part of a term can never, as plainly appears from the nature of an equation, become equal to nothing. This first property of all causes we call their *indestructibility*."[29] In other words, the main property of causes is their conservation; now, forces are causes (which is true as long as the term 'force' does *not* designate energy); hence forces are conserved (which is true provided the term 'force' *does* designate energy). Very likely this was not Mayer's original train of thought; but when writing it down he found it convenient to rely on the supposed authority of the causal principle, in the narrowest of its formulations. Moreover, Mayer went so far as to claim that his own discovery was but a consequence of the causal principle[30]—as if qualities could be causally connected with one another.

In short, whether understood as asserting quantitative constancy, or numerical equivalence, or proportionality, the principle *Causa aequat effectum* has often played a constructive role—or at least it has helped in disguising scientific novelty under a respectable appearance. The reason for that is simple: the maxim can be interpreted as affording a precise quantitative expression of the causal nexus. A more neutral and powerful principle, that of lawfulness, has nowadays taken the

[29] Mayer (1842), "Remarks on the Forces of Inorganic Nature", in W. F. Magie, ed., *A Source Book in Physics*, pp. 197 ff.

[30] *Ibid.*, p. 202: "We will close our disquisition, the propositions of which have resulted as necessary consequences from the principle '*causa aequat effectum*' . . ."

place of the hypothesis of the quantitative invariance of causes and effects; invariant relations have replaced unchanging events.

As to the qualitative invariability entailed by the principle *Causa aequat effectum* (an invariance which Mayer rightly rejected),[31] its positive content is the following. In the first place, although causalism may produce in us a bias against novelty, it warns us not to be deluded by apparent novelty; and this is in most cases a sound rule, for genuinc novelty is infrequent. As Milne Edwards said, "nature is prodigal in variety but niggard in innovation".[32] Secondly, to say *Causa aequat effectum* amounts to saying, for instance, that flies, not dogs or TV sets, nor even new species of insects, will come out of flies; it definitely rules out the possibility of transformation of this animal genus—which is also correct to a first approximation, since such things do not happen daily. The qualitative immutability asserted by causalism means that one cannot expect to obtain, from a given set of causes, whatever effect one's fancy invents, but only those effects that the given conditions can possibly bring out. So far, this is included in the general principle according to which nothing is produced *ex nihilo* and in arbitrary ways—which we regarded as the essence of general determinism (see Sec. 1.5). Not only pantheism but also miracles are thereby ruled out.

A third meaning which the maxim *Causa aequat effectum* may be assigned is that of structural invariance. This is, indeed, one of the many ways in which Russell has understood the causal link; let us hear him: "Take, say, broadcasting: a man speaks, and his speech is a certain structure of sounds; the sounds are followed by events in the microphone which are presumably not sounds; these, in turn, are followed by electromagnetic waves, and these, in their turn, are transformed back into sounds, which, by a masterpiece of ingenuity, are closely similar to those emitted by the speaker. The intermediate links

[31] Mayer (1842), "Remarks on the Forces of Inorganic Nature", in W. F. Magie, ed., *A Source Book in Physics*, p. 198: the second essential property of all causes is the "capability of assuming various forms . . . Taking both properties together, we may say, causes are (quantitatively) *indestructible* and (qualitatively) *convertible* objects".

[32] Cited by Darwin, *The Origin of Species*, p. 204.

in this causal chain, however, do not, so far as we know, resemble the sounds emitted by the speaker except in structure . . . It appears generally that, if A and B are two complex structures and A can cause B, then there must be some degree of identity of structure between A and B".[33] If biunique correspondence is regarded as a defining mark of the causal connection (see Sec. 2.4.1), then the last statement is tautologous, because structural similarity is nothing but biunique correspondence, that is, coördination of sets of elements of arbitrary kinds. Even so, Russell's is a useful statement, so far as it emphasizes that the causal principle does not involve quantitative or qualitative invariance, but just the identity of structure between the effect and its cause; it is only *causalism* that leaves no room for novelty.

8.3. Summary and Conclusions

In summary, strict causalism may account for novelty in number and in quantity alone; by declaring that actuals are either the mere manifestation or the quantitative development of possibles, causalism definitely excludes qualitative novelty. We are thus faced with the strange fact that the doctrine of causality, which is supposed to account for change, ends up by denying radical change, that is, that variety of change involving the emergence of new qualities. This paradox led Meyerson—who granted the possibility of novelty but saw no other form of determinism than its causal variety—to the conclusion that science leaves always an irrational remainder.[34] The same paradox has led others to fancy that novelty either is impossible or shows the failure of determinism—which would be right if causalism exhausted determinism.

[33] Russell (1948), *Human Knowledge: Its Scope and Limits*, pp. 485–486. Russell states also the converse, that "given two identical structures, it is probable that they have a causal connexion" (p. 486). The exceptions, however, seem to be too numerous, as shown by the fact that one and the same mathematical form can be correlated with a potential infinity of concrete objects with the same structure.

[34] Meyerson maintained that to explain (and, in particular, to explain causally) means to identify, to reduce an initial diversity to a final identity, to reduce change to permanence; and whatever cannot be thus reduced is an irrational residue. See *Identité et réalité* (1908), *De l'explication dans les sciences* (1921), 2 vols., and *Du cheminement de la pensée* (1931), 3 vols.

Novelty is not the insoluble *caput mortuum* imagined by irrationalism; it is explainable, but with the help of all the categories of determination. Moreover, the causal principle can be helpful in the detection of the very novelty causalism denies. In fact, whenever the principle "Same causes same effects" does not seem to be fulfilled, we tend to assume that the cause has *not* been the same in all cases, that is, that something *new* crept in unnoticed. As Bernard said, "given a natural phenomenon, whatever it may be, an experimenter should never admit that there is a variation in the expression of that phenomenon, without at the same time *new* conditions appearing in its manifestation".[35] That is, just because causalism is insufficient to explain novelty, it affords a criterion for disclosing the emergence of newness, namely, the failure of causal laws. The same holds for every conservation law; whenever such a law seems to fail we are led to assume either that the concrete object in question is not as isolated as it was supposed to be or that something new has emerged. This is the type of inference we perform when we find something missing at home, or when we assume the emission of a neutrino; in either case a *new* entity (burglar, neutrino) is assumed to exist, which restores conservation of something (property, energy).

But if the *doctrine* of causality is too narrow to account for every sort of change, on the other hand the causal *principle* is consistent with radical change, and causation itself seems to take part in the emergence of every novelty. Ideal schemes of change may be purely causal, purely random, purely self-determined, and so on; real changes, on the other hand, are always a mixture or, better, a combination of several types of becoming; their description should consequently include various categories of determination—if only because real changes happen to many-sided objects that bear a number of connections with other objects. Causal bonds may not constitute

[35] Bernard (1865), *Introduction à l'étude de la médecine expérimentale*, 2nd ed., p. 86. From this Bernard inferred that, if a phenomenon does not fit this scheme, if it cannot be explained in a causal way, the scientist must plainly reject it: "reason should reject the fact as an unscientific fact . . . for the admission of an uncaused fact . . . is no more and no less than the negation of science" (pp. 87–88).

the main connections in all cases, they may even be irrelevant to a given transformation, but it seems safe to assume that they somehow have a share in actual change.

In short, causation participates in the production of novelty although it does not exhaust it; and, although causalism is a conservative doctrine, the principle of causation is consistent with the emergence of newness. This is why the causal principle has a place in science, though not to the exclusion of other principles of determination. To find that place will be the concern of the remaining investigation.

PART IV

The Function of the Causal Principle in Science

9

Causality and Rational Knowledge

In this, the last part of the present work, we shall inquire into the place held by the causal principle in modern science. That is, we shall investigate the function of a philosophical principle in scientific research. Should it turn out that the principle does play a role, our conclusion would support the general thesis that philosophy neither towers above science nor stands at its basis, but is instead part of the very stuff of scientific research.

Before such analysis is begun, it will be convenient to make it clear what is here meant by scientific knowledge, or science, which is in turn a province of rational knowledge. In German-speaking countries, every *serious* (but not necessarily meaningful, coherent, and verifiable) discourse is entitled to be labeled scientific; thus, for instance, Husserl described phenomenology as *strenge Wissenschaft* (rigorous science). In other languages of the European continent, on the other hand, every discipline making *verifiable* statements—or at least striving to make them—whether requiring empirical confirmation (like those of biology) or not (like those of mathematics), is usually called a science. Finally, in English we find the paradox that mathematics and logic, though both admittedly *scientific* in the highest degree, are not usually included in science; indeed, in that language the word 'science' ordinarily covers the study of nature, spiritual activity, and society, to the exclusion of the disciplines dealing with the forms of thought. (Sometimes the term 'empirical science' is employed to denote all sciences in which observation or experiment or both play a decisive part, but this term is not convenient because factual sciences are not more empirical than rational, and because it seems to entail a commitment to empiricist philosophy.)

Although the English nomenclature is not entirely satisfactory, it will be adopted in what follows; that is, we shall include under science any discourse that is both rational (meaningful and coherent) and at least partly empirically confirmable—though not necessarily demonstrated, that is, not necessarily true with a high probability.

9.1. Is Causality Characteristic of Modern Science?

According to a rather popular belief, causality typifies modern science from its beginning till the birth of quantum mechanics, that is, roughly from the middle of the 16th century to our day. But most philosophers, and some scientists, know that the causal principle has survived the birth of the quantum theory (see Sec. 1.2.5), and that causal thought is much older than modern science. Explanation by causes is, indeed, as old as the phenomenological description of sheer time sequences. Moreover, the reduction of determination to causation is found in rather backward stages of knowledge, though probably not in the most primitive of all. (Thus, the Trobriand islanders are said to lack words to denote causation.) It seems, in fact, characteristic of primitive mentality,[1] at least at a certain stage of its evolution, to assign a cause to everything that is, begins to be, or passes away, and, particularly, to invent myths for explaining causally the origin of what we now regard as self-existent, unengendered, uncaused, namely, the universe as a whole; thus many cosmogonies, whether religious or not, besides fulfilling a social function satisfy the urge for causal explanations. A second typical characteristic of primitive mentality is the ignorance of chance, the refusal to believe in mere conjunctions and fortuitous coincidences, and the complementary belief that all events are causally connected, whether in an overt or in a hidden (magical) way. This belief in the universal causal interconnection—a belief probably born in

[1] Lévy Bruhl, *Les fonctions mentales dans les sociétés inférieures* (1910); *La mentalité primitive* (1922).

prehistoric times—was adopted in antiquity by Stoicism and is nowadays held by the continuers of prehistoric thought.[2]

Admittedly, it is a far cry from the primitive to the modern stage of causal thought; neither primitive nor archaic (pre-classic) thought had our view of an *impersonal*, *lawful*, and in principle *controllable* causation.[3] Causal thought, chimerical in most cases but often correct when it was a question of survival, had been one of the first ways of explaining becoming and even being. By requiring a ground for every existent, causal thought has prompted the search for objective connections; on the other hand, by being unable to conceive that something may exist per se, on its own account, without having been created, and by exaggerating the connectedness of the world, causal thought has also stimulated the elaboration of mythical and religious world views, quite apart from their well-known social roots.

Causal thought was codified by Aristotle (see Sec. 2.1). But, unlike primitives and children, who ignore chance, the Stagirite admitted it both at the ontic and at the epistemological levels, that is, both as objective contingency and as a name for our ignorance of the real causes. Only, he refused to admit chance as an object of scientific knowledge, since by definition "we cannot know the truth apart from the cause".[4] Every object, with the sole exception of the Unmoved Mover, exists and eventually changes in virtue of causes of several kinds; nothing but the First Cause is self-caused. The causal teaching of Aristotle was revived and worked out by the best of scholastics, one of whose watchwords was *Scire per causas*—To know by causes. The scholastic period was the golden age of causality; everything was then assigned a cause, and even chimeras were explained in causal terms in every detail. *Cognoscere causas rerum* —To know the causes of things—notwithstanding that it is the

[2] See Clymer (1938), "The Hermetic Teachings", in *A Compendium of Occult Law*, p. 116: "That which is manifest to us in the universe or the Macrocosm is the thought of God—the activity of the Law of Causation". (The author was Sovereign Grand Master of the Confederation of Initiates.) See also Amadou (1950), *L'occultisme: Esquisse d'un monde vivant*, chap. i.

[3] See Frankfort *et al.* (1946), *Before Philosophy*, 2nd ed., pp. 19, 24, and *passim*.

[4] Aristotle, *Metaphysics*, book II, chap. i, 993b.

motto of modern institutions of learning, is not a mark of modernity but rather a mark of a Peripatetic outlook—and, indeed, a mark of Christian philosophy in both its Aristotelian and its Platonic trends.[5]

What is characteristic of modern science, in connection with causality, is rather:

(*a*) the restriction of causation to *natural* causation (naturalism);

(*b*) the further restriction of all varieties of natural causation to *efficient* causation;

(*c*) the endeavor to reduce efficient causes to *physical* ones (mechanism);

(*d*) the requirement of *testing* causal hypotheses by means of repeated observations and, whenever possible, through reproduction in controllable experiments;

(*e*) an extreme *cautiousness* in the assignment of causes and a ceaseless striving toward the minimization of the number of allegedly ultimate natural causes (parsimony);

(*f*) the focusing on the search for *laws*, whether causal or not;

(*g*) the *mathematical* translation of causal connections.

Even so, the first characteristic—that is, the restriction of causes to natural factors—although practised by many scientists in antiquity, was not firmly established until the 18th century; before that, supernatural causes, as well as reasons, were admitted among causal determinants.

9.2. Cause and Reason

The principle *There is a reason for everything* has usually been regarded as the epistemological partner of the ontological principle according to which *Everything has a cause.* Moreover, the two have been welded for thousands of years. The identity of explanation with the disclosing of causes is even rooted in the Greek language, in which *aition* and *logos* are almost interchangeable, since both mean cause and reason. The confusion

[5] See Gilson (1940), *The Spirit of Mediaeval Philosophy*, chap. xviii; Nigris (1939), *Crisi nella scienza*, pp. 54 ff. There are, of course, exceptions; one of them is the defense of phenomenalism by Duhem and his followers, among them Fathers Gorce and Bergounioux (1938), *Science moderne et philosophie médiévale.*

of cause with reason, and that of effect with consequent are, moreover, common in our own everyday speech.

The identity of reason and cause was consecrated by Aristotle, to whom we owe the distinction between demonstrative (or empirical) and explanatory (or theoretical) disciplines; the former he held in lower esteem than the latter, which he regarded as pointing out the causes of things. Those who know by experience know only the *how* of things (the scholastic *quia*), whereas those who possess the art attain the understanding of the *why* (the *propter quid* of the schools). And "men do not think that they know a thing till they have grasped the 'why' of it (which is to grasp its primary cause)".[6] In the case of the Stagirite, this identification of cause and reason was not a mere confusion, although it may have been influenced by language; it was, I believe, an instance of his deliberate identification of logic and ontology—an identity that is still blocking the understanding of both. A similar identification of reason and cause, and consequently of explanation with causal explanation, was taken for granted by most Peripatetic schoolmen. (Albertus Magnus was among the very few who distinguished the physical *causa* from the logical *ratio*.) This shows, once again, that the identity of explanation and causal explanation is far from being peculiar to modern science, the moral being that we need not fear being called unscientific whenever we resort to noncausal explanations.

The 17th-century rationalists adopted the traditional identity of cause and reason, but they *inverted* the terms: causes were now reasons and, most often, reasons were of a mathematical kind. A mathematical proposition, not a physical agent, was regarded as the sufficient or determinant reason, not only of another idea but of material facts as well—as if things and ideas were on

[6] Aristotle, *Physics*, book II, chap. iii, 194b. See also *Posterior Analytics*, book I, chap. ii, 71b: "We suppose ourselves to possess unqualified scientific knowledge of a thing, as opposed to knowing it in the accidental way in which the sophist knows, when we think that we know the cause on which the fact depends, as the cause of that fact and of no other, and, further, that the fact could not be other than it is". In *De coelo*, book II, chap. xi, Sec. 2, Aristotle states that nature does nothing without a reasonable ground or in vain; in this case he identified reason and final cause.

the same level or as if the former hung from the latter. This reduction of *causal* explanation to *rational* explanation was typical of Kepler—at any rate of the young Kepler, who took no empirical fact for granted but, in a characteristically rationalist and Renaissance mood, tried to explain why the then known planets were just six. (He did it, as is well known, in terms of the five regular polyhedrons and the spheres containing them.) To young Kepler, the ultimate cause of things is some mathematical "harmony"; as Burtt explains, Kepler thinks that "he has reached a new conception of *causality*, that is, *he thinks of the underlying mathematical harmony discoverable in the observed facts as the cause of the latter, the reason, as he usually puts it, why they are as they are*".[7]

That rational explanations of this sort are in a sense the very *opposite* of causal explanations in terms of physical agents could not, of course, be realized by any rationalist who believed that nature is mathematically built, whence the use of mathematics, more than a mere practical convenience (as pragmatism holds), and much more than the sole adequate language of natural science, conveys the very essence of things. For, as Galileo put it in well-known words, the book of nature "is written in the language of mathematics".[8] The conviction that a mathematical object not only may be helpful in the description and explanation of nature but is in a sense the *cause* of it—or even that the fact *consists* in a given mathematical structure—is not a relic of the past; it can be found, for example, among those physicists who assert that the quantum properties of matter consist in, or follow from, the noncommutative multiplication of certain operators.

[7] Burtt (1932), *The Metaphysical Foundations of Modern Physical Science*, p. 53

[8] Galileo (1623), *Il Saggiatore*, in *Opere*, vol. VI, p. 232. This has little to do with Platonic idealism, although it is usually taken for it. Neither Galileo nor Kepler were Platonists, although they did seek the support of Plato's authority against the authority of Aristotle. To them, mathematical objects were not Platonic Ideas standing *apart* from natural things in a realm of their own, but were rather the kernel of the world. Thus "quantity", which was at that time regarded as the very essence of mathematics, was conceived by Kepler as the fundamental feature of things and as prior to other categories. The separateness of things and ideas made knowledge of nature a matter of opinion for Plato; the mathematical essence of things was for Kepler and Galileo the very ground of natural science.

9.3. Causation and the Principle of Sufficient Reason

No wonder, therefore, that to orthodox rationalism, whether medieval, Renaissance, or contemporary, the principle of causality is analytic and, moreover, has to be stated in the form of the principle of sufficient reason. According to the latter, which was Leibniz's pride and joy, *Nothing happens without a sufficient reason.*[9] This proposition has actually been regarded as a ground of being, as a *principium essendi* and not only as a principle of becoming or *principium fiendi.* In point of fact, a strict rationalist should require that a *reason* be given not only for that which *begins* to be or *ceases* to be, but also for every existent, including the sum total of existents, for "there can be found no fact that is true or existent, or any true proposition, without there being a sufficient reason for its being so and not otherwise, although we cannot know these reasons in most cases".[10]

But the confusion of the causal principle, which has an ontological status, with the principle of sufficient reason, which is an epistemological rule of procedure (although it has often been counted among the principles of logic), is not the monopoly of rationalists. This confusion of cause and reason is not peculiar to Kepler, Spinoza, or Leibniz; it has also been indulged in by some empiricists[11]—which should not be too surprising in the context of a doctrine holding that causation belongs to the sphere of the experiencing and knowing subject. Such a confusion

[9] Leibniz (1714), *Principles of Nature and of Grace, Founded on Reason,* Sec. 7. Though implicitly employed for centuries, and stated explicitly by Spinoza, *Ethics* (c. 1666), I, prop. xi, this principle was first regarded as a universal panacea by Leibniz. For a short history of this principle, see Schopenhauer (1813), *Über die vierfache Wurzel des Satzes vom zureichenden Grunde,* secs. 6–14, and Lovejoy (1936), *The Great Chain of Being,* chap. v. It seems, however, that the history of this important principle in connection with scientific thought remains to be written.

[10] Leibniz (1714), *Monadology,* sec. 32, in *Philosophical Papers and Letters,* ed. L. E. Loemker, p. 1049. See also *Theodicy,* sec. 44. Long before, in his *Réponses aux deuxièmes objections* (of Mersenne), Descartes had laid down the following proposition, calling it Axiom I: "Nothing exists of which it is not possible to ask what is the cause why it exists." If it is remembered that Descartes also identified 'cause' and 'reason', it becomes difficult to understand why Leibniz came to regard that principle as his own.

[11] See Neumann (1932), *Mathematische Grundlagen der Quantenmechanik,* p. 160; Rapoport (1954), *Operational Philosophy,* chap. 5 of which is entitled "The Problem of Causality: 'Why is X?'".

(or, as the case may be, deliberate identification) parallels the confusion of material reality with its reconstruction in thought; it has suggested the doctrine of preëstablished harmony between reason and reality, and it has fostered the notion that the result is contained in the cause in the same way as the consequence of a ratiocination is logically entailed by its premises, whence nothing new can emerge in the world—unless novelty is injected from outside the world itself.

The confusion of cause and reason in favor of the latter was useful for science as long as it rested on an unbounded faith in the rationalizable, intelligible structure of reality, and on confidence in the power of reason to disclose it. Moreover, in the beginnings of modern science this confusion carried with it a great enlargement of the scope of scientific research. Indeed, since looking for the causes of things was interpreted as advancing reasons and, whenever possible, reasons of a mathematical kind, instead of looking for one-sided cause–effect bonds among separate events, instead of searching for the ultimate, hidden causes of phenomena, and of formulating integral laws with the schematism required by the narrow frame of causality, and, above all, instead of inventing fictitious causes in order to comply with causalism, modern science began to frame *all* sorts of reasons, establishing all kinds of connections, whether among coexistents or successive qualities of the same object, or among aspects of different objects.

It was fortunate for science that Kepler and Descartes should regard rational (and, particularly, mathematical) explanation as ultimate, and that Galileo regarded it as indispensable. But rational explanation in terms of mathematical propositions could not satisfy physicists for ever. It did not satisfy Galileo, who, besides mathematical reasons, sought, without finding them, the hidden physical causes (which he identified with mechanical forces). It did not satisfy Hobbes either, who tried without success to combine ancient atomism with Galileo's mechanics. And it did not satisfy Newton; moreover, Newton was the first who succeeded in setting forth a *physical* (as contrasted to a *rational*) explanation of the motion of bodies, and in framing it in the language of mathematics. The rationalist program advanced by Descartes and consisting in the derivation of all physical phenomena from self-evident *mathematical* principles[12] was, in fact, definitely turned upside down by Newton who, in opposition to the Cartesians (though without mentioning them explicitly) asserted that "whatever is not deduced from the phenomena is to be called an hypothesis; and hypotheses, whether metaphysical or physical, whether of occult qualities or mechanical, have no place in experimental philosophy. In this philosophy particular propositions are inferred from the phenomena, and afterwards rendered general by induction".[13]

From that time on science has been explaining the world around and inside us with increasing success—especially to the extent to which determinism has unawares been enriched with the addition of new, noncausal types of scientific law. Modern science has, at least in this respect, complied with Bacon's requirement[14] of disclosing not so much the causes of things as their laws—something that cannot be understood if the identity of cause and reason is retained.

The principle of sufficient reason is extensively used in all fields of knowledge, though only exceptionally as synonymous with the causal principle. Giving reasons is no longer regarded as assigning causes; in science, it means to combine particular propositions about facts with hypotheses, laws, axioms, and definitions, some of which may but need not involve the cause concept. *In general there is no correspondence between sufficient reason*

[12] Descartes (1644), *Les Principes de la Philosophie*, part II, 64, in *Oeuvres*, vol. 3, pp. 178–179. As is well known, Descartes's secret philosophy, expounded in *Le monde* (1633), which he did not dare publish, was almost entirely materialistic; it is only his public philosophy (which ultimately prevailed) that was rationalistic to the extent to which reason did not conflict with dogma.

[13] Newton (1687), *Principia*, ed. Cajori, book III, gen. schol., p. 547. Needless to say, the value of this famous statement is that it rejects the aprioristic method of Descartes. But it is far from being a faithful account of the method of scientific theory actually practised by Newton, which hinges not so much on induction as on hypotheses, only not on arbitrary but on testable ones.

[14] Bacon (1620), *Novum Organum*, book II, sec. 2. Laws of nature were sometimes called *forms* by Bacon, who held that it was the concern of metaphysics to discover the formal and final causes of things, whereas physics was to deal with the efficient and the material causes, which he regarded as "slight and superficial, and [which] contribute little, if anything, to true and active causes".

and causation: it suffices to recall that mathematics, which does employ the principle of sufficient reason, remains outside the scope of the causal principle, which has an ontological status. Consequently, the principle of sufficient reason cannot be regarded, as it usually is, as "the mental aspect of causality".[15] And, generally, scientific knowledge does not consist in a mere reflection of material reality, but in a real reconstruction employing materials (images, concepts, ideas, ratiocinations) of its own, obeying laws of their own (those of formal and "inductive" logic) and often lacking a factual (empirical or material) correlate.

9.4. Limits of the Principle of Sufficient Reason in Connection with Theoretical Systems

9.4.1. *Should Everything Be Rationalized?*

One of the most amusing shows provided by rationalism was Christian Wolff's attempt to *demonstrate* the principle of sufficient reason—which attempt was, of course, grounded on the assumption of the validity of the very principle he tried to demonstrate. The principle in question has a wide field of application, but its range of validity is not the whole universe of scientific and philosophic discourse; first of all, it obviously cannot be applied to itself but only—as philosophical semantics teaches us—to propositions belonging to a lower level of language.

That not everything should or could be defined, explained, or demonstrated (in short, rationalized) *in a given context* or level, that some starting point is always required, was dimly perceived in antiquity by many philosophers and scientists, as is shown by the fact that they often developed their discourse starting from first principles that were either taken for granted as obvious or regarded as results of induction. Thus, for example, Leucippus and Democritus explained everything in terms of moving atoms—not simply atoms, nor solely motion,

[15] Enriques (1941), *Causalité et déterminisme dans la philosophie et l'histoire des sciences*, p. 106. The quoted statement contradicts Enriques' own empiricist assertion that the causal principle is essentially a methodological rule, not an ontological proposition.

but ceaselessly moving ultimate particles of matter; atoms were regarded by them as self-sufficient, ultimate, uncaused entities, which had been in motion from all time.[16] Some ancient atomists may have suspected that such atoms *need* not be explained, since they are the roots (*rhizōmata*) or principles of things. Moreover, it is likely that some of them had even realized that atoms *must* not be explained (although they can of course be described), if they are to function as ultimate explainers of everything that is and that comes into being and goes out of being.

Aristotle was not satisfied with this procedure, which he regarded as lazy; he asked about the *cause* of the atoms' motions,[17] which he did not regard as self-evident. (Only self-evident truths could be accepted as starting points according to Aristotle—which was obviously not the case with the propositions of the atomists.) As a consequence, he did not grant atomic motion the status of a first principle, *archē*,[18] or unengendered root.

Neither Aristotle nor his followers seem to have been aware of the *logical* necessity of admitting, in every context, a set of unexplained or primitive concepts and ideas in order to avoid reasoning in a circle. They did not regard first principles as *hypotheses* to be eventually justified a posteriori, and to be

[16] See Bailey (1928), *The Greek Atomists and Epicurus*, chap. iii, sec. 3.

[17] Aristotle, *Metaphysics*, book XII, chap. vi, 1071b: "Some, *e.g.*, Leucippus and Plato, posit an eternal actuality, for they say that there is always motion; but why there is, and what it is, they do not say; nor, if it moves in this or that particular way, what the cause is." Democritus had held that the eternal motion to which Leucippus had referred was the rectilinear motion of free atoms in a void; the *what* (that is, the type of existent) was then answered, but not the *why*—and this was unsatisfactory to Aristotle. In his *Physics*, book VIII, chap. i, 252 a, b, Aristotle asserts that Democritus had reduced natural causes to the eternity of matter in motion, without believing that he ought to enquire into the principle or ground of that eternity; he himself, on the other hand, believes that a cause, hence a reason has to be afforded for everything that is and that becomes.

[18] Originally, *archē* meant first principle *in time*. This was also the way Plato used it in his *Phaedrus*, 245 c, d, e, where he stated that whatever comes to be must originate in a principle (*archē*) that is unengendered. On the other hand, it seems that Anaximander gave the concerned word the meaning of ground, of that which is at the basis. And Aristotle, *Metaphysics*, book V, chap. i, distinguished no fewer than seven meanings of this word; the common core of them all is "the first thing from which something either exists or comes into being or becomes known."

assumed because something must be granted in every discourse at the start; they regarded them as self-evident and undemonstrable verities that had to be received without dispute just because they were obviously true.[19]

The methodological prescription of modern science, according to which every discourse shall start with a handful of notions among which some are either undefined (but become actually defined, explained, or at least made intelligible a posteriori, by the function they perform) or are defined in a *different* context—this prescription, I say, seems to have risen to consciousness not before the 17th century, and this in connection with mathematics exclusively—and moreover in a few cases. I am thinking of Pascal, who was perhaps the first to realize that the penalty for not granting some undefined notions at the start is circular reasoning.[20] Even Leibniz, the most powerful logician of his time, wished to prove the axioms of Euclid's geometry.[21]

It is no longer seriously disputed, I believe, that logical consistency requires the admission, in each theoretical context, of a set of notions that are primitive *in it*. Their meaning is clarified by the very function they perform in the given context, and their adequacy and fruitfulness can be judged from the consequences that are deduced from them. Finally, definitions (whether explicit or implicit) of terms, and demonstrations of primitive or undefined propositions, are sometimes afforded in other contexts.

9.4.2. *The "Principle" of Insufficient Reason*

There is a rule, occurring in elementary treatises on probability, which at first sight contradicts the principle of sufficient

[19] Even Sextus Empiricus, despite his extreme skepticism regarding the very possibility of proof, demanded that nothing be accepted *ex hypothesi*, not even in the realm of mathematics; see *Against the Professors*, book III, in *Works*, vol. IV, pp. 245 ff.

[20] Pascal did not, however, apply consistently his own rule about "primitive words". See Jacques Hadamard, "La géométrie non euclidienne et les définitions axiomatiques", *La pensée*, No. 58, 74 (1954).

[21] Leibniz (1703), *Nouveaux essais*, book I, chap. ii, p. 62; chap. iii, p. 68. Leibniz granted the existence of undefined ideas and of principles "which cannot be proved and need no proof"—but only in the domain of metaphysics.

reason, namely, the so-called principle of indifference, or principle of insufficient reason. According to Jacob Bernoulli, who formulated this celebrated rule, chance events can be said to be equally probable (for instance, $p_1 = p_2 = \frac{1}{2}$, as in coin throwing) if we know of no *reason* why one of them should be expected in preference to the alternative events. Thus, in the case of a perfectly balanced coin, we may *assume* that the mutually exclusive alternatives "head" and "tail" are equally likely, because we have no ground or reason to expect the appearance of one of them rather than of the other. The "principle of indifference", as it is also called, has often been regarded as an equitable distribution of ignorance among mutually exclusive alternatives; besides, it is often thought to be a purely *logical* criterion for ascertaining a priori when events may be regarded as equally probable, and even as a logical definition of equal probability enabling us to circumvent the circularity involved in Laplace's definition of probability in terms of equally probable events.

Now, strictly speaking, the "principle of indifference" is not a principle of the calculus of probability but an assumption that is often made in its applications to specific situations, and that, far from being purely logical (formal), is usually supported by explicit or tacit symmetry considerations. That is to say, the hypothesis of equiprobability can often be grounded on, or explained in terms of, geometrical or physical symmetry properties; consequently, it is not an ultimate explainer, an *Urgrund*. The "principle of indifference" is just a hypothesis to be confirmed or invalidated by its testable consequences. Thus, we cannot assess a priori the probability of casting an ace when throwing a real die by merely applying the "principle of indifference". The estimate $p = \frac{1}{6}$ is valid for the ideal die; it is approximately true for a "true", unbiased real die, but definitely wrong for a loaded die. Moreover, such an estimate does not rely on a complete ignorance but on a definite knowledge of the symmetry properties of the cube—not to speak of our previous experience with dice throwing.

Besides, the "principle of indifference" does not shrink the domain of validity of the principle of sufficient reason, although it may occasionally *shift* the range of its applicability. In point of fact, even admitting that the lack of any sufficient reason may lead us to conjecture equiprobability in some cases, this will not prevent us from asking, in a *different* theoretical context, *why* the alternatives concerned are equally probable (in case the hypothesis of equiprobability is actually confirmed). To say it in a paradoxical way, we can always ask for the sufficient reason why we have no sufficient reason to expect that a certain chance event will happen rather than another (equally probable) event.

The answer to this question, asked on the ground of the principle of sufficient reason, belongs neither to logic nor to the mathematical theory of probability, but to the special science concerned with the class of events under consideration. Thus in statistical mechanics the ergodic theory is supposed to answer the question, Why are microscopic configurations equally probable in the equilibrium state? Analogously, the method of arbitrary functions applied to an ideal model of roulette enables us to *deduce* the equiprobability of "blacks" and "reds".[22] In other words, what is taken as a primitive in a given context (for example, equiprobability in the applications of the probability calculus) may turn out to be derived in a further context.

In short, the principle of sufficient reason is limited in every theoretical domain by the need of avoiding circularity; but most such limitations disappear if the entire field of relevant rational knowledge is taken into account.

But what about matters of fact? Are we to stop here, too, at a certain critical point, our asking questions about sufficient or determinant reasons?

9.5. Limits of the Principle of Sufficient Reason in Connection with Matters of Fact

In modern physics the following question has been asked: Why are the values of "empirical" constants what they are? And, in particular, What is the reason for the values of the so-called fundamental constants, like the gravitational constant

[22] See Poincaré (1912), *Calcul des probabilités*, 2nd ed., pp. 148 ff.

and the velocity of light in a vacuum? Bridgman, in a curious rationalistic slip, entertained "the hope that some day we may be able to give some sort of account of the numerical magnitude of these constants".[23] Other outstanding physicists have wondered about the particular numerical value of the so-called fine-structure constant, a dimensionless number ($\alpha = 2\pi e^2/hc$) relating Planck's constant h with the basic electromagnetic constants c (the speed of light) and e (the electronic charge). Some have gone so far in their Pythagorean faith as to believe that the "explanation" (theoretical determination) of this number might be the key for solving the riddles of the quantum theory, being consequently "the most important of the unsolved problems of modern physics".[24]

The search for an "explanation" of the fine-structure constant has resulted in a large number of papers out of which nothing seems to have come—as could not be otherwise, since there does not seem to exist a clear idea of what is meant by "explaining" the so-called *fundamental* constants or their various dimensionless combinations. It might be that the failure is just due to the failure to point out what sort of reason would be regarded as sufficient in this case. But it might also be that this happens to be a pseudo problem, there being nothing to be further explained or understood about *fundamental* constants—apart from an eventual derivation of such constants from other constants, to which "fundamentalness" has been shifted. Besides, it is not altogether true that "empirical" constants, such as the mass and charge of the electron, "mean nothing to us because we have to read them from experiment"—as is sometimes stated. That would be so if the values of the mass and charge of the electron were entirely

[23] Bridgman (1927), *The Logic of Modern Physics*, p. 185. See also Einstein (1949), *Autobiographical Notes*, in Schilpp, ed., *Albert Einstein: Philosopher-Scientist*, p. 63: "I would like to state a theorem which at present can not be based upon anything more than upon a faith in the simplicity, i.e., intelligibility, of nature: there are no *arbitrary* constants of this kind; that is to say, nature is so constituted that it is possible logically to lay down such strongly determined laws that within these laws only rationally completely determined constants occur (not constants, therefore, whose numerical value could be changed without destroying the theory)".

[24] Pauli (1949), "Einstein's Contributions to Quantum Theory", in Schilpp, p. 158.

isolated pieces of information; but they mean something to us just because they fit into the whole framework of experimental and theoretical electron physics—and they fit so tightly that their values can be obtained in a number of seemingly disconnected ways.

To turn to more general factual questions, it should be pointed out that there are certain basic principles concerning material reality whose admission *qua* starting points not only does not entail any irrationality but enables us to avoid a lot of nonsense. Thus the principle that matter in motion, and the universe as a whole, are self-existent, uncaused, or causes of themselves rules out in a radical manner the validity of inquiries into the sufficient reason for the existence of moving matter—which inquiries usually transcend the boundaries of rational knowledge. Moreover, if the principle is granted that the universe is uncaused, and that matter is self-existent, then the existence of the material world, rather than having to be justified by human reason, becomes the very possibility for affording the sufficient reason (or, rather, the determining ground) of other things—among them, of reason itself.

The admission of this axiom dispenses us from performing the impossible task of explaining the existence of the sum total of existents—a task that is thereby rendered an apparent question. It does not dispense us, however, from the duty of explaining the *changes* in the world or the existence of the various *parts* of it—which can be explained, at least in principle, as the outcome of processes. Thereby the old and odd question, Why is there something rather than nothing? which has haunted so many thinkers[25] and existentialists,[26] becomes deprived of every sufficient reason.

[25] Leibniz (1714), *Principles of Nature and of Grace*, sec. 7, in *Philosophical Papers and Letters*, ed. L. E. Loemker, p. 1038: the principle of sufficient reason having been stated, "the first question which we have a right to ask will be, 'Why is there something rather than nothing?' For nothing is simpler and easier than something. Further, assuming that things must exist, it must be possible to give a reason *why they should exist as they do* and not otherwise". The final reason of things is found by Leibniz in his own, highly highbrow God.

[26] Heidegger (1929), *Was ist Metaphysik?* ends thus: "The basic question of metaphysics, to which Naught itself impels us, is—*Why is there Being rather than Naught?*"

In brief, the unrestricted application of the principle of sufficient reason, both in theoretical and in factual questions, may lead to the very destruction of reason.

9.6. On the Formalization of Causal Statements

9.6.1. *Logical Equivalents, or Logical Correlates of the Causal Connection?*

If there is some truth in what has been said in the preceding sections, textbooks on logic are not the proper place to deal with the causal problem—nor with any other ontological question. The reason for this is that the causal connection is not a logical relation, that is, it is not an abstract relation of the type $x \mathrm{R} y$ among abstract objects. The causal problem is an ontological, not a logical, question, for it is supposed to refer to a trait of reality, and consequently cannot be settled a priori by purely logical means; it can be analyzed with the help of logic, but cannot be *reduced* to logical terms.[27]

An elementary proof that the causal problem does not belong to logic is that laws of nature, whether causal or not, are by no means *logically* necessary; they are not the sole conceivable ones, and they are not necessitated by the laws of logic. In other words, the law *statements* are contingent truths (in Leibniz's sense of the term) because they lack the certainty that is supposed to characterize analytic statements (necessary truths). This does not mean, of course, that the laws themselves (the objective patterns of being and becoming) are contingent. (The assignment of contingency to the laws of nature is often nothing but a consequence of their mistaken interpretation as factual singular truths.) Nor does it mean that the establishment of law statements is a purely empirical matter; some laws of nature have indeed been deduced theoretically, but always on

[27] For a characteristic reduction of the causal problem to its logical aspect, see Popper (1950), *The Open Society and its Enemies*, rev. ed., p. 720: the terms 'cause' and 'effect' are defined with the help of Tarski's concept of truth in the following way: "Event *A* is the cause of event *B*, and event *B* the effect of event *A*, if and only if there exists a language in which we can formulate three propositions, *u*, *a*, *b*, such that *u* is a true universal law, *a* describes *A*, and *b* is the logical consequence of *u* and *a*".

the ground of empirical knowledge, however scanty it might be. Moreover, the objective validity of such theoretical results cannot be ascertained by purely logical means; only experiment can help to decide (though never irrevocably) about their condition as natural laws.

Conversely, what is *logically* possible need not be causally possible. Thus, it is logically possible for something to emerge out of nothing (in particular, to be noncaused); in other words, no rule of logic prevents us from conceiving of such a possibility, even though it contradicts the known laws of nature. Thus, nothing is easier than framing nonconservation (that is, creation) hypotheses, like $\partial T_{\mu\nu}/\partial x^{\nu} \neq 0$, contradicting the principle of productivity; it is only theoretical reflection upon experimental data, and not pure reason, that suggests putting $\partial T_{\mu\nu}/\partial x^{\nu} = 0$ instead of the previous inequality. Again, it is logically possible for the whole world to cease existing, thereby frustrating all causes; but this conceptual possibility not only contradicts the known laws of nature but is unverifiable, that is, it is not a scientific hypothesis.

The fact that the causal problem is of an ontological *nature* is not inconsistent with its having both a logical and an epistemological *aspect*—as every other philosophical problem does. The causal problem is an ontological question because it refers to a pervasive trait of reality that, rightly or wrongly, is often supposed to appear on all levels of the material and the cultural world, with the exception of the level of abstract objects; that is, the question itself is an ontological one, because its referent belongs to the ontal sphere. But, like every other philosophic and scientific question, the causal problem raises epistemological problems (exemplified by the question of the verification of the causal principle) and logical problems. The logical side of the causal problem essentially consists in the logical structure of the propositions by means of which causal statements are formulated.

Thus the logician is interested in studying the sentence *If C then E always*, by abstracting from the nature of the entities designated by 'C' and 'E', as well as from the specific character of the connection between C and E. He will probably say that, from the point of view of logic, the sentence can be regarded in any of the following alternative ways: (*a*) as a relation of material implication $p_C \supset p_E$ among propositions; (*b*) as an inclusion relation $C \subset E$ among classes; (*c*) as a dyadic relation $x\mathrm{R}y$ among members x and y of the classes C and E. In either case, the logician will divest the statement of its eventual ontological meaning, transforming it into a *logical* form—a form shared by an unlimited number of statements. And this is just what formal logicians are supposed to do, namely, disclose the most general forms, by dispensing with the contents.

Something similar will happen, a fortiori, with the eventual formalization of the more complex sentence *If C happens, then E is always produced by it*—which we have regarded as an adequate formulation of the causal principle (see Secs. 2.5 ff.). Any process of formalization would turn this synthetic statement into a *noncausal* statement, which, however, would be a *logical correlate* or transcription of it. It should not be expected, therefore, that the formalization of the propositions expressing ontological statements, such as the law of causation, will ever exhaust their content or meaning. It is not the business of formal logic to lift the veils hiding the face of the world, but rather to sharpen the rational tools with whose help such a task is performed by the sciences.

On the other hand it is, of course, perfectly legitimate and interesting to inquire about the adequate *logical correlates* of the various verbal statements of the causal principle, as well as of the combinations of causal statements. I speak of logical *correlates* rather than of logical *equivalents*, implying that causation is a form of generation or production, and that statements (contrary to their physical marks, such as spoken sentences) cannot produce anything, so that they cannot stand in a causal relation to one another. This problem of the formalization of causal statements has, of course, been approached, but, so far as I know, with no particular success. Let us take a glance at the difficulties besetting this task, too often hindered by the confusion of logical relation with factual connection—as well as by the Humean tenet that there *are* no factual connections.

9.6.2. *Causation and Implication* (*Material, Strict, and Causal*); *the Relational Approach*

Consider the conditional *If p, then q*, which is often (and erroneously) regarded as the essential scheme of the correct formalization of the causal principle. In the calculus of propositions, the material implication $p \supset q$ (sometimes also symbolized in the form $p \rightarrow q$) is usually regarded as the adequate formalization of the sentence *If p, then q*, p and q being propositions.

But the logical connective '$\supset$' is far from representing the causal connection at the level of logic. In fact, there is general consensus that the logical correlate of the causal connection is logical *entailment*, or the relation of premise to conclusion; that is, the projection of causation on the logical plane is deduction —which does not mean that causation is a logical category, or that deducibility is an ontological category. Now, contrary to logical entailment or deducibility (Carnap's L-implication), the sense of $p \supset q$ is not that, given p, q will *unambiguously follow* from it.[28] In fact, in the first place, a true proposition q is implied by *any* proposition p, whether true or false; particularly, anything "follows" from a false proposition (*ex falso sequitur quodlibet*: $\sim p \supset (p \supset q)$). If the antecedent p were interpreted as the cause and the consequent q as the effect, this theorem would roughly read: "A cause, if absent, entails any effect". Secondly, every proposition implies itself ($p \supset p$). Interpreted in causal terms, this could be taken as meaning "Anything is self-caused"; on the other hand, the causal connection is essentially nonreflexive (nothing being self-caused according to causalism). Thirdly, the relation of implication need not be irreversible or asymmetrical; on the other hand, the cause–effect link is essentially asymmetrical. The lack of uniqueness, irreflexivity, and asymmetry characterizing material implication renders this logical connective a very inadequate logical transcription or correlate of the cause–effect link.

In ordinary language, one says that q follows from p and q is

[28] See Hilbert and Ackermann (1938), *Principles of Mathematical Logic*, p. 4.

deducible from p in an unambiguous (unique) way; we also say that p *entails* q, or that p *implies* q—though not in the above-mentioned, technical acceptation of the word 'implication'. As far back as 1912, C. I. Lewis suggested restoring the correspondence between the ordinary and the technical meanings of 'implication'; to this end, he introduced a new notion of logical implication, to wit, true or *strict* implication, $p \supset_{st} q$, which is read 'it is not possible that both p and $\sim q$ be true';[29] that is, strict implication among propositions holds whenever q is uniquely deducible from p. However, strict implication has not solved the "paradoxes" of material implication[30] and has moreover "paradoxes" of its own; it does not entirely overlap with deducibility, whence it cannot be regarded as the adequate logical correlate of the causal bond.

A third kind of implication has been put forward in an attempt to build a closer logical correlate of causation, namely, the *causal* implication[31] of propositions, symbolized by '$p\ c\ q$' or, in an analyzed form, '$(x)\ (Ex\ c\ Dx)$'. This pioneer work is valuable, if only because it is not an exclusively extensional approach. However, it is ridden by paradoxes and it is open to technical criticism.[32] Besides, the following objections can be raised against it: (*a*) it seems to establish (by convention) the rationalistic tenet that every logically necessary statement is causally necessary; (*b*) 'causal implication' seems to refer to all types of biunique connections, whether involving a genetic

[29] Strict implication is a connective in modal logic. It is defined with the help of the (primitive) concept $\Diamond p$, which is interpreted 'it is possible that p'. The definition of strict implication reads thus:

$$p \supset_{st} q =_{df} \sim \Diamond(p \cdot \sim q),$$

which is the analogue of the definition of material implication in terms of negation and conjunction:

$$p \supset q =_{df} \sim (p \cdot \sim q).$$

[30] See Rosenbloom (1950), *The Elements of Mathematical Logic*, pp. 59–60. Actually there is nothing paradoxical in material implication except the name; the propositional calculus is purely *extensional*, being unconcerned about relations of meaning (content), such as those involved in ordinary entailment. See Langer (1953), *An Introduction to Symbolic Logic*, 2nd ed., 1953, pp. 276 ff.

[31] Arthur W. Burks, "The Logic of Causal Propositions", *Mind* (n.s.) *60*, 363 (1951).

[32] See G. P. Henderson, "Causal Implication", *Mind* (n.s.) *63*, 504 (1954).

relation or not—hence it does not account for what is peculiarly causal.

No satisfactory logical correlate of the causal nexus seems so far to have been proposed; but work is being continued in this field.[33] In my opinion, an adequate verbal formulation of the causal connection (on which there is no consensus) should precede every attempt to formalize causal statements; besides, it should be clearly pointed out from the outset that what is required is not an extension of formal (extensional) relations, but the determination of a type of semantic connection among terms that are relevant to each other: *the logical aspect of the causal problem is semantical rather than syntactical.* Finally, there is no reason why logical analyses of the causal bond should continue to be restricted to the context of the theory of propositions, or even to the calculus of one-place functions; a *relational* approach might be more rewarding. In fact, it allows us to disclose the following formal properties of simple causation: (*a*) it is a *dyadic* relation $x\mathrm{R}y$ holding among elements interpretable as events; (*b*) it is *irreflexive*, that is, $(x) \sim (x\mathrm{R}x)$—which can be read *nihil est causa sui*; (*c*) it is *transitive* (chainlike), that is, $(x)\ (y)\ (z)\ [xRy\ \&\ yRz \supset xRz]$; (*d*) it is *asymmetrical*, that is, $(x)\ (y)\ [xRy \supset \sim yRx]$. Notice, however, that this set of properties does not specify the causal relation: they are common to all ordered sequences, such as the succession of days and nights; they specify the topology of the series, not the nature of its terms.

No formalization should be expected to *solve* the causal problem, although it may contribute to its clarification—especially to the extent to which the causal problem is thereby shown *not* to be of a logical (and, particularly, extensional) nature. Moreover, an exaggerated insistence on the logical aspect of the causal problem at the expense of its ontological content might lead us back to pre-Humean times, when the term 'cause' was so often identified with 'reason', and when so many philosophers thought that the validity of any particular causal statement, and of the causal principle in general, could be settled a priori—as if it were a logical question. (It was a

[33] See Herbert Simon, "On the Definition of the Causal Relation", *Journal of Philosophy* 49, 517 (1952).

merit of Hume to insist that causation should be stripped of the rationalist element of logical necessity; it was, I believe, a mistake of his, as well as of his followers, to deny that the causal problem is an ontological one, and to claim instead that the concept of causation can be *exhausted* with purely logical means.)[34]

To sum up: the causal problem cannot be solved in the domain of pure logic, because it is about the nonlogical (or nonlinguistic) world. Like every other ontological problem, the causal question should be approached on the basis of the so-called empirical sciences, for they are concerned, among other tasks, with the study of particular causal connections. As to the formalization of causal sentences, the fact that no satisfactory solution seems to have been attained may partly be due to the very starting point of most logicians, which is some conventional highly-controvertible formulation of the causal law.

The formalization of causal propositions is, in conclusion, an interesting open problem the importance of which should, however, not be exaggerated, especially since the whole content or meaning of the causation category cannot be displayed with a few symbols. Finally, it must be borne in mind that, to the extent to which the logical side of the causal problem is solved, it ceases being causal.

9.7. Recapitulation and Conclusions

Even by the end of the Age of Enlightenment Lazare Carnot found it necessary to explain, almost apologetically, why he had not deduced or explained the *principles* of his famous *Essai sur les machines en général*: "A detailed explanation of these principles was not in the plan of the present work, and might result only in confusing things; in fact, the sciences are like a beautiful river, the course of which is easy to follow after it has acquired a certain regularity; but if one wishes to trace it to its source, one does not find the source anywhere, because it is nowhere; the source is in a way scattered over the whole surface

[34] See Reichenbach (1929), *Ziele und Wege der physikalischen Erkenntnis*, in Geiger and Scheel, ed., *Handbuch der Physik*, vol. IV, p. 59.

of the earth. Likewise, if one wishes to go back to the origin of the sciences, one finds nothing but obscurity, vague ideas, vicious circles—and one gets lost in primitive ideas".[35] Like almost everybody, Carnot was confusing the historical development of a science with its logical structure at a given time.

The admission of a few ultimates in every scientific discourse, as contrasted with the positing of material ultimates (that is, of unanalyzable material entities) entails no limit upon the intelligibility of the world. In fact, the meaning of such ultimate explainers is always clarified a posteriori, by the function they perform in the given context; that is, they are explained along the way (contextual definition). However, it should be recalled that in science rational ultimates (such as axioms) are not to be regarded as dogmas, or as indisputable self-evident verities, but simply as *hypotheses* to be justified both by the consequences to which they lead and by their consistency with other hypotheses regarded as established. Moreover, the possibility of a further explanation or demonstration of rational ultimates in a different context should not be ruled out a priori, especially in the case of the so-called empirical sciences.[36]

In summary, *the principle of sufficient reason may be applied to everything save to itself and to such elements of discourse as function as explainers in a given context.* Such a limitation of the range of the principle of sufficient reason, far from curtailing the program of attaining a rational understanding of the world, is rather a condition for its consistent fulfilment, for it avoids both vicious circles and the assignment of a fictitious "final reason of things".

What is true of reason in general is, of course, true of causal reasons, that is, of the reasons whose referents are causes, and which are invoked to explain facts of the concrete world outside and inside ourselves. *If not everything that exists has to be explained, then not everything need be assigned a cause.* On the other hand, the requirement that everything in the world be ex-

[35] L. N. M. Carnot (1783), *Essai sur les machines en général*, in *Oeuvres*, pp. 123–124.

[36] Agnostics, on the other hand, have held the absolute ultimacy of certain notions and even their utter unknowability or inconceivability. See Spencer (1862), *First Principles*, part I, chap. iii.

plained by causes necessitates the assumption that every causal chain has its ultimate source in an otherworldly final cause.[37] This helps to explain why the identification of science with the knowledge of causes has not always been beneficial for rational knowledge. While it is true that the description of the task of science as the disclosure of causes has led to the discovery of a number of specific connections and types of connection, it has also had the following negative sides. First, it has contributed to blocking the understanding that matter in motion exists per se and that the physical universe is uncaused, the material world being the ground of reason, rather than needing our rational justification for its existence. Secondly, the belief that science is coextensive with causality has fostered the invention of countless fictitious causal agents—among which the scholastic substantial forms have been prominent—and has stimulated a confusing multiplication of the varieties of causes. (The schoolmen of the 17th century went so far as to list about thirty kinds of cause.) Thirdly, the belief in question has made it easier for the enemies of rational knowledge to hail every instance of noncausal connection as a victory of indeterminism and irrationalism—a triumph determined, be it noted, by rational knowledge itself.

Such are some of the reasons why it is important to realize that causation does not exhaust determination, whence science cannot be restricted to the knowledge of causes.

[37] The deeply idealistic *Upaniṣads* are among the earliest theoretical documents showing awareness that consistent theology, to the extent to which it is possible, requires the postulate of the unrestricted operation of causation in the world of appearance—just in order to be able to posit an uncaused First Cause. In the *Upanisads* it is written that everything belongs to a causal series, nothing save Brahman being self-existent. See Radhakrishnan (1931), *Indian Philosophy*, vol. I, chap. v.

10

Causality and Scientific Law

The chief aims of the present chapter are to clarify the meaning of the term 'scientific law' and to show that the concept of causation need not be replaced by that of law but on the contrary often actually enters into it. A cursory examination will presently be made of both causal and noncausal types of scientific law; it will show that causal laws are only a species of the genus scientific law. This analysis should enable us to avoid excluding from science all those disciplines in which a certain type of law (for example, causal law) does not play a prominent role.

10.1. Law and Law Statement

Most scientists are prepared to grant that the chief theoretical (that is, nonpragmatic) aim of scientific research is to answer, in an intelligible, exact, and testable way, five kinds of question, namely, those beginning with *what* (or *how*), *where*, *when*, *whence*, and *why*. For the sake of brevity let us call them the Five W's of Science. (Only radical empiricists deny that science has an explanatory function, and restrict the task of scientific research to the description and prediction of observable phenomena.) Also, most scientists would agree that all five W's are gradually (and painfully) being answered through the establishment of scientific *laws*, that is, general hypotheses about patterns of being and becoming. Scientific laws enable us to answer the *what*, the *where*, the *when*, the *whence*, and the *why* of facts (events and processes), since with their help we can perform the most faithful description, the most accurate prediction (both forward and backward), and the most truthful explanation of natural and social facts that are possible at any given moment.

It will be necessary to distinguish between *laws* (whether of nature, thought, or society), and *law statements*: the former will be defined as the *immanent patterns of being and becoming*, the latter as the *conceptual reconstructions* thereof. Whenever a law statement is a general and testable (not necessarily tested) hypothesis, and whenever it has been forwarded in accordance with the (fluid) standards of scientific method, we may call it a *scientific law*. For the sake of brevity we shall often refer to $laws_1$ (objective patterns, or laws at the ontic level) and $laws_2$ (scientific laws, or laws at the epistemic level). Later on we shall have to distinguish a third level of meaning of 'law', namely, that of predictive nomological statement (see Sec. 12.1.2).

This distinction between objective laws (laws_1) and their tentative and improvable reconstruction in the human mind (laws_2) is not usually made, although it should be obvious to all nonmentalists—if only for the fact that the difference in question is involved in the very notion of perfectibility of the scientific account of facts, as contrasted with the assumed constancy of the patterns of facts. In fact, with but few exceptions—among them Ampère[1]—philosophers of science mean by 'law' what I am calling scientific law (law_2). Thus Mach asserted that a law of nature is nothing but "a rule for displaying all single predictions".[2] Ostwald[3] defined laws as predictive patterns of the form "If *A* is experienced, then we expect *B*". Carnap saw in laws of nature just "assertions with a general content".[4] Braithwaite, in his important treatise on scientific explanation, gave the following description: "What we call the laws of nature are conceptual devices by which we organize our empirical knowledge and predict the

[1] Ampère (1843), *Essai sur la philosophie des sciences*, vol. II, p. 28: to hold that laws of nature are nothing but *vues de notre esprit* is to take an idealistic stand not far from Berkeley's or Fichte's; we must distinguish the *concept* of the relation from the *relation itself*, as it exists before we have discovered it.

[2] Mach (1900), *Die Principien der Wärmelehre*, 2nd ed., p. 439.

[3] Ostwald (1908), *Grundriss der Naturphilosophie*, p. 57.

[4] Carnap (1926), *Physikalische Begriffsbildung*, p. 49.

future".[5] And Hutten, in his valuable work on the semantic aspect of metascience, states that "a law is merely a statement, or theorem, within a scientific theory".[6] These views are all, of course, in line with the purest empiricist and Kantian[7] tradition; and they are so widespread among philosophers that even Meyerson, despite his conviction that scientific statements have objective referents, could not help writing that "a law of nature that we ignore does not exist in the most rigorous sense of this term".[8]

On the other hand, what is here meant by a law of physical or social reality is not supposed to depend on our knowledge of it; we assume that $laws_1$ work objectively, that they are immanent in things, that they are the modes of being and changing of things. As a consequence, $laws_1$ are discoverable and not inventable—although they are certainly discovered with the help of a great many inventions, tools (both material and conceptual), and instruments, as well as constructs. We certainly do not grasp $laws_1$ in their virgin purity (we do not even grasp singular facts in themselves, without distortion). The process of scientific discovery is not a mere reflection of facts by way of induction, but arduous work of tentative reconstruction which, while penetrating deeper and deeper into the thing in itself, never attains it entirely. The constructs (hypotheses) called scientific laws (our $laws_2$) are the variable reconstruction of objective laws on the level of rational thought. If preferred, $laws_2$ are the (incomplete) projections of $laws_1$ on the conceptual plane. They never entirely overlap the objective laws—as is shown by the historical perfectibility of $laws_2$ in contrast with the assumed constancy of $laws_1$. This partial overlapping, or, better, this incomplete and ambiguous correlation between laws and scientific laws, is not only an instance of the distortions stemming from human frailty or immaturity, but also a consequence of the fact that scientific discovery is a *reconstruction* rather than a *reflection*, a process of building models

[5] Braithwaite (1953), *Scientific Explanation*, p. 339.
[6] Hutten (1956), *The Language of Modern Physics*, p. 222.
[7] Kant (1781, 1787), *Kritik der reinen Vernunft* (B), pp. 163 ff.
[8] Meyerson (1908), *Identité et réalité*, p. 17.

of reality rather than that of copying it—as is shown by the fact that science employs a host of abstract entities lacking a factual counterpart, and by the central role played in it by hypotheses.[9] In brief, whereas laws of nature, thought, and society are the structure of reality, the corresponding law statements belong to ideal models of reality, whence they apply (if at all) only approximately, never to any desired degree of accuracy.

Now, there are many kinds of scientific law: laws of class inclusion, of composition, and of structure; laws of motion in space, of force, and of conservation; laws of interdependence among qualities, of over-all trends, of stochastic connections, and so forth. There are inductive generalizations supported only by the instances they cover, and there are stronger scientific laws, belonging to hypothetical-deductive systems, that are supported not only by their instances but by their compatibility with other hypotheses as well. There are $laws_2$ stating almost everyday matters of fact and others containing elaborate theoretical concepts, while the indispensable form of others is mathematical (but not necessarily numerical or metrical). There are, in brief, many ways in which laws of nature, thought, and society can be looked upon from a meta-scientific viewpoint. But at this point we are interested in the problem of scientific law in connection with the question of causal determinacy. And this problem is essentially the following: Are there scientific laws that can be termed causal, and others that cannot?

At first sight, this is a trivial question. However, the influential positivistic school has claimed, since the days of Comte, that there is a certain *opposition* between causal explanation and scientific law, the business of science being to dispense altogether with the cause concept as well as with explanation, and to concentrate instead on the search for the so-called phenomenological or descriptive-predictive relation. Some anti-positivists, like Meyerson, have maintained on the other hand

[9] See Bunge (1959), *Metascientific Queries*, chaps. 3 and 5; "New Dialogues Between Hylas and Philonous", *Philosophy and Phenomenological Research 15*, 192 (1954).

that scientific laws are nothing but particular cases of the causal principle, and even that entire scientific theories, like analytical mechanics or the atomic theory, are but manifestations of the causal principle, since the chief aim of such theories is to explain phenomena, and explanation, according to an old tradition, is usually assumed to consist in or to be exhausted by causal explanation.

What follows is intended to show that scientific research, the chief aim of which is certainly the discovery of laws, does not dispense entirely with the cause concept, and that science contains both causal and noncausal laws, as well as laws having a causal range.

10.2. The Traditional Identification of Causality and Lawfulness

Following an ancient tradition recorded by Sextus Empiricus,[10] lawfulness—that is, agreement with law—is usually identified with causality, as if no other kinds of scientific law were possible than causal laws.[11] First-rate thinkers have concurred with this confusion even after the popularization of the discoveries of noncausal laws, such as the so-called phenomenological laws of geometric optics and of thermodynamics, the statistical laws of random populations, the teleological laws of living matter, or the dialectical laws of human history.

Thus in his old age Helmholtz recognized that in his famous youthful work on the conservation of energy he had let himself be influenced too strongly by Kant; only later did he realize that "the principle of causality is nothing but the assumption of the lawfulness [*Gesetzlichkeit*] of all natural phenomena".[12] Some years later, Mach stated a similar identity between lawfulness and causality: "The business of physical science is

[10] Sextus Empiricus, *Outlines of Pyrrhonism*, book III, chap. 5.

[11] A characteristic propounder of this identity of lawfulness with causality is Bernard (1865), *Introduction à l'étude de la médecine expérimentale*, p. 133: "We know natural phenomena only by their relation to the causes producing them. [Note this strange inversion of Newton's well-known statement, according to which we *infer* the causes from their observed effects, or phenomena.] Now, the *law* of the phenomena is but the numerical expression of this relation, stated in such a way as to enable us to predict the cause–effect relation in all given cases".

[12] Helmholtz (1881), additions to *Über die Erhaltung der Kraft*, p. 53.

the reconstruction [*Nachbildung*: copy rather than reconstruction] of facts in thought, or the abstract quantitative expression of facts. The rules which we form for these reconstructions are the laws of nature. In the conviction that such rules are possible lies the law of causality. The law of causality simply asserts that the phenomena of nature *are dependent* on one another".[13] From this chain of definitions it clearly appears that Mach did not regard laws as immanent in nature but as rules of procedure, as man-made rules devised in order to translate the "facts" (that is, the primary sense data) into thoughts. A strange paradox: on this view, laws of nature must be *contingent* relatively to nature, just as in the religious philosophies of nature.[14] The two identifications performed by Mach in the foregoing quotation—that of law with rule of procedure, and that of causality with lawfulness—have been retained by subsequent positivists. Thus Russell once gave the following definition: "By a 'causal law' I mean any general proposition in virtue of which it is possible to *infer* the existence of one thing or event from the existence of another or of a number of others".[15] The cryptic prophet of logical positivism wrote in his characteristic style: "If there were a law of causality, it might run: 'There are natural laws'. But that can clearly not be said: it shows itself".[16] The founder of the Vienna Circle[17] regarded laws of nature as *Vorschriften* (directions) and *Verhaltungsregeln* (rules of behavior or procedure) of the investigator —which is certainly true, but not the *whole* truth (see Sec. 10.1); and he also identified causality with lawfulness when he

[13] Mach (1883), *The Science of Mechanics*, p. 605.

[14] The thesis of the contingency of the laws of nature is involved in the positivistic view that they are bare conceptual shorthand or economical rules of procedure. But the metaphysician who most insistently expressed this doctrine was Émile Boutroux, one of the first spiritualists engaged in what Frenchmen call *la critique de la science*. Boutroux stated his main theses in his influential work, *De la contingence des lois de la nature* (1874) and repeated it in *De l'idée de loi naturelle* (1895), where he concluded that "what we call laws of nature is the set of methods that we have found in order to assimilate things to our intelligence, adapting them to the accomplishment of our wills" (pp. 142–143).

[15] Russell (1914), *Our Knowledge of the External World*, p. 216. Italics mine.

[16] Wittgenstein (1921), *Tractatus Logico-Philosophicus*, 6.36.

[17] Moritz Schlick, "Die Kausalität in der gegenwärtigen Physik", *Die Naturwissenschaften 19*, 145 (1931).

wrote that "The enquiry as to whether causality exists can only be interpreted as the enquiry whether a natural law exists. The principle of causality itself is not a law; it only expresses the fact that laws exist".[18] Even antipositivists, like Bergson[19] and Planck,[20] have retained the traditional identification of scientific law with causal law—though of course with different aims.

Although statistical physics is almost a century old, it is still possible to read sentences stating that there is an irreducible opposition between genuine laws of nature and statistical laws —as if the latter could not be laws of nature and indeed not laws at all, but "mere" over-all regularities, just because they do not enable us to predict *individual* events with *certainty*. Even Eddington[21] could not acknowledge the existence of a *law of chance*; although he was an indeterminist, the very notion of a *law* of chance conflicted with his Pythagorean rationalism. The same irreducible opposition between natural and statistical laws was held by two eminent empiricists[22]—presumably on the 18th-century thesis that chance is but a name for our ignorance of nature's "stern rules" of causation, and not an objective mode of being subject to a new type of law.

The mere existence of noncausal scientific laws, such as the laws of correlated variations in organisms, shows that scientific research does not consist exclusively—probably not even

[18] Schlick (1936), *Philosophy of Nature,* p. 57. See also "Causality in Everyday Life and in Recent Science" (1932), reprinted in Feigl and Sellars, ed., *Readings in Philosophical Analysis,* p. 523, and Carnap (1926), *Physikalische Begriffsbildung,* p. 57.

[19] Bergson (1888), *Essai sur les données immédiates de la conscience,* p. 150.

[20] Planck (1936), *The Philosophy of Physics,* chap. 2. Elsewhere, however, Planck drew the useful and now widespread distinction between *individual or dynamical law,* and *statistical or collective law*; see his *Introduction to Theoretical Physics,* vol. V, p. 225. He had, however, the hope that all statistical laws would ultimately be reduced to dynamical laws of the causal type (a hope as unwarranted as that of the ultimate reducibility of all types of law to statistical law). See *Where is Science Going?* (1933), pp. 146 ff.

[21] Eddington (1939), *The Philosophy of Physical Science,* p. 181. The usual (classical and popular) meaning of the word 'law', which Eddington seems to have had in mind, is that of a "stern rule" making no allowance for exceptions. An instance of general laws of nature making room for exceptions is the second law of thermodynamics as framed and interpreted by statistical mechanics.

[22] Borel (1914), *Le hasard,* chap. i; Enriques (1941), *Causalité et déterminisme dans la philosophie et l'histoire des sciences,* p. 68.

mainly—in the search for causal laws. Just a few types of noncausal scientific laws will be mentioned in following sections, with particular emphasis on taxonomic, kinematic, and statistical ones. They might all be forced under the heading of *exclusively* descriptive laws: but this classification, however widespread, is erroneous, since such laws serve explanatory purposes as well; only, as will be shown below, they do not enable us to formulate *causal* explanations.

10.3. Some Noncausal Types of Scientific Law

10.3.1. *Taxonomical and Morphological Laws*

All law statements *involve* the class concept, whether explicitly or not, since they all refer to kinds or classes of facts rather than to singular facts (see Sec. 2.2). But taxonomical laws are *just* classifications, that is, statements of class inclusion and therefore they are the simplest of all types of law statements. Linnaeus's epoch-making *Systema Naturae* (1737, 1768) was not concerned with change, hence it did not need any category of determination. This work introduced a method of artificial classification (by single characters) of *coexisting* living forms without any reference to lineage, and set forth a number of taxonomical laws. True, the Linnean classification was afterward supplemented with systems of phylogenetic classifications founded on aggregates of characters and based on hypotheses of genetic relations and evolution. That is, new arrangements of the available material were made with a predominantly (though not exclusively) descriptive purpose, on the basis of explanatory hypotheses. In other words, new descriptions were grounded on a new explanation. But Linnaeus's systematics was undoubtedly an indispensable preliminary stage in biology, at least in the science of plants; and this is easy to understand, for before knowing *that* species had been changing, and that genealogy accounted for parentage, it was necessary to know *what* might have evolved. The replacement of artificial by natural or genealogical systems of classification did not mean, however, a substitution of *causal* explanation for sheer description; in fact, to the extent to which evolutionary systematics accounts for biological progress, it goes beyond the scope of causality, which excludes higher novelty (see Sec. 8.2).

Another outstanding member of the class of noncausal and morphological theories is classical crystallography. The crystal types classified by this discipline may in turn be explained by the atomic theory, which does include a causal element. But the discipline concerning crystal forms is a science in its own right, and is moreover indispensable for *explaining* certain physical phenomena like piezoelectricity, which originates in crystallographic asymmetry. Generally speaking, the morphological branches of science need not contain causal elements, although they may turn out to be explainable in terms of theories containing causal elements. (For the explanation of laws in terms of higher-level laws, see Sec. 11.2.3.)

10.3.2. *Kinematical Laws*

Among kinematical theories, some of the most interesting and far-reaching are Ptolemy's and Copernicus' theories of planetary motion. They assigned no *causes* to the complex motions of the celestial vagabonds; however, they helped to explain and to predict *other* celestial phenomena, such as eclipses. That is to say, these kinematical theories were not—contrary to what is usually held—*purely* phenomenological theories; they were not restricted to describing phenomena without explaining any of them. Thus the aim of Ptolemy's astronomical theory was *to explain sensible appearances*, notably the apparently irregular motion of the planets (see Fig. 26), which was indeed irregular relative to the apparently circular motion of the stars. (Is it necessary to show that order and regularity are not absolute but relative?) The *Almagest* (c. A.D. 140) performed this task with the help of the hypothetical epicycles and eccentrics invented by Hipparchus (see Figs. 27 and 28), and which were curves to which Ptolemy and most of his followers seem to have assigned no reality. (The component motions functioned as *reasons* rather than as *causes*.) Ptolemy thereby succeeded in *explaining* the apparent motions of the planets, whose irritating irregularity was one of the most important sources of progress in astronomy; to this extent, his

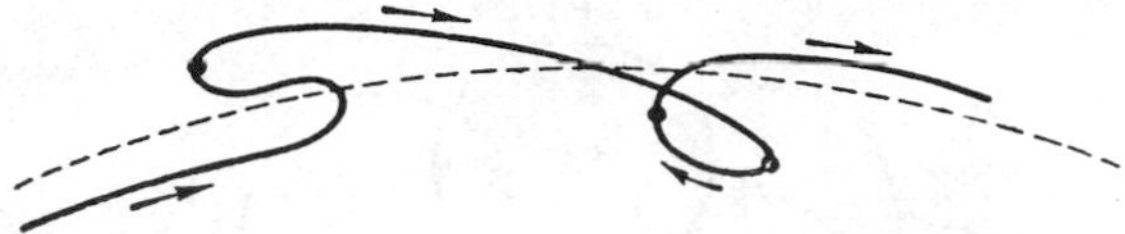

Fig. 26. Apparent motion of a planet. As seen from the earth, the advances, halts, and backward motions are irregular relative to the apparent motion of the fixed stars.

theory was not phenomenological—despite the assertion of phenomenalists.[23] What Ptolemy did not explain was why the component motions of planets were such as he assumed and not otherwise; in particular, his theory did not point to the *efficient causes* of planetary motion. His was a noncausal and predominantly descriptive, but partly also explanatory, theory.

Aristotelians could obviously not remain satisfied with such an astronomical system, and precisely because it was not a causal theory, hence not fully explanatory, on the Aristotelian doctrine that explaining is advancing reasons, which are in turn causes (see Sec. 9.2). By adopting and elaborating the astronomical ideas of Calippus and of Eudoxus of Cnidos, according to whom celestial bodies were dragged along by homocentric rotating spheres, Aristotle had completed a causal theory of heavenly motions. Every one of the numerous celestial spheres he invented was charged with the task of pushing a group of celestial bodies; the spheres were in turn pushed each by a First Mover. Unfortunately for the followers of the philosopher,

[23] See Duhem (1908), *Σώζειν τὰ φαινόμενα, Essai sur la notion de théorie physique.* Ptolemy had stated that the task of astronomy was restricted to saving (celestial) appearances (*sōzein ta phainomena, salvare apparentias*). Duhem took this statement as meaning that physical and astronomical theories should be *solely* descriptive, not explanatory. In my opinion, Ptolemy may have meant that astronomy was not then in a position to decide which one of the then known theories (which were equivalent as far as *description* was concerned) gave the truer account of reality. This abstention from judging was due to the discovery by Hipparchus (c. 146–126 B.C.) that with either system, the geocentric or the heliocentric, it was possible to account for the phenomena then known. New empirical data were needed to decide which of these systems was the true one; but this had to wait for Kepler and Galileo.

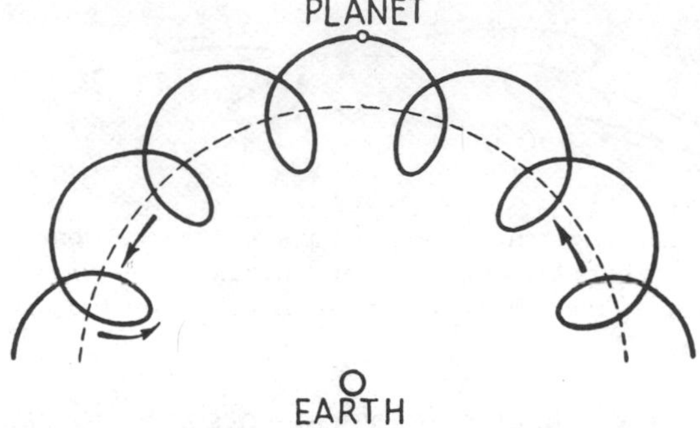

Fig. 27. Geocentric *explanation* of irregular planetary appearances (see Fig. 26) in terms of a regular (circular) motion along an epicycle. A more complete explanation requires the introduction of eccentrics.

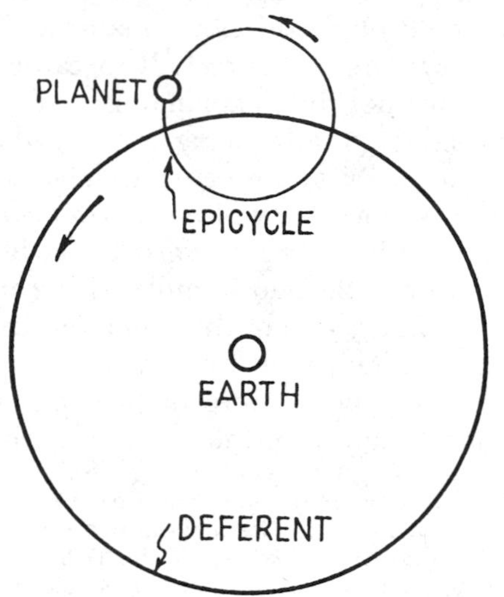

Fig. 28. Snapshot of process laid down in Fig. 27.

his astronomical theory did not match the results of observation. This is why Averroes, the *Commentator*, complained that the astronomy of his day did not exist, being a convenient means for computation, without however agreeing with what exists.

Many an Aristotelian must have wondered how it was possible for a noncausal theory like Ptolemy's to account for most of the then known celestial phenomena and even to afford an explanation of some of them (notably the awesome eclipses). However, nobody seems to have drawn the correct conclusion, namely, that a scientific theory need not be causal.

As is well known, Newton's theory enabled kinematic laws (such as Kepler's) to be explained; consequently, these need no longer be regarded as ultimates—though at the price of admitting higher-level ultimates, namely, Newton's second law of motion and the law of gravity, which in turn become derived statements in the wider context of the general theory of relativity. Thereby, astronomical theories have been built, which satisfied the rationalistic demands of Averroes, though not his Aristotelian requirement that reasons correspond to causes. Furthermore, modern dynamics made it possible to show that Kepler's kinematic laws of planetary motion are only approximate (owing essentially to the perturbations exerted by the remaining planets). Besides, it has made it possible to ascertain that the appearances which the geocentric theory took as unshakable starting points were wrong, the correct point of departure being an unperceived but intelligible and objective reality hiding behind the appearances.[24]

But the fact that the elementary kinematic laws of planetary motion can be explained by (that is, be deduced from) a higher-level theory containing an important causal element (Newton's theory) does not invalidate the assertion that the former are noncausal laws. For contrary to what Russell[25] and

[24] Not only believers, like Duhem, but also some positivists, like Frank, Reichenbach, and Dingle, have argued that the geocentric and the heliocentric systems are *equivalent descriptions* of the same set of phenomena (appearances). Two facts, among others, show the untenability of this claim. First, the heliocentric system, as perfected by Kepler and rationalized by Newton, accounts for *more* astronomical facts than the Ptolemaic system does. Second, the relativistic principle of equivalence (between gravitational fields and accelerated reference systems), which has been invoked in support of the alleged equivalence of theories, has no validity in connection with the solar system as a whole; in fact, gravitational fields can be "transformed away" by suitable coördinate transformations solely in differential volume elements of space-time—whence the geocentric description is *not* equivalent to the heliocentric one.

[25] Russell (1948), *Human Knowledge*, p. 344.

others have maintained, a statement such as "Planets move in ellipses" (essentially Kepler's first law) is not a causal law, for the simple reason that it does not refer to the *causes* of either motion or ellipticity.

10.3.3. *Further Noncausal Laws: Statistical Laws, Principles of Relativity, and Quantum Prohibitions*

Statistical laws are indisputably noncausal, even though some of them may be shown to be partly derivable from laws having a causal component. A typical illustration is afforded by the kinetic theory of heat, which explains thermal phenomena as emerging from lower-level mechanical motions, or, more exactly, as the collective result of a large number of individual mechanical motions. Three layers of theory and consequently three types of scientific law are involved in the statistical-mechanical explanation of molar, thermodynamic phenomena. They are: (i) *particle dynamics*, applying to single molecules; (ii) *statistical mechanics*, studying the over-all and long-run behavior of large collections of similar components; (iii) *thermodynamics*, treating physical systems as wholes the emergent properties of which are partially explained in terms of (i) and (ii).[26] The first level (particle dynamics) contains a strong causal component, summarized in the force equation; at the second level (statistical mechanics) the noncausal notion of random motion, or mutual irrelevance of trajectories, is introduced. (That is, while the individual motions are regarded as dynamically determined, every one of them is assumed to be contingent upon the remaining motions.) On the next level of theory (thermodynamics), both causation and chance have disappeared almost entirely; they have merged together, producing a theory dealing with concepts (referring to properties and processes) linked together by symmetrical functional

[26] For example, in the elementary derivation of the equation of state of the ideal gas, essentially the following statements are employed: (i) Newton's laws of motion are valid for each mass-point; (ii) The average kinetic energy (of translation) of a large assembly of freely moving mass-points is $\langle E_{\mathrm{kin}} \rangle_{\mathrm{Av}} = \frac{3}{2} pV$. These hypotheses are necessary (but not sufficient) for the derivation of the desired law statement, namely, (iii) $pV = nRT$, belonging to the third level of theory.

relations which cannot be polarized in an unambiguous way into cause–effect links.

But we need not resort to statistical physics in order to substantiate our contention that lawfulness is not exhausted by causality. The special theory of relativity, usually (and wrongly) regarded as an altogether causal theory, affords an outstanding example of theoretical structure having noncausal ingredients at its very basis. In fact, the basic axioms of this theory may be considered to be the special principle of relativity (stating the invariance of physical laws relative to the choice of inertial systems of reference), and the principle of the constancy of the velocity of light in a vacuum. It is only occasionally recognized[27] that neither of these principles is a causal law; that this is so can be verified by confronting them with any of the formulations of the causal principle (see Chapter 2). The same conclusion can be reached from a direct examination of the Lorentz transformation formulas, which can be derived exclusively from those axioms with some additions of a mathematical nature. Indeed, the Lorentz formulas,

$$x' = \frac{x - vt}{(1 - v^2/c^2)^{\frac{1}{2}}}, \quad y' = y, \quad z' = z, \quad t' = \frac{t - (v/c^2)x}{(1 - v^2/c^2)^{\frac{1}{2}}},$$

state a relation between inertial systems, not a genetic one-sided dependence, as the causal bond does; consequently, the Lorentz-Fitzgerald "contraction", following from these formulas, is *uncaused*. The realization of the noncausal nature of the Lorentz transformation formulas should alone be enough to discourage inquiries into the *cause* of the "contraction", which inquiries have contributed to clouding the interpretation of relativity, by hiding the relational character of both space and time.[28]

Finally, let us mention a rather odd group of law statements seldom, if ever, containing the causation category, namely, the

[27] Törnebohm (1952), *A Logical Analysis of the Theory of Relativity*, p. 47.

[28] At most, the Lorentz transformation formulas might be interpreted as expressing a *mutual* dependence among mechanical reference systems via the electromagnetic field; not, however, as stating a causal connection, since nothing is here reported to be engendered.

prohibitions or assertions of physical impossibility. The first two laws of thermodynamics may be framed as stating the impossibility of existence of perpetual movers of the first and second kinds respectively. Atomic and nuclear spectroscopy both contain a host of selection rules "prohibiting" certain transitions, that is, stating the impossibility of the existence of certain spectral lines or decay schemes. A further characteristic *Verbot* is Pauli's exclusion principle. These laws are neither causal nor statistical, but constitute a group of their own. (It is interesting to notice that both certain conservation theorems and selection rules—which are constants of the behavior of microsystems, and are derived from higher-level laws—do not refer to collections of systems but to *individual* objects; that is, they are supposed to be valid for every single microsystem, and not on the average or for a certain percentage of microsystems taken at random. This is important because it shows that quantum mechanics, in *either* of its interpretations, is far from being a purely *statistical* theory[29]—contrary to what is usually asserted.)

This will be enough, I hope, to show that scientific laws of nature need not be causal. I now turn to the thornier problem of sociohistorical laws.

10.4. Causality and Lawfulness in the Sociohistorical Sciences

10.4.1. *Are Sociology and History Scientific Disciplines?*

The reduction of lawfulness to causality is a mistake in scientific method and, like other mistakes of this sort, it is liable to have noxious consequences for science and for every general world outlook that claims to be based on science. In fact, such a restriction of the meaning of 'lawfulness' lends itself to the attacks of phenomenalism, indeterminism, irrationalism, and subjectivism upon science. This is only too clear in the case of atomic and nuclear physics, where noncausal lawfulness is often looked upon as a sign of indeterminacy, whereas it only shows

[29] See Eino Kaila (1949), "Beitrag zur Lösung der philosophischen Problematik der Quantenphysik", in Bayer, ed., *Nature des problèmes en philosophie*, vol. II, pp. 77 ff.

that causality has a limited range of validity. But the negative results of the restriction of lawfulness to causality are even clearer in the field of the sciences of man, where the exclusive operation of causation is far from obvious.

Essentially two arguments have been invoked in support of the claim that sociology and history are impossible *qua* sciences. The first is the uniqueness or unrepeatability of sociohistorical facts; the second is the noncausal nature of human affairs. If science were *defined* as the disclosure of the *causes* of *recurrent* (nonunique) events, then it would not be difficult to show that the sciences of society and history are impossible, since no two identical events are to be found in things human, and since noncausal categories of determination (such as statistical, teleological, and dialectical ones) are clearly involved in them. Two ways are open: either to acknowledge that the failure to include history and society in that definition shows only the inadequacy of the definition, or to maintain that the definition is correct but applies only to the natural, not to the cultural, sciences.

The latter course has been taken by the "humanistic" historical school, initiated toward the end of the 19th century by Dilthey, Rickert, Windelband, and other members[30] of the *Geisteswissenschaften* movement of reaction against the extension of the scientific method to all fields of research—or, at least, against its application to problems where it might yield "undesirable" insights. These writers, particularly Dilthey, denied that the chief object of historical science is both the material conditions of existence and the social (largely impersonal) "forces" that ultimately mold the spiritual culture; they asserted, on the contrary, that the sole concern of the historian is to trace the exploits of an allegedly free Spirit, which they regarded as independent of the material culture. Moreover, they refused to grant that the ultimate aim of historical research is the disclosure of general laws. They held that the aim of the historian should be restricted to the literary description and artistic resurrection of altogether *singular* and *lawless* events of

[30] Croce might be classed in this group, but his work on the philosophy of history was written a generation later; further supporters of this doctrine have been the sociologists Max Weber and Friedrich Meinecke.

a *spiritual* nature—for which task a mysterious and rarely found inborn intuition (*das Verstehen*) is indispensable; this supposed *Ersatz* of the scientific method is, of course, alien to general norms enjoying an intersubjective validity. According to this regressive trend history is, in brief, not a science but an art, or at most a branch of "philosophy" (in the traditional sense of the word) requiring a special insight.[31] Now, it would be foolish to deny that some "insight" is needed in order to disentangle orderliness, to solve problems, or to invent fact-fitting hypotheses, whether in history or in any other science. What is disputed is (*a*) that there is an inborn historical insight, and (*b*) that "insight" (whatever psychologists may find out it is) is the guarantee of truth; being incapable of providing a *test* for historical hypotheses, the celebrated "method" of *Verstehen* is not a method at all and has no title to replace scientific method.

The claim that historical events are the feats of a Free Spirit need not concern us; two monstrous world wars and the pending menace of nuclear and bacteriological warfare should suffice to dispose of such nonsense, the sole use of which might be to consecrate acts of rapine and murder as pure, spiritual exploits. What is relevant to our concern is the claim that it is not possible to disclose *laws* of history, because historical events are both specific, unique, singular (*einzeln*), and essentially noncausal. Let us start by examining the assertion that historical events are unrepeatable, for if they turned out to be *completely* unique, if they were singular in *all* respects, then obviously no general law, whether causal or not, would be found to fit, since law statements do not refer to single objects that are unique in all respects, but to kinds or classes of objects. The presumed noncausal nature of historical events will concern

[31] These opinions are, of course, very old. With a few exceptions, such as Thucydides and Polybius in antiquity, and Ibn Khaldun in the Middle Ages, history has until recently usually been regarded as the work of some supernatural entity (Destiny, Providence, Spirit) or as the direct outcome of the ideas and deeds of great men. In either case historical processes were regarded as *lawless*, hence unintelligible and even stupid. The immanent lawfulness of sociohistorical processes behind the apparent maze of the whole was asserted by Hegel—but, then, as the rational *Prozess des Weltgeistes*.

us after we have dealt with the problem of uniqueness (in Sec. 10.4.5).

10.4.2. *The Uniqueness of Historical Events*

The unique, unrepeatable, specific character of sociohistorical events is hardly disputable. The question is whether this uniqueness or *Einmaligkeit* is peculiar to human affairs or is a trait of *all* concrete objects—though admittedly to a lesser extent on levels lower than that of society or culture. That such a uniqueness is peculiar to sociohistorical facts is easily granted by those who, from Schelling to the contemporary romantics, are not familiar enough with the natural sciences. Even outstanding methodologists who grant the possibility of building a science of history, and who assert that causal explanations of historical facts are possible, claim that the difference between history, an allegedly "particularizing" science, and the "generalizing" sciences (like physics) lies precisely in that "it is the *particular event* that interests them [many students of history and of its methods], and not any so-called universal historical laws. For from our point of view, *there can be no historical laws* . . . Those who are interested in laws must turn to the generalizing sciences (for example, to sociology)".[32] Two points in this characteristic statement are relevant to our problem. The first concerns the concept of particular event, the second regards the controvertible identification of historical laws (most of which might consist in over-all and long-run trends) with *universal* historical laws, valid for every type of society and covering every single historical event of a given class.

The nature of historical laws will concern us in the next section; as to the first question, let us begin by granting that complete identity, identity in all respects, such as is defined in logic,[33] is a theoretical fiction in connection with sociohistorical

[32] Popper (1950), *The Open Society and its Enemies*, rev. ed., p. 448. Italics mine. Notice the strange (but usual) cut between history and sociology, which would be understandable on the assumption that human history is *not* essentially the development of human society.

[33] The usual definition of identity in logic is this: x is identical with y if all the predicates of x are also predicates of y, that is, $(x = y) =_{\text{df}} (F)(Fx \supset Fy)$.

facts—but, then, it is a fiction in *all* other fields as well, with the single exception of logic and mathematics, which build their own identical objects. This is a matter of fact for every dynamical ontology, from Leibniz's to dialectical materialism and Whitehead's philosophy of organism. And it is increasingly being felt by scientists, too; thus no thoughtful physicist believes now that there are in the real world (in contrast to theoretical schemes) two bits of matter (whether endowed with mass or not) in *exactly* the same state and interacting with *exactly* the same fields. No two macroscopic events can be strictly identical, if only because the entropy of molar systems never remains the same. Every physical object (material system, event, or process) is an *unicum* in at least one respect. Contrary to what Hegel thought, man—not nature—is the privileged inventor of strict (ideal) identity and recurrence; man is consequently the exclusive proprietor of boredom. Identity of coexistents, and exact repetition of successives, are as ideal as perfect circles: they are found to obtain to within a non-vanishing error only; total identity of concrete objects is an appearance sooner or later corrected by finer observations and deeper theoretical analyses. Only partial identity will be found at the ontic level. If something is not unique in at least one respect, then it does not belong to the external world.

Consequently, no science of the concrete world can aspire to find complete repeatability, that is, repeatability in *all* respects, the type of identity we build at the level of abstract thought. Hence 'exception' is not an unscientific word, as is usually believed; exceptions are just the least frequent alternatives in a collection of facts. The idea that exact identity is not a relation among facts of the external world but a *construct* that is very often *approximately* true is supported by the examination of the act of measurement. The measurement of the simplest physical attribute, such as length or weight, almost never yields the same result twice, provided sufficiently accurate instruments are employed (see Fig. 29). This is not a result of the limitations of our power of observation; quite the contrary, as the sensitivity of the measuring instrument increases, unrepeatability increases as well. That is, the finer the observation,

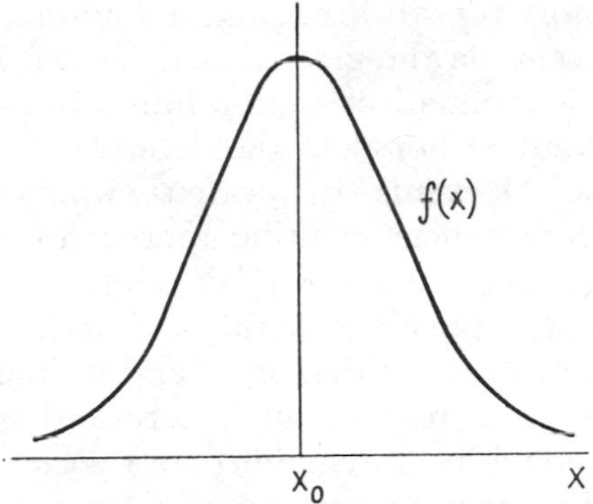

Fig. 29. The error curve. Measured values are designated by 'x', the mean (or "true") value by 'x_0'. The measurement "error" or discrepancy is distributed according to the law of Laplace and Gauss:

$$f(x) = \frac{1}{\sigma\sqrt{2\pi}} e^{-(x-x_0)^2/2\sigma^2}.$$

The probability that the error lies between $(x - x_0)$ and $(x - x_0) + dx$ is $f(x)dx$.

the more differences are detected; this is due to a deeper penetration in the fine structure of the object concerned, and to changes both in the object under observation and in its environment (in which the experimental setup must be included). In short, the finer the observation, the smaller the chance of finding two things exactly alike. (And this is just why finer observations are always desirable: not because they increase uniformity but because they permit the detection of a larger diversity.)

If such a specificity, uniqueness or unrepeatability of events in the world of matter does not make physicochemical sciences impossible, why should it render the social sciences impossible? Granted, sociohistorical events are enormously richer than those of the lower levels, and this lends them a *higher* degree of specificity or individuality. But this does not mean that sociohistorical events are *einmalig* (that is, happen but once) in *all* respects. To hold complete unrepeatability is as superficial as to repeat that history repeats itself, and it amounts to denying the existence of species on the ground that no two individuals are exactly alike—a nominalistic (and nihilistic) position criticized long ago by Kant;[34] nor does the diversity in question show that sociohistorical events are lawless—which would be the sole mark rendering their scientific approach fruitless. If sociohistorical events were all *essentially* different from one another, if no two similar happenings could be found, if uniqueness *in detail* were inconsistent with essential and partial identity or at least similarity, then nothing in society and in its historical development would be intelligible, and social scientists and historians could aspire to nothing but a more or less artistic chronicling of separate, unintelligible, hence useless facts. (Assuming that their mere description were possible—which assumption is far from obvious, since a new set of terms would have to be invented to narrate every event.) Presumably no intelligent person would be interested in such a description, even if it were possible.

Let us now turn to the question of whether there are historical laws.

10.4.3. *The Lawfulness of Historical Processes*

It can easily be granted that we know as yet very few *universal* historical laws, in the sense of patterns common to all periods and to all types of social organization. In the first place, most known historical laws are, as will be argued below, *statistical* laws, in the sense both that they involve numerous collections of men and that they do not apply in every single case but in a certain percentage of cases. In the second place, each type of social organization exists and evolves according to its own peculiar laws; thus the social life and the historical development of a community facing nature as a more or less integrated whole could not be the same as those of a social group ridden by struggles between antagonistic classes. Historical laws do not tower above men but are just the way in which men get along in time.

[34] Kant (1787), *Kritik der reinen Vernunft* (B), pp. 680 ff: a classical enlightening discussion of the need of applying simultaneously a "principle of homogeneity" and a "principle of specification" to account for identity in difference.

Notwithstanding, some social laws do remain quite constant through history, and the existence of such cultural invariants, or constant patterns of social behavior, is what entitles the anthropologist to speak of man, and the historian to speak of the history of man.[35] One such invariant is the ultimate or long-run predominance of material (biological and economic) conditions over spiritual culture, condensed in one of the few sayings that have both an individual and a social validity, namely, *primum vivere, deinde philosophari*. Another, by now quite obvious, sociohistorical law statement is the one according to which deep changes in the mode of production, such as the renewal or reorganization of technical equipment, in the long run elicit the renewal of the social structure—or at least render such a readjustment of social relations desirable to a section of society, which accordingly attempts to modify the prevailing social organization. A further universal historical law_2 is the statement that gathering and hunting precede plant, animal, and human domestication, which in turn precede civilization (at least in the case of autonomously evolving social communities).

Besides a handful of truly universal sociohistorical laws, which are grounded on general and nearly constant traits of mankind and are just the ones enabling us to group all men in the species *homo sapiens*, there are *general* laws peculiar to each type of social organization, or culture. That there are such laws is shown by the well-known fact that each type of society reacts in a typical way of its own to any given stimulus, whether external or internal. And is not the typical form of social behavior in the course of time what is meant by an historical law? However, it must be admitted that the social sciences are still characterized by too large a gap between the factual basis and the uppermost level of general principles. This gap between singular facts and higher-level law statements is gradually being filled with law statements (lower-level hypotheses) inferred from particulars—though not without the interference

[35] See Clyde Kluckhohn (1953), "Universal Categories of Culture", in Kroeber, ed., *Anthropology Today*; Gardiner (1952), *The Nature of Historical Explanation*, passim.

(favorable or otherwise) of general principles. To the extent to which this gap in the level of laws is filled, the valuation of both the factual basis and the top will undoubtedly change; different kinds of facts will be regarded as relevant, and new general principles will accordingly be proposed—but, now, not as mere hunches based on insufficient data, but as syntheses of whole bundles of particular law statements. (With regard to the scientific study of the existing social types, once this stage has been attained, and probably before the newly found general principles bear fully grown fruits—the discovery of new facts and new laws—the time may have come to alter the prevailing social structures "obeying" such principles—and so sociological knowledge will become historical knowledge.)

Granted, historical laws are enormously more complex than natural laws. In history we do not find it possible to apply easy schemes like the Hebrew prophets' law of increasing evil, or Spencer's law of unlimited progress, or the inevitable repetition asserted by Buckle and Schopenhauer, or the organismic rule of cyclical growth and decadence imagined by Vico and Spengler. Besides complexity, a further factor tends to obscure historical lawfulness, namely, the usually forgotten fact that single scientific laws, of whatever kind, do not hold for concrete, specific, individual instances; *scientific laws hold only for classes of facts*, and classes have a real existence in a limited number of respects alone, whereas concrete individual systems, events, or processes have an unlimited number of sides, so that they may belong to many different classes, their specification therefore requiring a number of laws. In order to make a group of concrete objects (systems, events, processes) fit into a class, we are forced to strip every one of them of peculiar marks that are irrelevant to the class in question. In a sense, by classifying events (and every application of law statements involves a previous classification), we turn concrete objects into abstract ones; but, if the classification is adequate—that is, if the attribute or set of attributes chosen for the grouping is actually possessed by every member of the class—then the universals concerned will have a real counterpart. In this case, the class, a product of intellectual activity, will also have an objective status; but the recognition of such an objectivity, far from being obvious, will demand a varying effort of abstraction.

Hence the uniqueness, specificity, individuality of every member in a collection does not render classifications illusory; the elements of the collection will all be the same *in certain respects*, never in all particulars—that is, if they are concrete objects. This applies, of course, not only to properties at any given instant but to their changes in time as well. Laws, whether of structure or of process, whether of society or of history, are then possible. Thus all wars consist of fights among armed people for some "reason" (usually a material interest); hence, although no two wars in the past have been exactly alike, they all share certain characteristics, they all fit a few elementary patterns. The existence of laws of war makes it possible and profitable to study strategy and tactics, using information about wars in the past in order to disclose and illustrate laws the knowledge of which may be useful in future butcheries.

As a consequence, historical particularizing is not inconsistent with scientific generalizing (establishment and use of laws), and is moreover performed on two successive levels: first, in the narrative of true histories; second, in the attempt to make these histories fit a comprehensive theoretical framework involving explanatory laws. The point is that the explanation of a unique historical event cannot be attained with the help of a *single* law statement, that is, it cannot be derived as an instance of a *single* nomological statement (in conjunction with a set of items of information performing a function analogous to the initial conditions of physics). Concrete, hence unique, objects, such as historical events, may be viewed as *intersections* of a large (perhaps infinite) number of laws.

The following sketch of a logical analysis of the problem of uniqueness *vs.* legality may contribute to clear the issues. Let X_1 denote any characteristic (property), and let L_n designate any of the laws connecting X_1 with other properties, X_2, X_3, . . . In the simplest case, the law L_n will have the form of a dyadic relation; that is, in this case $L_n =_{\text{df}} (\imath R_n)(X_1 R_n X_n)$. Let us call F_n the field of the nth law—roughly, the class of facts for which

L_n holds or, better, the set of elements x_1 for which the law L_n holds. And let us finally build the intersection of the fields of the laws which the property X_1 "obeys", namely, $F = F_1 \cdot F_2 \cdots F_n \ldots$ This will be, by definition, the domain in which all of the laws in question hold in conjunction. Such a field or range may be referred to as the *locus of the set of laws* $\{L_n\}$. My assumption is that every real fact (event, phenomenon) in which X_1 is involved can be regarded as *the sole element of the locus of a certain number of laws*; if preferred, unique facts are those loci of a given set of laws that reduce to a single "point". Even better: on this assumption singular facts are described in terms of general laws as the single-member loci of a set of laws, thus: $(\iota x)x\epsilon F_1 \cdot F_2 \cdots F_n \ldots$

It is both ignorance of presumably important sociohistorical laws and their large number that usually prevent historians from going beyond the level of narrative. In this regard, the situation in the historical sciences is not essentially different from the one that still prevails in medicine, where case histories are indispensable, not only as the raw material for the eventual inference of laws, but also as a substitute for as yet ignored, but presumed, laws. The present relative backwardness of medical theory does not, however, warrant the conclusion that there are no laws_1 of human pathology. Why should not the same apply to history?

10.4.4. *Obstacles to Disclosure of Historical Laws*

One of the difficulties besetting the disclosure of laws in any field is the low frequency of *observed* regularities. (Was not the apparent disorder in the phenomenal world one of the pretexts Plato gave for denying the possibility of a science of the material world?) Regularities are seldom *observed*; they are usually *inferred* or reconstructed, that is, first assumed and then tested—and this, just because regularities pertain not to individual, specific objects, but to classes of objects. In order to find regularities we must make an extensive use of abstraction, which alone can go beyond the maze of appearances into the essence of things; we must, in brief, impoverish the immediately given before we are able to recognize its orderliness. In other words, the establishment of laws requires the framing of *theoretical models*—and this will not become a habit as long as the descriptive attitude prevails in the sciences of man.[36]

Philosophers of social sciences should not dogmatically deny that man is capable of learning some sociohistorical laws. At least, man is entitled to *seek* for laws of cultural development, being moreover vitally interested in finding them, just in order to improve his chances of succeeding in sociohistorical business. Philosophers of cultural sciences should also grant that our comparative ignorance of the laws of society and history is due not only to the great complexity of human affairs, but also to the very prejudice that there *are* no laws of history—a prejudice suspect of being allied to powerful social (or antisocial) interests that are vitally interested in preventing deep insights into the social mechanism. *Et pour cause!* People who are able to take the social mechanism apart in theory may wish to change it in practice, and—what is more dangerous for those who live on the persistence of fossil social forms—such men may even succeed in their attempt. This is essentially why the pragmatic view of science, summarized in the positivistic maxim "To know is to foresee; to foresee is to control", which enjoys such wide acceptance in connection with the sciences of nature, is so often forgotten in the sphere of the social sciences, where it is found more desirable to hold that human society is rationally unknowable, hence unpredictable, hence incurable.

To summarize, there are historical laws,[37] both universal and general, the former applying to mankind, the latter to definite types of social organization—but this does not mean that history repeats itself. (Nor does the lawfulness of nature exclude novelty: see Sec. 8.2.5.) There are general patterns of collective behavior and of its changes; but most of them, being peculiar

[36] For a survey of recent mathematical theories of economic and social behavior, see Kenneth J. Arrow, "Mathematical Models in the Social Sciences", *General Systems 1*, 29 (1956).

[37] See Carl G. Hempel, "The Function of General Laws in History", *The Journal of Philosophy 39*, 35 (1942), reprinted in Feigl and Sellars, ed., *Readings in Philosophical Analysis*. A nonirrationalistic negation of historical lawfulness is found in Arthur C. Danto, "On Explanations in History", *Philosophy of Science 23*, 15 (1956).

to the type of social structure, change in turn with every radical social transformation. Finally, as will be argued in the next section, sociohistorical laws are not exclusively, perhaps not even predominantly, patterns of causation.

10.4.5. *Noncausal Features of Sociohistorical Events*

Many philosophers, but I believe few historians, would deny that most social processes of "historical importance" are at the same time strongly *self-determined* (internally determined by the structure of the social group itself), definitely *dialectical* (consisting in or brought about by struggles of human groups), partly *teleological* (striving, though mostly unconsciously, toward the attainment of definite goals), and typically *statistical* (the collective result of different individual actions largely independent of one another)—that sociohistorical events are, in short, eminently noncausal.

The self-determination of sociohistorical events is here understood in the sense that factors external to the human group concerned (natural environment and contacts with neighbor cultures) are effective solely insofar as they succeed in changing the essential inner processes, which proceed in accordance with immanent laws—those of material production, social relations, and spiritual life. (These laws are all certainly *rooted* in the animal basis of man, but they do not therefore *reduce* to biological laws. Thus, the laws of the production and circulation of commodities in a given society are rooted in the common animal needs of man, but they are economic, not biological, laws; this is why they change considerably in the course of history, even though man has changed but little, biologically speaking, during the last several millennia.) The dialectical nature of the sociohistorical events having some historical relief is related to that very self-determination, and originates in the lack of social homogeneity—an inhomogeneity that exists whether or not the social group concerned is divided into social classes, but that is characteristically intensified (to the point of eliciting destructive antagonisms) wherever the appropriation of material goods is individual. It is, in fact, no longer a secret —although it is still widely held to be a seditious heresy—that the *main* source of social change is the (pacific or warlike) clash of material and cultural interests.

As to the partially teleological or goal-seeking character of sociohistorical processes, it hardly needs to be proved, since it is a truism that men act not only because they are driven by present inward and outward pressures but also in view of the future—sometimes even in view of a nonexistent personal future. (Human behavior is of course not always *conscious*, and it is seldom intelligently planned, although it would be foolish to deny that the conscious planning of social life has been steadily increasing since the dawn of civilization.) Moreover, human action is too often unsuccessful—which only shows that teleology should not be understood in a fatalistic but in a statistical sense. However, in contrast to processes at the level of inanimate matter, human behavior is largely directed to the satisfaction of future needs and to the future attainment of freedom from want. Nevertheless, this does not happen in every individual instance; teleological laws obtain only in a large percentage of cases, that is, they are statistical.

The statistical nature of social events was recognized, though only in a restricted manner, early in the 17th century by Dutch statesmen and statisticians who realized the usefulness of statistics for governmental affairs. But the rise of this discovery, from a technique for the collection of taxes and estimate of budgets to a principle of the philosophy of the cultural sciences, had to wait till Condorcet's *mathématique sociale* (1795), Quételet's *physique sociale* (1836), and historical materialism (1850's). In the statistical determinacy of social events, historical materialism saw a confirmation of its thesis that the laws of society are as objective, and operate as blindly, as the laws of nature. Most economic, social, cultural, and historical laws do not consist in fixed recurrent patterns concerning the *totality* of a class of individual events; they are, instead, *tendencies*, that is, over-all and long-run trends. They do not apply without exception to every individual event of a certain kind, but in a given percentage, and hold for kinds of *large* collections of events; they are, in short, collective regularities. (The term 'probability law' is often used as a synonym of statistical law, but such a designation is misleading, since probability is a mathematical concept; only the propositions of the theory of probability should be termed probability laws.) Consequently, at the pragmatic level cultural laws are not infallible precepts but, at their very best, rough maxims that are found successful in a high percentage of cases.

The fact that most sociohistorical laws have a statistical character does not mean that individual events are lawless; it means only that they cannot be expected to "obey" *single* laws, but rather groups of laws operating in conjunction (see Sec. 10.4.3). And they are not only "governed" by sociohistorical laws, but by lower-order laws as well; that is, singular facts at the social level are the intersection of a number of laws belonging to both the social and the lower integrative levels. Thus the decisions of the individuals participating in a given historical event are motivated by biological, psychological, intellectual, and other factors—but they will be effective solely provided they fit a social scheme. And the statistical nature of most sociohistorical laws does not mean that the social sciences are solely interested in the statistical resultants of many individual deeds without caring for the latter in themselves. Not only the most frequent events are important, hence interesting to the historian; also large deviations from the mean, uninteresting to the statistician, are sometimes eminently interesting to the historian, for they may be associated with deep qualitative changes. An obvious illustration of such infrequent (unprobable) events is provided by the profound changes that comparatively small sections of mankind (and even exceptional single individuals) succeed in eliciting from time to time, namely, social and political readjustments. But even such exceptional events, far from being lawless, are the outcome of the laws of social growth; besides, even if novel in the highest degree, they will share *something* with similar events in the past—and this partial identity is what renders the study of history useful for both conservatives and reformers.

10.4.6. *Scientific Exactness: Not Exhausted by Numerical Accuracy*

The claim is often heard that sociology cannot apply statistical methods in all of its sections, owing to the existence of social *imponderabilia*, of allegedly unmeasurable characteristics such as taste, fashion, devotion to a cause, enthusiasm, and so forth. While it is true that the assessment of the weight of every cultural factor is far from easy, experience seems to show that there is no difficulty in principle in making it. The problem reduces essentially to devising the adequate conceptual tool and appropriate technique; the alleged *imponderabilia* are unmeasur*ed* rather than unmeasur*able* characteristics—and they usually finish by being weighted in some way as soon as it is found that they have an economic importance. No object in the external world is *inherently* nonquantitative; what is usually nonquantitative, but just classificatory or comparative, is *our approach* to concrete objects.

Besides, the aim and result of the "measurement" of social features need not always be a set of numbers fitting functions, vectors, tensors, matrices, and so on. Sociology is certainly interested in measuring the intensities of socioeconomic factors (the concern of sociometrics); but sociology is also interested in "topological" or order qualities and relations that can be described with an equivalent accuracy. Thus, the inquiry into the structure of a social group is a sociotopological, not a sociometrical, problem: although it is approached on the basis of definite numerical data—such as income distributions, in the case of certain stable societies—it requires further, nonquantitative data, such as those regarding the place occupied by the members of the particular group in material production and in the social hierarchy, if any. In short, the metrical or numerical approach to social problems is indispensable, but it must be supplemented with a likewise exact, but comparative or topological, analysis. However, this is not peculiar to the sciences of man; exactness is not exhausted by numerical accuracy.

An exclusively statistical approach to sociohistorical facts, and moreover an approach focusing on the measurement of petty details but overlooking the main stream, may be as misleading as a purely qualitative approach. However, the error does not lie in the method itself; it may lie in the selection of facts, a selection that is always guided (overtly or not) by general principles. Exclusively statistical accounts tend to overlook both the main trends of social development and the inner mechanism bringing about the given statistical results. This microsociological approach, so typical of contemporary American sociology, is unable to foresee the major social changes, although it may afford accurate predictions regarding the forthcoming sales of a cigarette brand. But this shortsightedness is not, let me repeat, inherent in the statistical approach; it is usually the result of a superficial selection of facts, and of lack of theoretical models in the service of whose test the data should be gathered. The microsociological approach must be supplemented with the inquiry into the inner mechanism of social evolution; the latter can both make use of and guide microsociological research. At any rate, the intimate mechanism producing the observable statistical results in the social field is not essentially causal—although it is perfectly determinate.

10.4.7. *The Defense of Scientific Method in the Sociohistorical Sciences*

The claim of the *Geisteswissenschaften* movement, that the study of history cannot be scientific because it has not been possible to find *causal* historical laws, is justified to the extent that (*a*) historical writings are usually less than causal analyses, remaining mostly at the level of chronicling or narrative; (*b*) a strictly causal analysis of historical facts would of necessity overlook an enormous number of facts and connections not fitting any bundle of causal threads; (*c*) the causal approach to historical problems has led to the isolation of causal chains that can easily be destroyed by organismic and functionalist philosophies of history (see Sec. 5.2.2). The identification of *scientific* analysis with *causal* analysis thus facilitates the view that history cannot be a science—a view that very nicely matches the desire to prevent people from understanding what they are doing and from learning how to participate consciously and effectively in historical events.

However, defenders of a scientific approach to human affairs often write that the chief aim of social science is to attain a knowledge of *causal* connections among social factors,[38] and that the historian's main task is the disclosure of *causal* bonds among events.[39] I think that such statements should not be taken too literally; what they mean is not always a systematic application of the causal principle but rather the application of the principle of determinacy, that is, of the lawful genetic connection (see Sec. 1.5.3). What most contemporary defenders of causality actually mean is, I believe, that events do not happen capriciously but according to law, and that they do not spring out of nothing but emerge from preëxisting conditions; hence, where they write *causal* connections one should perhaps read *some sort of lawful connection*, whether causal or not.

In order to decrease the prevailing confusion about historical method and its ontological presuppositions, it is desirable to dispense with a nomenclature attached to an outdated philosophy of science, namely, that asserting the coextensiveness of science and causality. A way of defending the application of the scientific method in sociology and history is to acknowledge the limitations of causality, and to demand the exploration of the whole range of types of determination. Being a particular nexus, causation is defensible as a trait of reality and is even indispensable as an approximate reconstruction of becoming; but it is indefensible as a category of unlimited extension. Lawfulness, on the other hand, is not only the chief result of scientific research but also, in a way, its very presupposition. No scientific approach to reality would be possible if at least research itself were not susceptible of a systematic organization in accordance with rules of its own; and only on a transcendentalist

[38] See MacIver (1942), *Social Causation*; Alan Gewirth, "Can Men Change Laws of Social Science?", *Philosophy of Science 21*, 229 (1954).

[39] See Maurice Mandelbaum, "Causal Analysis in History", *Journal of the History of Ideas 3*, 30 (1942). This is a valuable defense of scientific method in history; it entails, however, the controvertible theory (held by Meyerson) that causal explanations must *underlie* scientific laws. For a criticism of this theory, see Gardiner (1952), *The Nature of Historical Explanation*, pp. 83 ff.

anthropology, that is, only by supposing that man is altogether alien to the reality he studies, might this reality be imagined to be lawless.

10.5. Conclusions

Positivists have demanded that the concept of causation be *replaced* by that of law; they have required that science forsake what Comte called "*la vaine recherche des causes*"; and they have imagined that science has already eliminated the common sense notion of causal influence, "substituting the entirely different *denkmittel* of 'law'".[40] Some antipositivists, on the other hand, have held that lawfulness is insufficient, a causal connection having to be sought *behind* every law. An examination of the nature of scientific law has shown us, however, that science has followed neither advice, since it contains both causal and noncausal laws, as well as laws in which various categories of determination concur.

The trend of recent science points neither to the decausation preached by positivism in favor of purely descriptive statements of uniformity, nor to a return to traditional pancausalism. Present trends show, rather, a *diversification* of the types of scientific law, alongside of an increasing realization that several categories of determination contribute to the production of every real event.

The stage of formulating causal laws is not the last, or even one of the most advanced, of research; it is usually followed by stages in which questions are answered in terms not only of causes, but of several other categories of determination as well. Theoretical physics became firmly established the day Newton rejected the scholastic precept of looking for causal explanations only, and realized instead that the chief, though not the sole, task of scientific theory is the discovery of the laws of phenomena—whether or not the causal aspect is prominent in them. But this did not mean the replacement of causality by lawfulness; it meant instead an *enrichment* of determinism; causal determinism became sublated in general determinism.

[40] James (1907), *Pragmatism*, p. 119.

Finally, scientific laws afford not only an economical, abridged description of phenomena enabling us to dispense with the repetition of observation or experiment, as Comte and Mach believed. Scientific laws, or law statements, being approximate ideal reconstructions of the immanent forms of structure and process, not only enable us to answer what-, where-, when-, and whence-questions, but also provide perfectible answers to whys; they are the chief tools of the scientific explanation of nature, thought, and society. But the problem of explanation will be approached in the next chapter.

11

Causality and Scientific Explanation

Scientific explanation has traditionally been regarded as causal explanation; the explanation of a fact was not usually deemed to be scientific unless its proximate and ultimate causes were assigned. Positivism reacted against this mistaken identification, which ultimately rested on the assumed identity of reason and cause. However, the reaction of positivism against causal explanation was not grounded on the limited range of causality but on the rejection of *every* sort of explanation, in favor of description.

The thesis to be here defended is, again, neither the traditional nor the positivistic one, but a third position, namely, that answers to why-questions need not be causal in order to be scientific, although causal explanation does constitute an important ingredient of scientific explanation in many cases. To this end, a cursory examination will be made of some types of explanation that are actually being used in science. But, before this is attempted, the question whether science should have an explanatory function at all will be treated.

11.1. Is Science Explanatory?

It is worth while to recall briefly the story of the movement of reaction against the panexplanatory endeavor of rationalism, especially since that movement is far from simple, and is connected with the attacks on causality from a regressive standpoint. The explanatory function of science has been questioned from two opposite quarters. On the one hand, religious writers have usually held that scientific knowledge is powerless to attain the primary or ultimate causes of things, whence it is vain. Skeptics, agnostics, and early positivists, on the other hand, have regarded speculations about transempirical and "ultimate" causes as vain either because, as consistent empiricists, they believed the whole reality to be such as it appears to us, or because they deemed the essence of things to be unattainable, or, finally, because, in harmony with the *Zeitgeist* of an industrial era, they held an instrumental conception of science, seeing no profit in the search for ultimates. The "positive" character of science, as contrasted to the wild, obscure, and largely profitless speculations of medieval schoolmen and ulterior philosophers of nature (from the Renaissance down to our own day) consisted in accurate descriptions and predictions; the search for answers to why-questions, regarded as an impossible or at least useless enterprise, was explicitly proscribed by the various kinds of phenomenalism. According to phenomenalism, the task of scientific research is to describe how things appear to us to happen, not why they happen thus and not otherwise. Faithful to this rule, Comte rejected the wave theory of light—an epoch-making scientific invention of his own time—on the bare ground that it provided an explanation of optical phenomena (more exactly, an explanation of the laws of geometric optics in terms of higher-level laws).

There is not a single kind of phenomenalism—or at least not a single intention that may nourish phenomenalism. Thus Francisco Sánchez was a representative of progressive phenomenalism and agnosticism in the face of scholastic essentialism and dogmatism. On the other hand, Berkeley was a clear instance of regressive phenomenalism associated with dogmatism—and that not in the Renaissance, like Sánchez, but at the beginning of the Enlightenment. In Hume, the young Kant, and the young Comte, the progressive aim of discrediting supernaturalism and unscientific speculation about matters that were successfully being handled by science was already mingled with the regressive attempt to put a priori limits on the scope of rational enquiry in general, and on scientific research in particular. With their strict adherence to the phenomenalist program that circumscribes the domain of scientific research to the "positive" description of what is "immediately given", Comte and his followers have earned the name of negativists. By limiting the reach of research, curtailing the function of reason, and refusing to answer the most interesting questions, phenomenalism allied itself willy-nilly with the obscurantist trends. A tacit alliance was signed against the claims of rationalism and realism; the end of the long warfare between theology and science was proclaimed, and a division of the sphere of spiritual culture into nonoverlapping domains was established. From the last decades of the 19th century on, in wider and wider circles science began to be looked upon, not as essentially the pursuit of truth and the attainment of understanding, but just as a means of control over nature and over man, as a summary of facts and as a technique for prediction—hence as an aid to action—on the basis of sheer, uncommitted description. Science could be tolerated as long as it was assigned solely an instrumental, not a cultural value; a *modus vivendi* was possible as long as the reply to why-questions was declared to be beyond the reach of science.

The pragmatic attitude toward scientific research, the irrationalistic devaluation of theoretical knowledge—facilitated by the ignorance of the role of hypotheses in science—had captivated so many outstanding minds toward the end of the last century, that Mach could declare that "only the *relation among facts* [die *Beziehung des Thatsächlichen zu Thatsächlichen*] is valuable—and this is exhausted by description".[1] Le Dantec stated that the best "language" is the one involving the fewest hypotheses and affording the fewest explanations.[2] Duhem, the Catholic positivist, wrote bluntly that the explanatory part of science is "parasitic".[3] Petzold, the German follower of Mach, repeated the contention that "there is no other explanation than a complete and most simple description".[4] Pearson, the British follower of Mach, went so far as to state that "nobody believes now that science *explains* anything; we all look

[1] Mach (1900), *Die Principien der Wärmelehre*, 2nd ed., p. 437.

[2] Le Dantec (1904), *Les lois naturelles*, chap. xvi, esp. p. 114.

[3] Duhem (1906), *La théorie physique: son objet et sa structure*, Part I, chap. iii.

[4] Petzold (1906), *Das Weltproblem von positivistischen Standpunkte aus*, p. 147.

upon it as shorthand description, as an economy of thought".[5] More recently, Wittgenstein demanded that this attitude be taken by philosophy as well: "There must not be anything hypothetical in our considerations. We must do away with all *explanation*, and description alone must take its place."[6] The partisans of the *Prinzip der Denk-Ökonomie*—put forward by Ockham and reinvented by Mach[7]—practised it to such an extent that they failed to notice that the most economical of all procedures is not to think at all.[8] It was William James who, with his deep insight, realized that an outcome of the pragmatic outlook was the revindication or at least the toleration of "religious experience", alongside other kinds of experience.

Although rank-and-file followers of positivism still believe that scientific explanation is nothing but unbiased, uninterpreted, hypothesis-free description, the explanatory function of science has become so obvious with the quick development of theoretical science that the most distinguished and open-minded neopositivist philosophers have recently been led to acknowledge that it had been a mistake to uphold a "hypertrophied operationism" entailing "a radically antitheoretical attitude" and looking upon explanation as a "metaphysical misfit".[9] But those logical empiricists who now admit the legitimacy of explanation usually *restrict* the philosophical problem of scientific explanation to that of its *logical* structure,

[5] Pearson (1911), *The Grammar of Science*, 3rd ed., p. v.

[6] Wittgenstein (1945), *Philosophical Investigations*, part I, p. 47.

[7] Mach (1883), *The Science of Mechanics*, pp. 578 ff. While Ockham regarded his own celebrated maxim (*Entia praeter necessitatem non esse multiplicanda*) as a methodological rule for the search and exposition of truth, Mach seems to have regarded the economy of thinking as the very aim of science, and even as the essence of what he meant by laws of nature.

[8] Born (1949), *Natural Philosophy of Cause and Chance*, p. 207: "If we want to economize thinking the best way would be to stop thinking at all."

[9] Herbert Feigl (1945), "Operationism and Scientific Method", in Feigl and Sellars, ed., *Readings in Philosophical Analysis*, p. 503. See also Feigl, "The Philosophy of Science of Logical Empiricism", in International Union for the Philosophy of Science (1955), *Proceedings of the Second International Congress*, vol. I, in which he holds that the "hypocritical realism of phenomenalism must be supplanted by a hypercritical realism" (p. 110), and announces "the abandonment of reductive phenomenalism and of ultra-operationism in favor of a more constructive realism" (p. 114).

leaving entirely aside the epistemological and ontological sides which this, like every other philosophical question, has—and to which I now turn.

11.2. Some Aspects of the Problem of Scientific Explanation

11.2.1. *Conditions for an Explanation to Be Scientific*

The aim of scientific explanation is to render facts intelligible, that is, to rationalize reality—not, of course, in the psychoanalytic sense of 'rationalization'. But obviously not every explanation affording what is loosely called intellectual satisfaction is scientific; only general, meaningful, and verifiable ideas are involved in genuine scientific explanation. A factual hypothesis—that is, an assumption referring to facts—can be regarded as tested, verified, or confirmed (not *proved* beyond doubt, but verified to within a vague margin of uncertainty) if it meets at least the following requirements:

(i) the *rational* or logical condition of *consistency*, that is, of compatibility with the remaining propositions of the same theoretical system. Only *partial* consistency should be required, not consistency with the whole corpus of available knowledge, for the proposition (hypothesis) in question may happen to refer to a discovery invalidating some of the received ideas. Consistency, a mark of rational success, is also a temporary halt in the advancement of knowledge—which, as Socrates is said to have discovered long ago, proceeds chiefly through the clash of (internally consistent!) ideas and systems of ideas;

(ii) the *material*, factual, or empirical condition of satisfactory (not necessarily nor possibly perfect) *adequacy* to satisfactorily certified facts (sometimes called scientific facts, or facts under experimental control, in contradistinction to "brute" facts). Adequacy of factual hypotheses is *tested* by experiment and observations; but adequacy does not *consist* in empirical verifiability: adequacy consists in correspondence (however incomplete) between propositions and facts. The condition of adequacy is to be fulfilled by the particular consequences of the principles (basic hypotheses), not by the principles themselves, which may have an objective counterpart but need not have an empirical counterpart at the time they are formulated and, being general, are not directly verifiable. Both the certification of facts and the adequacy of a hypothesis to a class of facts are of course perfectible and even altogether infirmable—the contradiction between ideas and the facts to which they eventually refer being a further source of scientific progress.

At least some of the explanans-propositions involved in any scientific explanation must be capable of being submitted to an empirical test—if possible through prediction and, even better, through reproduction. But in order to be honored with the title 'scientific', a hypothesis need not have been verified already; what is peculiar to scientific hypotheses is neither truth (which some scientific hypotheses are assumed to share with some vulgar assumptions) nor infallibility (which is rather a mark of dogmatism) but logical consistency and empirical verifiability in principle. Verifiability means possibility of verification, and this entails both refutability and perfectibility (improvement of consistency and adequacy, and eventually extension of the domain of validity).

The foregoing remarks should suffice, for the time being, as a general description of scientific factual statements.[10] I shall now proceed to a brief characterization of the logical structure and epistemological import of the explanatory statements of science.

11.2.2. *The Logical Structure and Epistemological Meaning of Scientific Explanation*

Etymologically, to explain means to unfold, expose, or develop. The commonly received doctrine on the logical structure of scientific explanation is in agreement with the original meaning of the word; in fact, that doctrine holds that to explain a fact is nothing more than to show that the proposition(s) stating the fact is or are a particular consequence of one or more propositions of greater generality (eventually in conjunction with some items of specific information about the particular case in question). The concept of cause need not be involved in scientific explanation. Thus consider Leonardo's

[10] A fuller account will be found in Bunge (1959), *Metascientific Queries*, chap. 2.

explanation of the downward direction of the fall of bodies: "Every natural action is made in the shortest way; this is why the free descent of the heavy body is made towards the centre of the world: because it is the shortest space between the movable thing and the lowest depth of the universe".[11] The explanans-proposition is here a minimum principle (known from Hellenistic science), by no means a causal proposition.

Now, the general propositions acting as explainers or premises in science—that is, the explanans-propositions—are usually of that particular kind of hypotheses named law statements (our $laws_2$); hence, the scientific explanation of a fact consists, from a *logical* point of view, in showing that it is an *instance* of a general law.[12] In turn, the scientific explanation of a uniformity or regularity will, when available, consist in its deduction from a higher-level law, that is, in its subsumption under a statement of greater generality.[13] Thus, for example, Kepler's laws of planetary motion were shown by Newton to follow from the conjunction of his own law of motion and the particular law of force known as the inverse-square law of gravity—both of which have a range of validity wider than Kepler's laws. Hence, from the standpoint of logic, to explain is to show the existence of an entailment or implication of the particular by the general; the object to be explained is thereby shown to be a member of a certain group, the idea to be accounted for is thereby shown to be somehow "included" in a previously admitted set of assumptions. Consequently, to the extent to which deduction may be regarded as tautological, explanation will be tautological as well; that is, from a strictly formal-logical point of view, explanation produces nothing new,

[11] Leonardo, *Notebooks*, G 75 r, ed. MacCurdy, p. 551.

[12] Mill (1843, 1872), *A System of Logic*, book III, chap. xii, sec. i, defended a similar theory of explanation in one of his frequent rationalistic whims. His mistake in this respect was to believe that an account of the logical framework of explanation could exhaust the problem.

[13] For the logic of explanation, see C. G. Hempel and P. Oppenheim (1948), "Studies in the Logic of Explanation", reprinted in Feigl and Brodbeck, ed., *Readings in the Philosophy of Science*. Braithwaite (1953), *Scientific Explanation*, has studied the level structure of scientific hypothetical-deductive systems; the highest-level hypotheses are the premises from which the "theorems" (lower-level generalizations) are deduced.

nothing that was not in some way "contained" in a previously admitted system of ideas. In brief, the logical aspect of explanation consists in demonstration.

Unfortunately or not, logic does not tell the whole story of knowledge. Explanation, like deduction in general, does add to knowledge, because actually the object to be explained was not previously *contained* in its class (or in its law statement) from the start; it was put by us in it a posteriori. The explanation operation is not a mere drawing of an element out of a given collection; from an epistemological point of view, explanation does not consist in the mere identification of an element of a class that is overtly displaying its characteristics before us: explanation consists, rather, in the *inclusion* of the given object (fact or idea) in its class. And this is a *constructive*, synthetic operation, requiring the previous schematization of the given object, comparison of it with other objects, and so on. Now, at the level of formal logic, change does not enter; hence processes, like the epistemological one involved in explanation, have no place in deductive logic, which dispenses with the time concept and treats the actual processes of thought as laid down in an eternal present—just in order to avoid contradictions between successives. In other words, what from an epistemological viewpoint is a real transition from ignorance to knowledge appears in formal logic as a purely analytic relation. Deduction, and in particular explanation, always entails a novelty in knowledge—and this is why we care to perform it. To put aside the nonlogical aspects of explanation, by concentrating exclusively on its logical structure—as is usual among contemporary empiricists[14]—is, I believe, a token of one-sidedness.

Like every other exaggeration of logicism, the reduction of

[14] See Schlick (1936), *Philosophy of Nature*, chap. 3; Reichenbach (1951), *The Rise of Scientific Philosophy*, chap. 2; Popper (1950), *The Open Society and its Enemies*, rev. ed., p. 445: "To give a causal explanation of a certain event means to derive deductively a statement (it will be called *prognosis*) which describes that event, using as premises of the deduction some *universal laws* together with certain singular or specific sentences which we may call *initial conditions*". The objectionable identifications of reason and cause, and of cause and law, are involved in this statement; otherwise it is an unobjectionable description of the logical aspect of scientific explanation, whether causal or not.

explanation to deduction has strange ontological and epistemological consequences. (In philosophy, as in everyday life, to abstain from looking at facts does not eliminate them.) Indeed, the view that scientific explanation is *nothing but* the disclosure of identity in the midst of difference, the discovery of like in unlike, the deduction of the particular from the general—such a view suggests a metaphysics holding the monistic belief that there must be one or at most a few universal essences, ultimate substances, fundamental laws, basic languages, or what not, to which everything must "ultimately" be reduced, freely trespassing on the boundaries between levels. It also suggests the odd theory that the advancement of knowledge consists in a progressive unification reducing the apparent multiverse to an extremely impoverished universe depictable by a highly economical unified science. Such views have actually been held not only by traditional metaphysicians but also by modern positivists.[15] To say the least, that is a one-sided view of the matter; the history of science shows that the number of law statements and of categories *increases*, that the number of fields of research *increases* as well, that the number of irreducible but eventually connectible concepts *increases* accordingly. A progressive *differentiation*, instead of a progressive reduction, is the sign of scientific advancement. But, just as in the case of biological and cultural progress, the very process of differentiation calls for an increasing *integration* of science, both ontological and methodological. In point of fact, there is evidence of an increasing *unity of method*—not excluding the increasing diversification of particular techniques—and of a ceaseless discovery of new *connections* among fields that initially looked

[15] Schlick (1936), *Philosophy of Nature*: "The first step towards a knowledge of nature consists in the description of nature which is equivalent to the establishment of the facts . . . The next step towards a knowledge of nature—explanation—is characterized by the fact that a symbol (concept) which is employed in the description of nature is replaced by a combination of symbols which have already been used in another context. In point of fact, progress in knowledge consists in the discovery that a substitution of this kind is possible" (p. 17). It follows that "it is obvious that in the progress of knowledge, the number of concepts necessary for a description of nature, will become increasingly reduced; so that what is denoted by the term 'world-picture' will become more and more unified. The world will become a '*Uni*-verse'" (p. 18).

separate—connections which need not erase qualitative differences. The advancement of knowledge follows Bernard's law of biological progress, namely, "*Plus l'organisme est élévé, plus il y a de variété dans plus d'unité*". The increasing integration (or "unity") of science should not prevent us from realizing its increasing diversification.

In conclusion, the logical structure of scientific explanation consists in deduction from scientific laws in conjunction with particular items of information; or, what amounts to the same thing, the logical kernel of explanation is generalization, the process of showing that the fact in question fits a general pattern. But logic does not exhaust the analysis of scientific knowledge—nor is it the concern of (formal) logic to perform such an analysis. In order to complete the picture of scientific explanation we must take a look at its ontological aspect, especially since a peculiar characteristic of scientific explanation is that some of the explanans-propositions in it are law statements that are supposed to have an ontological status, that is, $laws_2$ that are assumed to be correlated to $laws_1$.

11.2.3. *The Ontological Basis of Scientific Explanation. Explanation of Facts and Explanation of Laws*

Usually a scientist does not "feel" that he has understood a demonstration if he has grasped *only* the logical mechanism of it, that is, the derivation of the result from the assumption(s). He may, on the other hand, "feel" that he has understood a derivation even if he has not looked into the detail of the logical chain, that is, even if he has neglected a number of steps in the process—which he often does purposely just to avoid getting lost in the formal net. Moreover, whenever a scientific explanation is summarized, the details of its logical structure are left aside; it is merely taken for granted that they have been taken care of, and what is required is a short presentation of the very "stuff" of the explanation. This actual procedure of scientists suggests that, in the so-called empirical sciences, explanation raises at least two philosophical problems: (i) that of the *nature* of the materials (explanans-terms) out of which the explanation is built; (ii) that of the *logical relations* among these materials. The reduction of explanation to deduction (or generalization) overlooks the very materials of explanation; the logical analysis of a given scientific explanation is able to ascertain whether it is formally (logically) valid, not whether it is materially plausible; the logical analysis of explanation is incomplete, because the explanans-terms are left unanalyzed in it. These explanans-terms, regarded as primitives at the logical level, constitute the object of an ontological analysis, to which I now turn.

The object of scientific explanation may consist of (*a*) classes of *facts* (natural, mental, social), or (*b*) scientific *laws* themselves (for instance, empirical generalizations may be subsumed under law statements containing elaborate theoretical concepts). In either case scientific explanation, in contrast to popular explanation, is performed in terms of *laws*. That this sort of explanation has an ontological aspect should be obvious except to those holding that both facts and laws are constructs.

As an illustration of scientific explanation of the first kind—where the explicandum is a class of facts—consider the case of the recoil undergone by guns upon firing. This class of facts can be explained in two different ways, which are at first sight essentially different from one another. On a first level of explanation we account for the recoil motion of the gun as being an effect of the equal pressure exerted in all directions by the gas generated in the chemical reaction started by the trigger. On a second level of explanation, the recoil is explained as being a mere instance of Newton's law of the equality of action and reaction. The former we are tempted to call a *causal* explanation, because it invokes an efficient cause, namely, the gas pressure; as to the second explanation, we are on the verge of terming it a *rational* explanation, as it openly consists in a deduction from a general principle. However, this dichotomy of explanation into causal and rational is—like every dichotomy of scientific statements into purely factual and purely rational, or purely synthetic and purely analytic—too simple to be adequate. In fact, our first-level explanation was incomplete, since it did not state that the cause (the gas pressure) is in its turn a result of the self-sustained, hence noncausal, chain reaction initiated by the trigger in the explosive; moreover, that explanation made not only a direct, explicit reference to a fact (the gas expansion), but also an *indirect* or implicit reference to a *law* immanent in that process, namely, the law of the isotropic distribution of gas pressure (in the absence of external fields). Hence, our first-level explanation was not simply an explanation of one kind of fact in terms of a further sort of fact, but included a reference to a law statement. In short, our first-level explanation is not complete insofar as it is exclusively causal; and it is as rational as the second-level explanation, since its logical structure is that of deduction. As to our second-level explanation, although it did not involve a *direct*, explicit reference to facts, it did appeal to the wide class of facts covered by Newton's third axiom—which is a physical, not a formal, principle. Our second-level explanation, which is manifestly a deduction, is not an explanation of facts in terms of pure reason; it is, again, an explanation in terms of a scientific *law*.

First-level explanations are psychologically more appealing, hence didactically more effective, than second-level explanations, because their *immediate* referent is a further fact, but they are misleading as long as the concealed law is not noticed. There is no essential difference between the two kinds of scientific explanation; at both levels reference to facts *and* laws "governing" the same or related facts is made. Only, in first-level explanations the reference to facts is explicit, whereas the appeal to laws is implicit—which obscures the role of inference in it. (Most explanations of historical facts seem to be first-level explanations in which the reference to law statements is concealed—which accounts partly for the current unawareness of historical lawfulness.[16]) And in second-level explanations, the roles of facts and of law statements are reversed relative to the former case—which creates, especially among logicians, the illusion that scientific explanation is a purely logical affair.

Let me now turn to the case in which the explicandum itself

[16] For the concealed appeal to universal hypotheses in sociohistorical explanation, see Carl G. Hempel (1942), "The Function of General Laws in History", in Feigl and Sellars, ed., *Readings in Philosophical Analysis*, esp. p. 464.

is a scientific law. Science seldom remains satisfied with the explanation of a class of facts in terms of a given set of laws; either different explanations are tried or, to the despair of ultimate-lovers, scientific laws are explained in terms of further, more general, laws. Moreover, there is no evidence that the process of finding the laws of laws may ever finish with the discovery of fundamental, ultimate laws—even less, with the disclosure of a single universal formula. Empirical generalizations are shown to follow from statements richer in theoretical concepts—as when Galileo's law of falling bodies was deduced from Newton's second axiom in conjunction with a specific assumption regarding the force function; and statements having a certain range of validity are subsumed under statements of greater generality—as when Newtonian mechanics was shown to be a limiting case of relativistic mechanics. Weakly explanatory (that is, mainly, though not exclusively, descriptive) laws are deduced from laws having a stronger explanatory component—as happened with geometric optics, which was deduced from the wave theory of light; conversely, strongly explanatory laws may be subsumed under weakly explanatory laws—as in the case of the deduction of the laws of motion from a variational principle. The laws characterizing a given integrative level are explained as emerging from lower-level laws, as was the case of the laws of chemical binding, which follow from the laws of quantum mechanics—and so on.

Science does not stop asking why laws are as they are; it does not, however, seek explanations *behind* laws but in terms of *further* laws—which, if believed to be "final" at a given moment in a given context, are called principles or axioms, that is, highest-level law statements. Science, then, does not remain contented with the phenomenalist dictum that there is nothing to be further understood about laws—as if laws were contingent, that is, as if they might have been different. (Laws are *logically* contingent, in the sense that there is no logical necessity in them if torn out of the theoretical system to which they belong; a change in the laws of nature or of society would not be *illogical*, that is, would not entail logical contradiction.) Science is always looking after the "sufficient reason" of laws, and finds it in still other laws. Nothing warrants the belief in ultimate first principles or absolutely fundamental laws; $laws_2$ may be *fundamental in a given context* or for a while, not absolutely or forever. By subsuming scientific laws under statements of greater generality, by deriving the laws characterizing a given level of reality from laws working at other levels, science tests the ontological hypothesis of the connectedness of reality, just as the variety and heterogeneity of the world is shown by the fact that an increasing number of qualities and systems of qualities, of laws and systems of laws, is continually being discovered.

But it is high time to approach the problem of the relevancy of the causal principle to scientific explanation, that is, the question of the extent to which scientific explanation is causal. To this end, the types of scientific explanation may be grouped in the following classes: (A) explanations that *may* but need not be causal, and (B) those that are characteristically *non*-causal, in the sense that they do not involve the causation category. Needless to say, no attempt will be made to achieve an impossible completeness.

11.3. Explanations That Can Be Causal

Let us examine some types of explanation in which the notion of causation may enter. (An explanation will be scientific *and* causal if, besides being coherent and adequate in the sense explained in Sec. 11.2.1, it invokes a causal law.)

(A, 1) *Inclusion in a Sequence of Events or of States*

Illustrations: (*a*) "The present month is September, because the last month was August". (The "law" here invoked is the uniformity of sequence "September comes after August".) (*b*) "Charlie's voice is now changing because he is entering puberty". (The regularity serving as explanans is here "The voices of almost all boys entering puberty change".)

The typical form of this explanation is "*B* is thus and thus because it was preceded by *A*—and whenever *A* happens it is known or assumed to be followed by *B*". No genetic link need be asserted in this sort of explanation, which invokes a mere *sequence*. This is a very rudimentary type of explanation indeed; but it is often sufficient, whether the antecedent can be regarded as a cause or not, that is, whether the explanation is *causal* or not. From the logical point of view, this kind of explanation often consists in an appeal to a *definition*, and consequently is tautological. Thus the definition "September is the month following August" (explanatory premise) entitles us to infer that the first day after August is over will be September 1. But we are not interested at this stage in the logical structure of explanation: we are concerned with its ontological aspect, which in this case focuses on the concept of *sequence* (not necessarily of the causal type), not on the logical concept of definition.

(A, 2) *Tracing of Genesis and Evolution*

Illustrations: (*a*) "*A* behaves the way he does because of his social origin and of his upbringing". (*b*) "Let us then suppose the mind to be, as we say, white paper, void of all characters, without any ideas; how comes it to be furnished? Whence comes it by that vast store, which the busy and boundless fancy of man has painted on it with an almost endless variety? Whence has it all the materials of reason and knowledge? To this I answer, in one word, From experience: in that all our knowledge is founded, and from that it ultimately derives itself".[17]

In this typically, though not exclusively, modern kind of explanation, the present traits of the explicandum are accounted for as a stage in its development; laws of evolution, or of emergence, play here the role of explainers. In contrast to the previous kind of explanation (consisting in the inclusion of an object in a sequence), a genetic connection is here assumed whereby the *raison d'être* of something is accounted for in terms of its production at a given time, and of its later development. As in the former case, however, a why-question is reduced to a whence-question, and the law to which reference (explicit or implicit) is made may but need not be causal.

[17] Locke (1690), *An Essay Concerning Human Understanding*, book II, chap. i, sec. 2.

(A, 3) *Connection with Different Facts*

Illustrations: (*a*) "Iron rusts when in contact with air and moisture". (*b*) "Electrically charged particles, when approaching our planet, move along a helix owing to the magnetic field of the earth".

In this case, the given fact is connected, not with other facts of the same series (as in the two previous kinds of explanation), but with facts of a *different order*; and these act as the determinants (or codeterminants) of the change in question. Explanation is here achieved by placing the event concerned in what is assumed to be its actual context, or nearly so—a system of interconnections that had been overlooked before the explanation. This sort of explanation (which Whewell called colligation) is common in the sociohistorical sciences, where the assumption of interconnections is instrumental in suggesting the search for events which available evidence has not yet established.[18] Often, this kind of explanation is characteristically causal, since it may invoke agents acting extrinsically and uniquely; but even more often, it revolves around the category of reciprocal action, or reciprocal causation, since it consists in showing the existence of interdependencies (see Chapter 6).

(A, 4) *Analysis of Complex into Simpler Facts of the Same Nature*

Illustrations: (*a*) Ampère's explanation of large-scale magnetic fields in terms of molecular currents acting as elementary magnets. (*b*) Explanation of the working of complex mechanisms in terms of the functioning of simple machines.

In this case, the given fact is explained as the aggregate or composition of simpler facts, usually on a smaller scale, which are either known or assumed to be known—or simply forwarded as an untested conjecture. An essential point is that the explicandum and the facts acting as explainers shall not be radically different from one another (when they are we get essentially noncausal explanations of types (B, 4) and (B, 5) below). A conspicuous subclass of this kind of account is that of mechanical explanations, consisting in decomposing complex events

[18] See Gardiner (1952), *The Nature of Historical Explanation*, part III.

into changes of place of mass-points (which may but need not be atoms).

In contrast to the traditional conception of explanation as reduction of the unknown to the known, of the new to the familiar, the scientific analysis of complex into simpler (not "simple", which involves an assumption about ultimacy) facts has sometimes just the opposite character, since it may be performed in terms of unperceivable hypothetical entities, sometimes called "theoretical entities"; such was the atomic theory before our own century. Scientific analysis need not reduce the new to the old, the unfamiliar to the familiar, as reductionism requires; what it does is to reduce the *complex nonunderstood* to the *simpler understood.*

11.4. Noncausal Explanations

(B, 1) *Recognition, Identification, or Inclusion in a Class*

Illustrations: (*a*) "This body sinks in water because it is made of iron, and iron is heavier than water". (*b*) "That animal does not sing because it is a dog, and dogs do not sing".

The typical form of this explanation is, "If a thing is F, it is G; now, a is F; hence a is G". This kind of explanation might be termed taxonomic, for it essentially boils down to a classification (that is, to the inclusion of the class F in the class G: $F \subset G$). Causes are definitely absent from this sort of explanation, although further explanations, at other levels, might introduce the causation concept.

(B, 2) *Description*

Illustrations: (*a*) The formation of images in optical instruments is explained by means of the laws of geometric optics. (*b*) The functioning of heat engines is explained with the help of the laws of thermodynamics.

By definition, the so-called phenomenological laws do not contain the causation category. Thus geometric optics regards light rays as stationary states rather than as processes; and the principles of conservation and of evolution, which stand at the basis of thermodynamics, definitely do not refer to any causal agent. Despite the traditional opposition of description and explanation, the so-called phenomenological laws do serve explanatory purposes; we saw this in the case of kinematic laws (see Sec. 10.3.2). In a sense, description is prior to explanation; in another sense it is *a sort of explanation*, though admittedly a superficial one. Sooner or later, truthful explanations enable us to frame more complete and more accurate descriptions; they may even suggest the existence of facts overlooked in descriptive statements. Pure description, "unhampered by theory", "unbiased by interpretation", entirely hypothesis-free, is a myth invented by traditional positivism, intuitionism, and phenomenology. Science ignores wholly uninterpreted facts; the very selection of facts is guided by theoretical principles (such as criteria of relevance) and by hypotheses of an explanatory nature. The interpretation of pointer readings is based on theoretical considerations embodied in the very building of the measuring apparatus; even the wording of the so-called protocol sentences entails a host of assumptions about the relevancy of the factors usually present in every concrete situation. No scientific statement is meaningful outside a theoretical system. In short, science is both descriptive and explanatory; and description can be distinguished from explanation but not separated from it.

(B, 3) *Explanation in Terms of Static Structure Laws*

Illustrations: (*a*) The *raison d'être* of every piece in a metal framework is accounted for by the place it holds in the whole. (*b*) The differences among isomeric molecules are explained by the relative positions of the component atoms.

In this sort of explanation the given object is analyzed into entities which are shown to hold a certain place in a static structure, which place accounts for the peculiarities and functions of the whole. Nothing is here said of changes;[19]

[19] Laws of structure have sometimes been denied. Thus Schlick (1936), *Philosophy of Nature*, p. 60: "Only laws of succession, but not of coexistence, exist". Thereby the laws of statics, systematics, morphology, and so forth, are all thrown away.

hence, the causation category is not involved in this kind of explanation, which may be regarded as a subclass of (A, 4)—just that for which the connections are not causal.

(B, 4) *Reference to a Lower Level*

Illustrations: (*a*) Molar properties of matter are explained as emerging from molecular behavior. (*b*) Laws of mental processes are in principle explainable in terms of physiological laws.

The object to be explained (the explicandum) is shown to be, or assumed to be, not the mere aggregate or sum of smaller scale facts (as in (A, 4)), but the outcome of qualitatively different events belonging to a lower level. A genetic but noncausal connection is thereby established among different domains. This sort of explanation is often referred to as a reduction.[20] This designation is misleading insofar as it suggests that the peculiar qualities of a given level are "explained away" by their connection with the properties of the underlying level. Every explanation is a reduction from a *logical* point of view, since it consists in the derivation of a particular from a general statement; but it is not always ontologically reductive: not, at least, unless the higher-level entities are shown to be mere aggregates of lower-level entities, without properties of their own. Genuine reductive explanations were dealt with above, under (A, 4).

(B, 5) *Reference to a Higher Level*

Illustrations: (*a*) The measured values of the atomic properties called 'observables' depend not only on the atomic objects themselves but also on their interaction with the macroscopic apparatus with which the measurement is performed. (*b*) The behavior of a member of a collection (molecule in a mass of liquid, limb or gland in an organism, individual in a community) depends strongly on the behavior of the whole.

[20] See Ernest Nagel (1949), "The Meaning of Reduction in the Natural Sciences", reprinted in Wiener, ed., *Readings in Philosophy of Science*; "Mechanistic Explanation and Organismic Biology", *Philosophy and Phenomenological Research* *11*, 327 (1951).

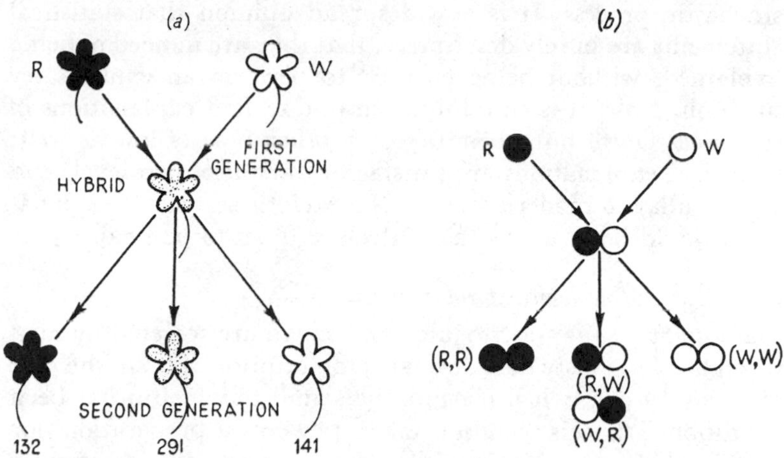

Fig. 30. (*a*) Simple hybridization: summary of facts. The figures (obtained by Correns) are in the proportion 132 : 291 : 141, or approximately 1 : 2 : 1. (*b*) Statistical *explanation* of cross-fertilization. The reproductive cells of the two kinds combine at random (hypothesis of randomness). The theoretical prediction of proportion is accordingly 1 : 2 : 1.

This sort of explanation (often called organismic) consists in disclosing the place of the given object in a whole and in showing the reaction of the whole on the part. The synthesis (reconstruction of the whole) results from an analysis of the interdependence of the parts. The part-whole relation, not the causation category, is here at stake.

(B, 6) *Statistical Explanations*

Illustrations: (*a*) The equation of state of the ideal gas is explained by means of mechanical and statistical hypotheses. (*b*) The distribution of hereditary characteristics in certain populations of living beings is explained by means of Mendel's laws (see Fig. 30).

Statistical explanations consist, essentially, in showing that the given object is a member of a statistical population or of a stochastic process. It is a widespread opinion that statistical statements are purely descriptive, that they are in need of being explained, without being entitled to perform an explanatory function. Now, it is certainly desirable to find explanations of statistical laws—but, then, of every other type of law as well; statistical explanations are satisfactory at their own level, and are peculiar to modern science. To exclude statistical explanations would be as unwise as to declare them to be final.

(B, 7) *Teleological Explanations*

Illustrations: (*a*) Gastric juice and saliva are secreted by man when he sees a tasty dish or hears a description of it (in the case of a Pavlov dog, when it hears the sound to which it has been conditioned); this secretion takes place as a preparation for (possible) future digestion. (*b*) Certain wars are consciously planned in order to prevent economic crises or placate social uneasiness.

No one seems to have afforded convincing grounds for doubting that human conscious behavior is purposeful or goal-directed; the controversial point is whether unconscious end-seeking functions and activities can be found at the level of life. What is not usually denied is that, in contrast to physico-chemical processes, many plant and animal functions and behaviors are not indifferent to the end result but occur instead as if they were somehow directed by it. (They are actually determined by the immediately previous states and by the whole past history of the organism, as well as by its environment; organs, functions and behaviors could not be determined by future, still nonexistent needs; they are presumably determined by past and present conditions and are *adapted* beforehand to coming conditions, though not with foresight or conscious planning, but as a result of a long and blind past history of successes and failures.)

In the case of the conditioned reflex producing secretion by certain glands, the sound (stimulus or efficient cause) initiates or triggers, but does not produce, the entire process; moreover, the process anticipates another function (digestion)—which, after all, may not take place. Whatever the real nature of the nexus stimulus (cause)–response (effect) may be, it certainly is not a *direct* causal link. And as soon as physiology and psychology attempt to go beyond simple schemes of the stimulus–response type, as soon as they try to explain the whole process between the observable cause and the observable effect, they transcend not only the frontiers of phenomenalism (for example, behaviorism), but also those of causality.

The largely noncausal character of biological processes is realized when it is understood that, within a wide (though limited) range, they are independent of the precise environmental conditions, and even of the particular means or "methods" employed to achieve goals, such as self-preservation, that are comparatively stable.[21] Life processes are eminently *self-determined* and *historically preadapted*; far from being the passive toy of its environment, the organism seems to select actively the most favorable conditions for the attainment of its ends (preservation, development, propagation). It should however be emphasized that teleological laws are *statistical*, in the vague sense that organisms do not always succeed in fulfilling their ends; teleological laws are consequently deprived of the necessity (uniqueness and regularity) claimed by vitalism, and characteristic of causal laws.

Whether teleological explanations will be retained in future biology and psychology or not, cannot (or should not) be foretold. The fact is that teleological explanations have not been eliminated from these branches of science. What modern science has done in connection with teleology is not to discard it but rather to deprive it of supernaturalistic overtones, to confine it to the higher integrative levels, and to show that even there it is insufficient—and, moreover, that teleology is often replaceable by other categories of determination. So far as I know, modern science has not proved that inner finality (or immanent finality, in contrast to transcendent design) is a

[21] See E. S. Russell (1945), *The Directiveness of Organic Activities*, p. 144: "What is distinctive is the active persistence of directive activity towards its goal, the use of alternative means towards the same end, the achievement of results in the face of difficulties". Besides this multiplicity of means, teleological processes are characterized by the stability of goals despite the variations (within limits) in the external circumstances.

myth. It may be a mistake to dispense with teleological explanation just because in the past it has been associated with anthropomorphism and theology; the unscientific attitude toward teleology is to regard goal-striving activity as ultimate, unexplainable, or supernaturally directed. The present task of science, in connection with teleology, does not seem to consist in denying it a priori but in trying to *explain* the laws of goal-adapted organs, functions, and behaviors, in terms of evolutionary laws, feed-back processes, and so forth; it consists, in short, in explaining teleological patterns in terms of other laws of nature, thereby extruding the notion of design from biology once and for all.

It is even likely that teleological laws will not be replaced but will be explained by other laws, that, for example, they will be shown to have emerged in the course of the evolution of organisms and associations of living beings. (Although the individual organism may not as yet have gone through the end state that is coincident with the attainment of the goal, its ancestors have, and this experience may have been preserved for the species by the mechanisms of selection and heredity.) Teleological laws will perhaps be explained as a new mode of behavior of material systems, resulting from a long past process of trials and errors in the adventure of adaptation, and stabilized by the mechanisms of heredity. Scientific philosophy does not require the extrusion of teleological explanations in connection with the higher integrative levels; it demands merely the avoidance of the obscurantist interpretations of teleological patterns in terms of immaterial and unintelligible entelechies, such as those imagined in the early days of biological and psychological research.

(B, 8) *Dialectical Explanations*

Illustrations: (*a*) The composition of atomic nuclei is the result of the conflict of two opposing tendencies: the equalization trend stemming from the conversion of neutrons into protons, and the reduction of the proton number owing to electrical repulsion. (*b*) "In the process of demonstration the thinker splits himself into two [*entzweit sich*]; he contradicts himself;

and only after having stood and overcome this inner opposition is he a demonstrator".[22]

Typically dialectical explanation consists in the disclosure of the inward and outward conflicts that keep certain (not all) processes running, or that bring about the emergence of entities endowed with new qualities. Although dialectical explanations employ several other categories of determination (such as interaction and causation), they have peculiar characteristics that cannot be explained away.

11.5. Conclusions

In order to ascertain whether a given explanation is causal or not its logical examination—that is, the analysis of its logical framework—is not only insufficient but also irrelevant. An explanation can be called causal provided the causation category lies at the center of it; and this is not a logical but an ontological question—or, more precisely, this can solely be decided upon an examination of the ontological referent of the explanans-proposition(s). Therefore the logical-empiricist analyses of causal explanation, restricted as they are (when they acknowledge explanation and causation at all) to the logical aspect of the question, are all insufficient; insofar as they are correct, they apply to *all* kinds of scientific explanation, and therefore they do not enable us to discern causal from noncausal explanation.

A succinct ontological examination of current types of scientific explanation has shown us that there are many ways of understanding, that is, of answering why-questions, the disclosure of causes being but one of such ways—or, rather, a conspicuous but not universal component of scientific explanation. Eight types of scientific explanation based on noncausal laws have been listed in the preceding section; but a closer examination should yield even further types of explanation, in which no reference is made to the causation category. To assume or to require beforehand that every scientific explanation be couched in causal (or in statistical, or in dialectical) terms does not seem much wiser than the procedure of the

[22] Feuerbach (1839), *Zur Kritik der Hegelschen Philosophie*, p. 42.

shaman who asks the complex question: "Which are the evil spirits dwelling in this sick man?"

As a consequence, it is not possible to retain the identity *causa sive ratio*; contrary to the traditional teaching, causality is not a sufficient condition for understanding reality, although it is often a component of scientific explanation. Moreover causal explanation, if scientific, will not be found *beyond* scientific laws[23] but *in* the laws themselves, as it is characteristic of scientific explanation in general to be made by means of laws, some of which have a causal component while others lack it. Scientific explanation is, in short, explanation by laws—not necessarily explanation by causes.

What lends causal explanation a delusive majority over the remaining types of explanation is that most explanations are framed in a causal *language*. Thus in English most explanatory statements contain the word be*cause* (by cause); they are, indeed, of the form 'q because p'—save teleological explanations, which usually have the form 'p in order that q'. To fall into linguistic traps like this one is a sign of naïveté; but it is also a mark of philosophical candor to conclude the bankruptcy of causality from the undeniable fact that the statements of *particular* scientific laws do not usually contain the words 'cause' and 'effect', which designate philosophical categories.

Now a test of scientific hypotheses, whether causal or not, is prediction. Prediction (backward or forward) is, indeed, a touchstone of the factual adequacy of theories, just as theorems are the touchstones of the logical consistency and fruitfulness of axioms. Let us then proceed to inquire into the place of the causal principle in scientific prediction.

[23] The exteriority of causes and laws, probably a remnant of religious thought, was maintained by William Whewell. See also Meyerson (1921), *De l'explication dans les sciences*, vol. I, p. 53: "Everyone knows that what the scientist seeks beyond law is frequently denoted by the word *cause*, which is in this sense more or less synonymous with the term *explanation*: when the cause or causes of the phenomenon are known, the latter will have been explained, and the mind will declare itself satisfied".

12

Causality and Scientific Prediction

We shall now examine the nature and function of scientific prediction, in order to ascertain whether successful forecast requires the knowledge of causal laws, as is usually assumed, and whether successful prediction is both the meaning and the criterion of causality, as is often maintained. It will be shown that there are several kinds of prediction, scientific prophecy by means of causal laws being only one of them—whence both the traditional and the positivistic solutions of the problem of the relevance of causality to prediction, and vice versa, are inadequate.

12.1. Nature and Functions of Scientific Prediction

12.1.1. *Nature of Prediction According to Law*

Scientific prediction may be defined as the deduction of propositions concerning as yet unknown or unexperienced facts, on the basis of general laws and of items of specific information. It has rightly been pointed out that the logical structure of scientific prediction is the same as that of scientific explanation; in fact, both are consequences of the conjunction of laws and particular pieces of information. A theory can predict to the extent to which it can describe and explain.

But the identity in *logical* structure between prediction and explanation does not entail an identity in nature or kind; prediction is *epistemologically* not the same as description and explanation because, as everyone knows, prediction is usually affected by a peculiar uncertainty of its own. Descriptions are never complete and explanations are never final, because an unlimited number of variables always escape us. The uncertainty of prediction stems partly from the incompleteness of description and explanation; but in prediction an *additional* uncertainty appears, namely, the one associated with the unexpected emergence of novelty. The most careful prediction may fail to foresee important novelties that would not remain unnoticed in an accurate description; as the saying goes, it is easy to be wise after the event. In other words, the difference between description and prediction is simply due to the fact that the event that is predicted *need not be the same* as the fact that is described. To put it even more briefly, there is knowledge but not foreknowledge.

True, the uncertainty that characterizes prediction could be removed or decreased if more complete knowledge of the relevant or significant variables, laws, and specific data were at hand. But the fact is that satisfactory descriptions and explanations *are* often attained with the help of a fixed and even small number of laws and data, which may however be found insufficient for an equally satisfactory prediction. In this sense, every specific problem in description is closed, whereas every prediction problem is *open*, just because of the (provisional) closure of description. Prediction is, then, epistemologically quite different from description and explanation, even though the logical structure of prediction is the same as that of explanation.

12.1.2. *Predictive Nomological Statement: A Third Level of Meaning of 'Law'*

The consideration of causal laws affords a clear illustration of the difference in kind between scientific explanation and scientific prediction. In point of fact, on the basis of causal laws (or partly causal laws) we can frame causal (or partly causal) explanations—but very seldom predictions that are "causal" to the same extent, for most predictions about the empirical verification of laws, whether causal or not, have a *statistical* ingredient (such as the statement of the probable error). Moreover, *quantitative* predictions, such as those of future positions on orbits, are *always* partly statistical, since the estimate is incomplete unless the probable error is assessed.[1]

As a compensation, statements of $laws_2$ for purposes of verification, prediction, or application *may have a causal component that was absent in the corresponding* $laws_2$. This is the case whenever we are able to *control* some of the variables related by the law under consideration—provided the relation is unambiguous. The set of variables under experimental control we may call the 'cause' if, by changing their values in a prescribed way, a given effect is invariably produced in a unique way, without in turn appreciably influencing the "cause". But this will not be enough to warrant that the relation concerned —that is, the law_2—is itself a causal one; to this end, we have to prove that, by choosing the complementary set of variables as parameters under experimental control—that is, by handling the former effect as a cause—a *different* connection will obtain. If the same connection obtains, it cannot be termed causal for, being reversible, it lacks the essential unidirectionality of causation. And such a proof will often fail; most connections are, in fact, symmetrical (hence noncausal) functional dependences, that is, they do not alter if the "cause" and the "effect" are interchanged. As Campbell[2] wrote in connection with the process of verification of a functional relation of the type $y = ax$, "the law states a relation which is not that of cause and effect, although it may be established by observing such a relation; there is a distinction between the meaning of the law and the evidence on which it is asserted".

The foregoing considerations suggest rendering explicit a *third* meaning of the term 'law', besides the two recorded in Sec. 10.1, namely, the form taken by law statements ($laws_2$) when used for predictive and other purposes connected with experience. We thus have three levels of meaning of 'law': objective pattern (law_1), conceptual reconstructions thereof ($laws_2$, or theoretical laws), and adaptations of $laws_2$ to their

[1] See Hans Reichenbach, "Das Kausalproblem in der Physik", *Die Naturwissenschaften 19*, 713 (1931), p. 715: "*Every causal proposition, when applied to the prediction of a natural event, takes on the form of a probability proposition*".

[2] Campbell (1921), *What is Science?*, p. 54.

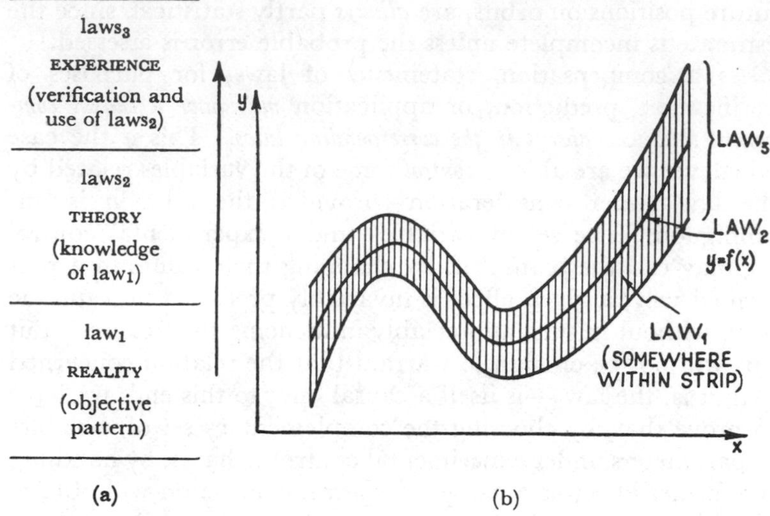

Fig. 31. (*a*) The levels of meaning of 'law'. (*b*) Illustration: a two-variable law.

empirical verification or practical use ($laws_3$). Reality, its theories, and the empirical verification of the latter are thereby accounted for (see Fig. 31). The mere fact that a neat distinction must be drawn between law statements ($laws_2$) and their adaptations to practical purposes ($laws_3$) is, by the way, a strong argument against the pragmatic (or operational) principle according to which the meaning of a proposition is the mode of its verification.

The nature of scientific prediction will be further clarified through an analysis of its functions, to which I now turn.

12.1.3. *Functions of Scientific Prediction*

The function of prediction is twofold: (*a*) it is a *forecast* of some kind (whether like those of astronomy or the statistical sort), and as such an indispensable aid to successful human action and control; (*b*) it is a *test of hypotheses*, for by means of prediction we confirm or disconfirm scientific assumptions: prognosis is a test of both the diagnosis (description) and the etiology (in the wide sense of explanation, whether causal or not).

Prediction is then both practically and theoretically important; moreover, these two features of the question are mutually dependent. In fact, a forecast will be successful to the extent to which it is made on the basis of sufficiently true law statements and specific information regarding the particular case in question. In turn, as pointed out above, a test of law statements (*and* of the items of information involved in the framing of the corresponding $laws_3$) will be the successful prognosis formulated with their help. Hence, although prediction is a very important task of science, the most cursory examination of its function shows us that it cannot be *more* important than either description or explanation, on which prediction rests and on which it reacts; prediction is *as* important as description and explanation. To assert that prediction is *the* most important task of science—as positivists and pragmatists have held—shows a strange lack of acquaintance with the life of science.

Successful prediction is of course verified (or refuted) by means of observation, experiment, and artificial reproduction. Reproduction, such as is performed in industry on the basis of technology, is the optimum test of prediction, since it requires the control of the most relevant variables. But, of course, reproduction is not always possible; moreover, it is hard to ascertain when reproduction may be regarded as crucial, that is, as affording a definitive confirmation or refutation of a given set of hypotheses—to the extent to which there is anything definitive in empirical matters. Furthermore, the fact that successful reproduction is the best test of prediction does not mean that making entails knowing, that is, it does not mean that, if we know how to produce or reproduce a given thing or phenomenon, we know all there is to know about it. The atom bomb was made with the coarsest nuclear theory. Making *is* not knowledge and does not entail exhaustive knowledge: making is a *test* of knowledge; it is not the sole test and it does not afford the last word—but it is, undoubtedly, the best test.

However, our concern is not prediction in general but rather the various kinds of scientific prediction.

12.2. Kinds of Prediction

According to a widespread prejudice, a single sort of forecast is entitled to be called scientific, namely, the highly accurate *quantitative* prediction of future events, such as those based on the laws of classical mechanics or classical electromagnetic theory.[3] This view is outdated; there are many kinds of prediction, actually as many as there are kinds of scientific law, on which scientific prophecy is based. Although the last statement is merely a consequence of the definition of prediction adopted in Sec. 12.1.1, it will be convenient to insist on this point just in order to show that our definition does cover the actual usage of the word 'prediction' in contemporary science. To this end, let us examine the kinds of forecast afforded by some types of scientific law—or, if preferred, the kinds of extrapolation allowed by them.

Consider, in the first place, taxonomic rules—the simplest of all from a logical point of view, since they assert "mere" generalizations. They are of the form "Every a is A", where A stands for a class and a denotes either an individual or a sub-class of A. An elementary illustration of this kind of law is "All birds are warm-blooded", that is, "The class of birds is included in the class of warm-blooded animals". This inductive statement enables us to *predict*, with probability near certainty,[4] that the *next* bird we catch will be warm-blooded. A more distinguished example of a taxonomic law enabling us to perform scientific predictions is Mendeleev's periodic table of the elements.

[3] Curiously enough, this function, now popularly attributed to science, was formerly regarded as an attribute of the deity: *pronoia* and *providentia* mean foresight.

[4] The word 'probability' is here used with a loose meaning; it just means that, the above-mentioned rule (taxonomic law) being an empirical generalization, it may be refuted by new empirical evidence, that is, by the finding of as yet unknown cold-blooded birds. Indeed, the "probability" in question is not the one dealt with by the calculus of probability, but merely a degree of rational belief based on empirical evidence. It cannot in our case be estimated on the basis of relative frequencies (all of which yield exactly the value 1 as long as no exception is found), and it cannot be calculated on the basis of available bird physiology; it is not, in brief, a quantitative concept.

Taxonomic rules enable us to predict, sometimes with a high probability of success, that every *next* member of the given class (or every next subclass) that we meet will have the qualities typifying that class. The word 'next' need not refer, in this context, to the objects themselves, that is, to a future state of theirs; it may refer solely to the predictor himself—as is the case with the empirical generalizations of paleontology, archaeology, history, and so on. This kind of forecast does not consist in the accurate prediction of any numerical values referring to future events; but it is none the less scientific, since, far from being a blind guessing, it is a grounded and testable inference.

Next come structure laws. For example, the existence of some light isotopes was predicted on the basis of the assumed structure of their nuclei. A certain set of hypotheses concerning this structure permits the formulation of the following rules, all of which can be used to predict the formation of isotopes: "All stable nuclei up to O^{16} are obtained by adding a neutron and a proton in succession, in the order *n–p*"; and "From O^{16} up to A^{36} the isotopes are obtained by repeating the sequence *n–n–p–p*". There are a few exceptions to this rule, such as He^{5}; this shows that something is wrong with the basic hypotheses concerning nuclear structure—but this lack of completeness and certainty is precisely a mark of scientific knowledge, as contrasted with revealed formulas.

Here again, prediction is not an inference from present to future, but *from known to unknown*. Structure hypotheses, as well as taxonomic rules, laws of colligation, and so forth, enable us to forecast the *existence* (or the absence) of certain things and properties, whether in the future, the present, or the past. In this sense they are much more important than many accurate quantitative predictions.

Then come the so-called phenomenological laws, such as those of geometric optics, thermodynamics, and pre-Newtonian astronomy. For example, if the refractive index n of a transparent body is known (either from previous measurement or from other physical characteristics linked to it in a known way), and the angle of incidence i of a light ray is measured, the Snell law $\sin i/\sin r = n$ enables us to predict the angle of refraction

r of the ray. Note that time does not play a role in geometric optics; nor does it in the thermodynamic theory of equilibrium states. This shows that, contrary to a popular belief, scientific prediction does not require law statements asserting that certain events will follow other events according to prescribed time patterns. Prediction on the basis of this particular kind of law is just a kind of scientific prophecy, not its prototype. Time does, however, enter the "phenomenological" laws of pre-Newtonian astronomy, such as Kepler's second and third laws. In none of these cases have *causal* laws been at stake, yet accurate predictions were shown to be possible.

Let us now approach prediction on the basis of time patterns —which is usually, and wrongly, believed to be the *sole* kind of scientific forecast or at least its paradigm. Laws of development in time, which may have a causal component or not, can sometimes be expressed in a mathematical form. The simplest types are:

$$\frac{\Delta x}{\Delta t} = f(t) \qquad \text{(finite-difference equation);}$$

$$\frac{dx}{dt} = f(t) \qquad \text{(ordinary differential equation);}$$

$$\frac{\partial u}{\partial t} = f(x, t) \qquad \text{(partial differential equation);}$$

$$f(t) = U(t, t_0)f(t_0) \qquad \text{(operator equation).}$$

Such formulas are needed in order to predict the *precise* future (or past) values of a certain variable *at a prescribed time.* (They are, together with knowledge of initial or boundary conditions, necessary for the formulation of this kind of prediction, but they are not sufficient. An additional condition is that the equation concerned actually has solutions for the prescribed time—and this is not obvious but must be demonstrated.[5])

[5] For instance, the general solution of the partial differential equation

$$\frac{\partial u}{\partial t} + i\frac{\partial u}{\partial x} = 0$$

is

$$u(x, t) = \sum c_n \exp\,(inx + nt),$$

and this series converges for $t \leqslant 0$ only, thus rendering retrodictions, but not predictions, possible.

However, as should be clear from the previous examination, this sort of prediction is the *exception* in science rather than the rule, despite which it is practically the sole one dealt with by philosophers of science.[6]

The reason why we are able to predict even *without* the help of law statements containing the time variable is a truism, namely, that in such cases it is our knowledge that undergoes a change in time—whether that knowledge refers to a process or not.

Let us now take a look at statistical prediction.

12.3. Statistical Prediction

12.3.1. *Insufficiency of Prediction of Individual Events*

The fact that *certain* astronomical predictions have been and are still regarded as *the* model of scientific prognosis (because of their numerical accuracy, and because they have been the sole reliable ones for millennia) has had two opposite effects. On the one hand, it has fostered research in other sciences; on the other, it has spread an erroneous notion of the nature and function of scientific prediction—and consequently an incorrect idea about what are the disciplines that can be honored with the title of science. Astronomical predictions based on laws of motion should not be regarded as *the* type of scientific prediction but rather as a highly exceptional case favored by the extreme mechanical isolation and stability of the solar system, as well as by the large interplanetary distances. These conditions are almost ideal; they are the furthest from the ordinary conditions met with in science, and they account partly for the early development of astronomy.

Even so, if we asked the exact position of our planet one billion years hence, no Astronomer Royal would venture to give us an accurate answer, even assuming that no qualitative changes were to occur in the meantime; tidal friction, random perturbations from other celestial bodies, and the very errors

[6] Thus Destouches (1948), *La mécanique ondulatoire*, chap. vii, and "Quelques aspects théoriques de la notion de complémentarité", *Dialectica 2*, 351 (1948) has built a "general theory of predictions" based exclusively on time patterns of the above-mentioned types.

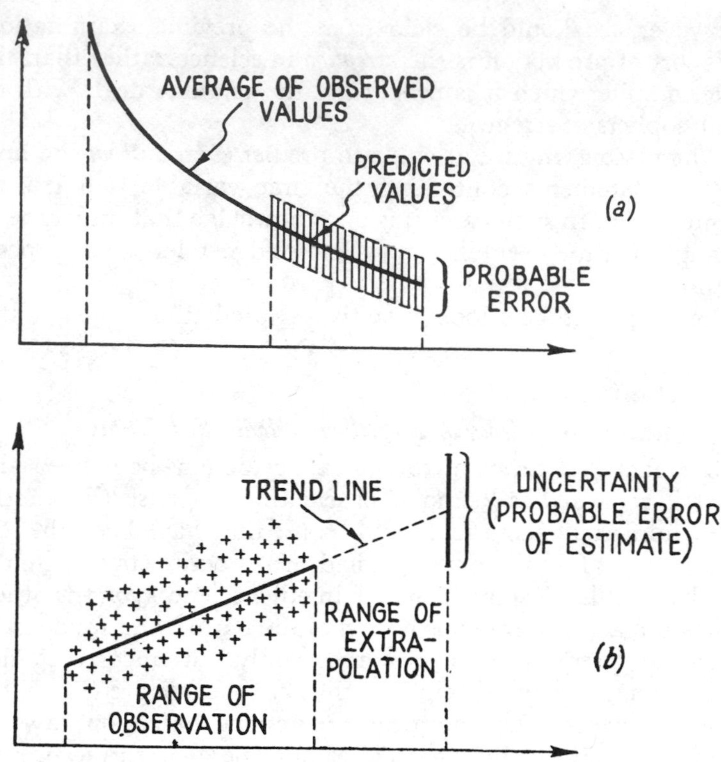

Fig. 32. (*a*) Prediction on the basis of a functional relation. (*b*) Prediction on the basis of a trend line (linear regression line).

in the estimate of the present values of the earth's position and velocity (initial conditions) would suffice to blur completely the future values. This problem is similar to that of asking about the exact result of a roulette wheel after a sufficiently large number of turns: the answer is that all final positions are equally likely.

Moreover, modern astronomy makes extensive use of statistical methods in the exploration and classification of heavenly bodies; it also uses the theory of probability in the study of large collections of stars *qua* collections, and even in problems connected with individual bodies, such as comets (which are then regarded as members of a potential collection, since the probability concept is a set property). As far back as 1816, Laplace published a study entitled *Sur les comètes*, in which, by means of probability considerations, he tried to account for the observed frequency of elliptic and hyperbolic comet orbits. This kind of study, based on definite physical and statistical hypotheses, enables us to explain observed facts (such as frequency ratios) and to estimate the probability that the next member of the collection will have such and such characteristics (for instance, that it will move along an elliptic orbit). But the chief use of statistical laws (established either empirically or on the basis of hypothetical models) is not the prediction of the probability of occurrence of *individual* events. Their chief importance, as far as prediction is concerned, lies in that they help in foreseeing collective properties, that is, properties of large collections of entities that are similar in some respects. Among other things, statistical laws enable us to predict probable frequencies (inferred from probabilities that are handled without any uncertainty), averages, mean deviations from the mean, degree of association (statistical correlation) among causally disconnected entities, the most probable trend of a course of events, and so on (see Fig. 32).

12.3.2. *The Statistical Predictions of the Sciences of Man*

A popular argument against the scientific character of anthropology, sociology, history, and related disciplines is that sociohistorical sciences are seldom capable of predicting. If it is implied that the required prediction be of *singular* events with *precise* indication of place and time, then it can easily be granted that such a sort of prediction is rare in the field of the so-called cultural sciences. It is also easy to see why this is so: most sociohistorical laws are statistical (see Sec. 10.4.5), and statistical laws are simply not supposed to permit the formulation of *precise* predictions about *singular* events. (The word 'statistical', in connection with sociohistorical laws, means not only that they cover mass phenomena, but refers also to the uncertain character of the predictions of singular facts attained on the basis of such laws.)

Thus one of the goals of anthropology (in the large sense of the word) is "to predict or indicate the *general directions* of change *likely* to be taken by the phenomena under consideration. The proper sphere of anthropology as a science is therefore to strive to secure historical, sociological, and psychological laws that describe over-all trends or processes among the peoples of all prehistoric and historic periods".[7] General predictions about over-all trends (which are typically statistical) can be and have been made on the basis of sociohistorical laws; on the other hand, it is next to impossible to predict particular historical events with accuracy. For instance, Marx's hopes for an early social revolution in England did not come true; on the other hand, he certainly succeeded in predicting some general features of future developments.[8] It is no wonder that he failed in the forecast of *singular* historical events while he was able to predict some of the main traits of the world we live in (large centralization of the means of production, increasing participation of the whole of society in production, decadence of colonialism, growth of socialism, and so forth). To this end, Marx relied on a few sketchy law statements concerning the economic structure of the capitalist society, which have "only" a statistical validity, so that they permit the prediction of what is *likely* to occur in the *long run* wherever *certain conditions* are met.

Likewise, the Mendelian laws of heredity do not enable us to foresee with certainty the appearance of a given hereditary trait in a *prescribed* offspring of peas at a *prescribed* time—and this, just because they do not refer to individual events and because they do not contain the time variable. Scientific prediction cannot transcend the scope of the law statements on which it is based, and it cannot be more precise than the specific information that it uses. This is precisely what essentially distinguishes scientific forecast from popular prophecy. A further difference between prophecies and predictions is that the former are unconditional and can consequently be couched in *categorical* statements, such as "Peace will break out"; scientific predictions, on the other hand, are formulated as

[7] Jacobs and Stern (1947), *General Anthropology*, p. 5. Italics mine.

[8] Bernal (1949), *The Freedom of Necessity*, p. 413.

hypothetical statements exhibiting the conditions required for the occurrence of the event under consideration, as in "Peace will break out *if* such and such conditions are fulfilled". The ground of this logical difference is that whereas prophecies do not rely on laws, scientific predictions do rest on them. In short, scientific prediction is lawful; consequently, if scientific laws are employed in an incorrect manner for predictive purposes, the user of them and not the laws themselves must be blamed.

12.3.3. *Are Statistical Predictions Less Complete Than Others?*

The complaint is often heard that statistical laws are *incomplete*, in the sense that they do not allow us to infer with certainty what *individual* events will happen at a prescribed place and time. Thus, from the birth rate in a given country we cannot derive the birth date of a given individual. That statistical laws are incomplete in this sense, and that they should consequently be supplemented with laws of a different type, is easy to grant. Yet, statistical laws are no *more* incomplete than other kinds of scientific law; only, their incompleteness is of a different kind—and this is all. In fact, all kinds of law statements (laws$_2$) permit the framing of predictions (instances of laws$_3$) that are both qualitatively incomplete and quantitatively inaccurate. The difference between the sort of prediction afforded by the various types of scientific law is a difference in *kind* rather than in degree of completeness or accuracy. Thus Newton's laws of motion, usually regarded as the paradigm of scientific law, will not tell us much about the over-all behavior of a group of, say, 1000 stars, whereas a statistical-mechanical treatment of a star cluster (that is, a model of it in which every star is regarded as if it were a molecule in a mass of gas) may lead to certain accurate prognoses; for example, the disintegration rate of a star cluster may be derived by applying Maxwell's statistical law on the distribution of velocities. Statistical predictions can be, generally speaking, as certain (or uncertain) as every other kind of prediction; they do not refer with certainty to individual events, but as a compensation they refer to collective properties, that is, to over-all behavior or long-run characteristics.

The claim that statistical laws, in contrast to other kinds of scientific law, are incomplete, hence provisional, is largely a matter of metascientific inertia; it is a heritage of the 18th and 19th centuries, when it was commonly thought that the progress of science consisted in a close imitation of mechanics and astronomy, the predictions of which were deemed to be the sole decent kind of prophecy that science should afford. In contemporary science and technology, and even in everyday life, we often ask questions that simply cannot be answered on any individual or dynamical laws, questions requiring a statistical approach and analysis. And this is so because they refer to events that occur, to a large extent, *no matter what the precise behavior of the individual components of the collection under consideration is*. Thus an insurance company is not interested in the precise date of the death of every one of its clients, for the whole insurance business is based on the knowledge of certain statistical laws with the help of which the premiums are calculated—a calculus that amounts to a statistical forecast. Even if complete information is available, statistics is not a dispensable luxury;[9] statistical laws are patterns of collective behavior, and such behavior is not rendered any the less real and peculiar when the detailed behavior of the parts is known.

If we are able to answer questions about statistical populations that individual or dynamical laws cannot answer, why should we continue to describe statistical laws as less complete than other kinds of law, which do not allow us to formulate predictions about over-all or long-run behavior?

12.4. Degrees of Certainty in Prediction

12.4.1. *Uncertainty with Causal Laws and Quasi Certainty with Statistical Laws*

It is often thought that causal laws, in contradistinction to other kinds of laws, allow the formulation of predictions with certainty. This is not true, for, even supposing that our law

[9] Neumann (1932), *Mathematische Grundlagen der Quantenmechanik*, p. 107, stated, on the other hand, that the statistical treatment of classical phenomena is "*ein Luxus, eine Zutat*" if the exact values of the initial position and momentum of every particle in an ensemble can in principle be known.

statements were a faithful reflection of reality (and they never are), they would not be enough to formulate predictions, whether accurate or not; law statements have to be supplemented with specific information such as initial or boundary values, past history over a stretch of time, constraints, and so on; and the limited precision with which this specific information is known is always a source of error (see Sec. 12.4.3). Consequently, the causal nature of a scientific law does not warrant the certainty of the predictions attained with its help.

On the other hand, *certain statistical laws allow us to attain near certainty*. Thus, the second law of thermodynamics may be interpreted as stating that it is *almost impossible* for the entropy of a closed molar system to decrease. And the analysis of so simple a mechanical system as the one consisting of two uncoupled linear oscillators with phases standing in an arbitrary ratio to each other shows that it is likewise *almost impossible* that a given configuration of the system be repeated—whence it may be concluded that eternal recurrence is nearly impossible for an aggregate with a finite number of independent (or nearly independent) components[10]—even if they constitute a mechanical system, in which no qualitative change can occur.

An event of probability unity is *almost* certain to happen, that is, it will happen almost always; and an event of zero probability will *almost* never happen, although it is not impossible. (In this context, the word 'almost' means that there may be exceptions and even infinitely many, but such that they do not accumulate and are so rare compared with the alternative events that they have no appreciable *quantitative* influence. Such exceptions do not affect the stable values of the relative frequencies, although they may give rise to *qualitatively* new things—a problem that does not concern the theory of probability. In technical language, that is expressed by saying that the exceptions constitute a zero-measure set, because they form a denumerable sequence. This lack of overlapping of the frequency (which may not vanish even if the probability is exactly zero) and the probability is but an illustration of the incompleteness of the correspondence between the world of

[10] See D'Abro (1939), *The Decline of Mechanism in Modern Physics*, pp. 243 ff.

experience, to which frequencies refer, and its reconstruction in thought, a conceptual system in which the concept of probability may take part.)

In short, in the extreme cases of zero and unity probability, statistical laws lead where causal laws lead only exceptionally, namely, to prediction *almost with certainty*.

12.4.2. *Almost Necessary Truths of Fact*

In some cases, the degree of certainty corresponding to zero and unity probability (which are interpreted respectively as near certainty of failure and near certainty of occurrence) can be attained without resorting to either causal or statistical laws. Thus, on the basis of the functional and teleological laws of human physiology we can predict with almost complete certainty that no child lacking a heart will survive its birth. This prediction is not made on the basis of empirical evidence regarding past cases; if it were, it would be open to the usual objections against inductive generalizations. The empirical knowledge of the zero frequency of children without hearts surviving their birth rather *supports a posteriori* the prediction. In fact, the prediction is a result of our knowledge of the central role of the heart; it is the understanding of the heart's function, rather than any empirical generalization, that permits us to attain such a certainty. Before Harvey's discovery of the nature of the heart's function, the eventual empirical finding of the zero-frequency ratio of surviving children without hearts would *not* have guaranteed the prediction of the death of *every* next such baby.

We can often ascertain a priori, not so much the necessary outcome of an event, as its being next to impossible, on the ground that its occurrence would violate certain well-established (though still hypothetical) laws of nature. For example, the laws of the strength of materials enable us to predict that no arbitrarily high column can be built—although the height at which collapse would take place would only be estimated to within a certain error. And the physiology of the nervous system allows us to infer, with the same certainty, that nervous, hence psychic, functions cease with bodily death, so that survival of the soul is out of the question. Do these instances of *almost necessary synthetic truths* refute the traditional claim of both Platonic idealism and skeptical empiricism, that no certainty can be attained in factual and, particularly, in empirical matters? That is, do these examples refute the thesis that every general proposition having a factual (or an empirical) content is at best a probable hypothesis? No, they do not refute it, but they qualify it considerably, as we shall see presently.

Skepticism about the possibility of attaining anything near certainty in empirical affairs was amply justifiable at a time when most reliable scientific knowledge of the external world was little more than a heap of *empirical generalizations*, that is, general propositions supported solely by their favorable instances. Moreover, such skepticism played a progressive role in the fight against nativism, which claimed that everyone has innate true ideas with factual referents, that is, self-evident truths of fact having a necessary character. That time ended in 1687 with Newton's *Principia*, in which experience and reason were firmly united for the first time, and in such a way that it became possible, so to speak, to calculate facts. From that time on the ideal type of scientific statement is no longer the empirical generalization but the universal law containing some elaborate theoretical concepts and supported, not only by its instances, but by *other* $laws_2$ as well, so that it stands or falls together with a whole bundle of hypotheses. In fact, hypotheses belonging to theoretical systems are not confirmed or invalidated as if they were isolated statements: the test of them strengthens or weakens entire frameworks of ideas referring to facts. This does not change the fact that all general propositions with a factual content are more or less probable hypotheses, but it introduces the important novelty that universal law statements are not all *equally* probable; some of them are *almost certainly* true and others *almost certainly* false.

Contrary to what Kant held, this certainty (actually, quasi certainty) in empirical matters is not grounded on a priori principles (whether logical or intuitive), but upon the rational reconstruction of experience; it is theory (including the theory of experiment), rather than either pure speculation or pure empirical generalization, that enables us to set forth some almost necessary truths of fact or, if preferred, some *rational truths of fact*—a name that in a way typifies the most advanced sections of modern science, which have gone beyond both rationalism and empiricism. Neither experience by itself nor pure mathematics has led us to attain certainty in factual matters; in this sense, the quest for empirical certainty has come to naught. But the theoretical sciences of nature (and, let us hope, of society as well) are rendering the quest for *near certainty* meaningful, dispelling the paralyzing ghost of equally probable hypotheses.

12.4.3. *Grounds for the Failure of Specific Predictions*

Scientific laws do not enable us to draw *completely* safe inferences about the future course of physical or cultural events —and this, quite independently of the presence of the causation category in the law statements concerned. The reasons why a scientific prognosis may fail fall rather naturally into three groups: they may be connected with the law statements themselves, or with the complementary specific information, or with the process of inference. In more detail, the sources of uncertainty in prediction are essentially the following:

(i) *Errors in the Law Statements or in the Choice of Them*

(*a*) The statements of the laws are incomplete or inaccurate (incompleteness or sketchiness referring to qualities, inaccuracy to quantitative precision); that is, our knowledge of the objective laws that are relevant to the fact(s) under consideration is too sketchy or inaccurate.

(*b*) Some of the laws themselves fail, that is, they cease to apply owing to the emergence of qualitatively new features during the process; in other words, unexpected new laws begin to operate. Complete predictability would require not only a thorough knowledge of laws, but also a knowledge of the laws of variation of those laws.

(ii) *Errors in Specific Information*

(*a*) The information regarding the nature of the system is neither complete nor accurate enough, which leads to a wrong

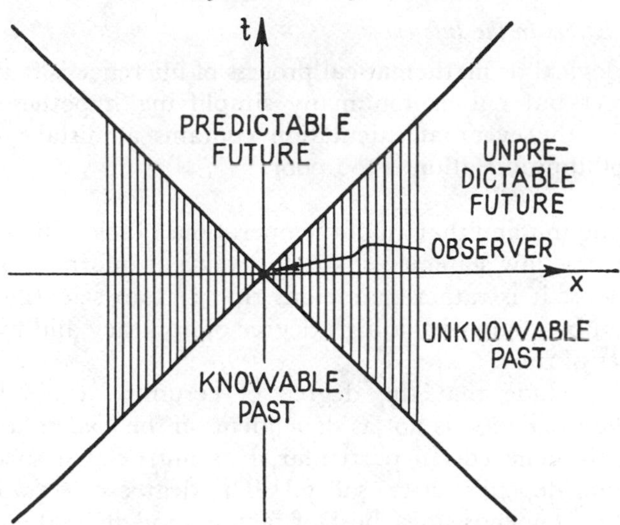

Fig. 33. Limitation of possibility of information gathering owing to the finite velocity of light messengers. Regions of space-time other than the one controlled by the observer in the figure can be scanned by observers (or observing devices) placed at other positions.

choice of the relevant law statements; in particular, the information may not report that the system under consideration is inadequately isolated, so that unexpected interactions with external systems occur.

(*b*) Specific information depending on observation and measurement of quantitative characteristics (such as initial conditions, constants, and so on) is assessed with an excessive error.

(*c*) Accessible information is not enough, either because of technical limitations or because the nature of reality is such that it prevents us from gathering all the items of information needed in order to apply the relevant law statements. This may be due, for example, to the finite propagation velocity of physical signals (such as light rays); this limit upon the speed of message carriers confines the knowable past within the backward light cone and the predictable future within the forward light cone of the observer (see Fig. 33).

(iii) *Errors in the Inference*

The logical or mathematical process of inference is faulty, or is correct but entails too many simplifying hypotheses; for example, the eventual calculation contains a mistake, or its order of approximation is too poor.

Bearing in mind that all of the operations of scientific prediction involve law statements, specific items of information, and deductions, it is rather marvelous that certain scientific forecasts can be made with a high degree of accuracy and for long stretches of time.

We conclude that the degree of certainty attainable in scientific prediction is not as dependent on the *kind* of law as is frequently believed. In particular, it is untrue that successful prediction depends on causality.[11] The degree of certainty in prediction depends on a host of factors, and it is almost impossible to predict with certainty the conditions that will make one prediction more accurate than another.

12.5. Should Causality be Defined in Terms of Predictability?

12.5.1. *The Positivist Criterion of Causality*

The traditional doctrine that prediction is possible with the help of causal laws alone can be found in the belief that predictability is a symptom of causality—or even that causation can be *defined* in terms of predictability. A characteristic statement in this connection is the following by Feigl: "The clarified (purified) concept of causation is defined in terms of *predictability according to a law* (or, more adequately, according to a set of laws)".[12] The definition of causal law as a predictive

[11] This belief goes as far back as Aristotelian scholastics. See Aquinas (1272), *Summa theologiae*, part I, q. LXXXVI, a. 4.

[12] Herbert Feigl (1953), "Notes on Causality", in Feigl and Brodbeck, ed., *Readings in the Philosophy of Science*, p. 408. See also Frank (1937), *Le principe de causalité et ses limites*, pp. 47, 200, and passim; Reichenbach (1944), *Philosophic Foundations of Quantum Mechanics*, pp. 2–3; Hutten (1956), *The Language of Modern Physics*, p. 222. Schlick (1936), *Philosophy of Nature*, p. 58, while asserting that predictability is the *criterion* of causality, refused to regard it as "suitable for a logical formulation of the principle of causality".

instrument led Russell[13] to regard generalizations such as "dogs bark" or "lions are fierce" as causal laws just because they enable us to make predictions, even though they assert nothing but invariable associations.[14]

The following points should be noted in connection with the equation of causality and predictability. First, it involves the identity of *scientific* law with *causal* law—an identity which we found outdated (see Chapter 10). Second, contemporary scientists know that the degree of certainty in prediction does not depend solely on the kind of law, but that a host of circumstances determine the completeness and accuracy of prediction (see Sec. 12.4.3); inaccurate predictions can be made on the basis of causal laws, while statistical laws sometimes permit the attainment of quasi certainty (see Sec. 12.4.1). Third, a trait of reality (causation) should not be identified with a criterion (predictability) for the empirical test of scientific hypotheses that may but need not contain the causation category; in other words, predictability is not the *meaning* of causation but is a *criterion* of truth of both causal and noncausal hypotheses.

Unlike causation, which is an ontological category, predictability is an epistemological category with an obviously historical status; indeed, what seemed formerly to be unpredictable in principle is now seen to be predictable in some sense—and vice versa. Both the kind and the probability of every prediction depend on our knowledge of general laws and of specific situations. On the other hand, causation is a mode of behavior of things in the real world. Predictability, a variable human ability, does not mirror causation at the level of knowledge; it is grounded in *all* types of lawful determination, including causation.

12.5.2. *Uncertainty and Causality in Quantum Mechanics*

The equation of causality with predictability is, of course, common among the upholders of the positivistic interpretation

[13] Russell (1948), *Human Knowledge*, p. 327 and passim.

[14] Even Planck defined causation in an anthropomorphic way in *The Philosophy of Physics* (1936), p. 45: "An event is causally conditioned if it can be foretold with certainty".

of the quantum theory. Thus Heisenberg,[15] in a famous paper, held that what prevents modern physics from retaining causality is the physical impossibility of measuring simultaneous exact values of conjugate variables, such as the position and the momentum of a "particle"; and this, in turn, prevents us from formulating accurate predictions about the future states of the "particle". In fact, "in the rigorous formulation of the causal law, 'If we know the present exactly we may calculate the future', not the inference but the premise is false. We *cannot* in principle know the present with every detail". Heisenberg's "rigorous formulation of the causal law" is at least striking, since it happens to be the formulation of an *epistemological consequence* of the hypothesis of invariable sequence in time—whether causal or not. It is incorrect to hold that the causal principle assumes the premise "We can know the present exactly", because the law of causation is a hypothesis referring to a kind of connection among events as they happen whether under experimental control or not. It is only in order to *test* hypotheses concerning connections among phenomena that we have to know the antecedents from which predictable consequents will evolve.

To say the least, it is hasty to proclaim the failure of determinism on the bare ground that it is not possible to gather all the information that is needed to *test* the causal principle in a particular domain (such as atomic physics). Only the erroneous equations Determinism = Causality (ontological narrowness) and Causality = Predictability (anthropomorphic conceit) can be made the ground of indeterminism. Uncertainty in knowledge is far from being the unequivocal sign of physical indeterminacy or haziness. Moreover, the empirical indeterminacy characterizing the usual interpretation of the quantum theory is a consequence of its idealistic presuppositions regarding the nonexistence of autonomous physical objects.[16]

[15] Werner Heisenberg, "Über den anschaulichen Inhalt der quantentheoretische Kinematik und Mechanik", *Zeitschrift für Physik 43*, 172 (1927).

[16] See Niels Bohr, "Kausalität und Komplementarität", *Erkenntnis 6*, 293 (1936): "We have been forced to forego the ideal of causality in atomic physics solely because, as a consequence of the unavoidable interaction between the object of experiment and the measuring instruments . . . we cannot speak any longer of

The so-called "observables" of the quantum theory are in general only *statistically predictable*; more exactly, with the exception of pure cases it is not possible to predict with certainty what values a given variable will assume upon measurement. This comparative uncertainty in the *prediction* of the results of *measurement* is an empirical indeterminacy (an uncertainty) which by no means warrants the validity of indeterminism in an ontological sense. Moreover, it is not an unbound indeterminacy, for definite probability distributions are predictable, that is, the probability of finding every single value is predicted. Furthermore, the uncertainty under consideration is not the reflection of an objective indeterminateness, of a lack of precise connection between the successive states of microphysical systems, since it openly refers to results of measurement —hence to the interactions of the object with its macroscopic environment—and not to things in themselves, as they behave in the absence of interactions with macroscopic devices. In other words, the empirical indeterminacy in question refers only to physical variables represented by operators; on the other hand, the so-called hidden parameters, or intrinsic variables, by means of which de Broglie and Bohm have succeeded in setting forth the causal interpretation of quantum mechanics, are not affected by any inherent uncertainty.

Those who wish to invoke the quantum theory in favor of some of the fashionable indeterministic reveries must decide to forget, (*a*) that not even the usual formulation and interpretation of quantum mechanics foregoes determinism in general, but questions only causality (see Sec. 1.2.5), the universal validity of which had been questioned in philosophy long before the advent of the quantum theory; (*b*) that from 1952 on various consistent and empirically equivalent interpretations of quantum mechanics have been proposed, in which a good deal of causal determinism is reinstated, though not to the exclusion of statistical determinacy.

---

an autonomous behaviour of the physical object". See also Weizsäcker (1943), *Zum Weltbild der Physik*, pp. 30, 42, 76, and passim, where the Kantian thesis is defended that there are causal chains but not independent of the experimenter.

12.5.3. *Uncertainty and Indeterminacy. Is Ontological Determinism Inconsistent with Epistemological Probabilism?*

Successful prediction enables us to confirm or invalidate scientific law statements, whether causal or not. Prediction and practical application are tests of all the general hypotheses of science and technology, whatever their causal rate may be. The systematic failure to predict by means of a given law with a strong causal component is a strong presumption against the truth of that law statement in the range under investigation; but it does not invalidate the law in every range, and it does not constitute a refutation of the causal principle. The failure of particular causal hypotheses—and even the eventual failure of causality in entire fields of research—proves only that the causal principle has not a universal validity. In particular, the statistical ingredient in all the statements which we have named $laws_3$ (such as the statements of $laws_2$ for predictive purposes) shows that at least one domain, that of the subject-object relation, is not exhausted by causal connections.

Causality cannot consequently be defined in terms of predictability according to law; nor is it correct to define chance as inability to predict. Prediction is a human—both fallible and perfectible—ability; predictability is an epistemological consequence of the existence and knowledge of laws of any kind, both causal and noncausal. Uncertainty therefore does not entail indeterminacy. Complete determinacy, univocally fixed by the totality of laws, is a necessary but not a sufficient condition for attaining complete predictability; it would not be fair to blame the world for our own shortcomings. Ontological determinism is therefore consistent with epistemological probabilism—the doctrine according to which only probable general propositions can be asserted regarding empirical matters (the degree of confirmation being treated as a probability), though not all equally likely.

12.6. Conclusions

In contrast to divination and guessing, scientific prophecy does not claim to attain complete certainty. Very few facts in the concrete world are predictable with near certainty, and none in all detail, because scientific forecast is based on the knowledge of laws and of specific information regarding singular facts, neither of which is ever complete and exact. Still, scientific prediction is often satisfactory; moreover, it is perfectible—which is to say that it is fallible, since the failure of every prediction calls for the improvement of the formulation of the premises on which it rests. The conviction of scientists that satisfactory prediction is possible whenever the relevant laws and specific information are at hand with a minimum accuracy is a form of the confidence in the lawfulness of reality. But prediction, being together with reproduction the decisive test of law statements, does not exhaust lawfulness and its function. Comte was certainly right when he put forward his famous slogan, "*Science d'où prévoyance d'où action*". But he was wrong in denying the explanatory function of science, which not only satisfies in a rational way the inborn instinct of investigation, but also is instrumental in suggesting new, as yet uncertified facts. What? where? when? whence? why?—these essential questions that scientific research tries to answer in an intelligible and verifiable way imply each other. To answer one of them is not more important for science than answering any of the remaining questions—at least in the long run.

Causation is not identical with predictability. Hence, successful prediction by means of noncausal laws refutes the belief that causal legality is coextensive with science. Secondly, failures in prediction on the ground of causal laws do not entail the failure of determinism in general. In particular, inability to predict radical (qualitative) novelty, far from confirming indeterminism, only suggests that the law of emergence of the new feature under consideration has not yet been found. (Moreover, complete predictability, as contrasted to complete determinacy, is in a sense inconsistent with our recognition of radical novelty; a being who was endowed with the power of predicting every novelty would be unable to recognize it as such when it occurred—he would always have the feeling of *déjà vu*.)

In a word, neither causal laws nor laws of evolution in time are indispensable to scientific prediction; conversely, successful prediction with the help of a set of laws is not a test of their causal nature. A pragmatic criterion, like predictability, cannot decide about the *meaning* of $laws_2$ or about the *nature* of $laws_1$. Predictability and artificial reproducibility are empirical *criteria* for testing the truth of law statements; the attempt to derive all the meaning of law statements from their use and from the procedures of their verification amounts to confusing truth with one of its criteria, semantics with pragmatics. Only a philosophical analysis of scientific laws can decide upon their meaning.

The kind of laws needed for prediction will depend on the kind of prediction that is being asked for; conversely, the kind of prediction that can be obtained will depend on the available laws (and specific information). *Any* kind of scientific law will, if true enough, allow us to perform scientific predictions of *some* sort; in contrast to divination, scientific forecast is just foresight grounded on an insight including a knowledge of the objective patterns of being and becoming. In brief, there is no necessary relation between causality and prediction, any more than there is between causality and explanation.

13

The Place of the Causal Principle in Modern Science

13.1. Causality: Neither Myth nor Panacea

The problem of causation divides philosophers into roughly three camps: causalists or panaitists, who may be regarded as the conservative party; acausalists or anaitists, who exhibit the nihilistic tendency; and semicausalists or hemiaitists, whom I take pleasure in imagining to be representatives of the progressive or constructive trend. Causalism is the traditional attitude of disowning all noncausal categories of determination, holding dogmatically that every connection in the world is causal. The nihilistic party, on the other hand, declares that the concept of causal nexus is a "fetish" (Pearson[1]), an "analogical fiction" (Vaihinger[2]), a "superstition" (Wittgenstein[3]) or a "myth" (Toulmin[4]). This view is usually accompanied by the phenomenalist rejection of *every* kind of explanation, including of course causal explanation, in favor of description.

Needless to say, the denial of the existence of genetic bonds among events is vital for every kind of subjectivism; in the case of laic empiricism, the sole admissible link among events is the experiencing subject, whereas in the case of Neoplatonist,

[1] Pearson (1911), *The Grammar of Science*, 3rd ed., p. vi. Other "discarded fundamentals" are, according to Pearson, matter and force.

[2] Vaihinger (1920), *Die Philosophie des Als Ob*, 4th ed., passim. "Things with properties, and causes that act, are myths" (p. 44).

[3] Wittgenstein (1922), *Tractatus Logico-Philosophicus*, 5.1361.

[4] Toulmin (1953), *The Philosophy of Science*, p. 161.

Malebranchian, or Berkeleyan idealism, there can be no other bond than God. In either case a person, human or divine, is assigned the role of glue among facts that would otherwise be disconnected or simply nonexistent. The entirely negative attitude taken by indeterminism and by empiricism toward the principle of causation is inconsistent with the very goal of science, which is the search for the objective forms of determination and interconnection. To declare that the sole verifiable relations are those obtaining among sense data, concepts, and judgments, and to hold that it is vain to try to disclose autonomous interconnections and real modes of production, is an anthropomorphic attitude blocking scientific advance; it is a regressive attitude, even if most of its upholders sincerely believe that they are in the van of modern thought, just because they substitute modern dogmas for traditional ones. On the other hand, the recognition that causation is more than a psychological category akin to habit, the acknowledgment that genetic links, among them those of the causal type, do exist in the external world, is an attitude that, far from being anthropomorphic, helps to avoid the pitfalls of subjectivism, whether sensationistic or spiritualistic.

Many who wish to resist the phenomenalist and indeterminist attack on rational knowledge have found no better way than to repeat with obstinacy the traditional view according to which no scientific knowledge worth its name is possible apart from causal lawfulness and apart from explanation and prediction on the basis of laws which, like Newton's or Maxwell's, are wrongly supposed to be purely causal laws just because they are not statistical—the truth being that they have a causal component combined with doses of self-movement and reciprocal action. This conservative attitude, often rooted in a sound desire to preserve the right to understand the world—a right denied by indeterminists—has proved powerless in the face of philosophical criticism and the increasing realization of the importance of other types of determinacy, such as reciprocal causation, self-determination, and random interplay. The course of science has not confirmed the conservative hope that all the noncausal types of determination would finally be seen to be reducible to causation. Quite the contrary, a richer variety of types of determination is being recognized.

We are not, then, faced with the dilemma indeterminism–causalism. Just as in the problem of moral freedom, in intellectual affairs it is not always a question of choosing among two given alternatives; sometimes the act of choice is replaced by the creation of a third alternative. The right way of resisting the combined attack of phenomenalism and indeterminism upon rational and objective knowledge is not, I believe, to take refuge in the past by dogmatically disowning all noncausal types of scientific law, explanation, and prediction, regarding them as mere temporary contrivances—as is so often done in connection with statistical determinacy. The right and progressive attitude is to face the agreeable fact that science has advanced to such a point that, without dispensing entirely with the causal principle, it has assigned it a place in the broader context of general determinism—a role that is neither the principal nor the meanest, neither that of "main pillar of inductive sciences" (Mill) nor that of a "superstition" (Wittgenstein). The causal principle is one of the various valuable guides of scientific research and, like most of them, it enjoys an approximate validity in limited ranges; it is a general hypothesis with a high heuristic value—a fact suggesting that, in certain domains, it does correspond rather closely to reality.

13.2. The Domain of Causal Determinacy

13.2.1. *Conditions of Applicability of Causal Hypotheses*

The statement that the causal principle has a limited range of validity raises at once the following questions: What is the causal range? and When is it permissible to apply causal hypotheses? The first question refers to the objective operation of causation; the second concerns the conditions under which the use of causal ideas is valid. The answer to the first question depends on the answer to the second one, as the scientific way of drawing the strip demarcating (rather vaguely indeed) the domain of causal determinacy is to ascertain the degree of adequacy of our causal ideas about the world, that is, to ascertain to what extent they are confirmed, a task that is of course performed by science in every particular case.

It is a consequence of what has been said in the foregoing chapters, and especially in Part III, that some of the conditions for the applicability of specific statements fitting the necessary-production formula of causation are the following:

(i) *That the main changes under consideration be produced by external factors*, that is, when the system is largely (never entirely) at the mercy of its environment, so that inner processes are not the main sources of the change in question—although external conditions will be efficient solely to the extent to which they succeed in modifying those inner processes. The predominance of external over internal factors finds frequent illustrations in technique and in industry—which, as Bacon would have said, are just concerned with turning the *natura libera* into a *natura vexata*.

(ii) *That the process in question can be regarded as isolated*, that is, when it is permissible to regard the given process as torn out from its actual interconnections—which are usually numerous, but often irrelevant to the aspect that is being investigated; in other words, when such an isolation does not affect essentially the feature that is being investigated, or (what amounts to it from a pragmatic viewpoint), when the perturbations can be corrected for. This is often possible for limited intervals of time.

(iii) *That interactions can be approximated by agent–patient relations*, that is, when reciprocal actions either do not exist or, far from being symmetrical, are such that the action is considerably more important than the reaction; in other words, when reactions either are absent, or are negligible for all practical purposes. It is, again, typical of human production and technology to regard the raw material as a patient on which human work is exerted.

(iv) *That the antecedents and the consequents be uniquely connected to each other*, that is, when each effect can be considered as following (not necessarily in time) uniquely from a fixed cause (simple causation, as contrasted to multiple causation). This is particularly the case when the relevant causes are not all equally important in the concerned respect, but may on the contrary be arranged in a hierarchical gradation (chief cause, first-order perturbation, and so on).

It should be noticed that all of these conditions concern facts rather than our cognitive "grasp" (reconstruction) of them; in other words, the conditions for the validity of causal ideas depend primarily on the nature of the object (or on what is assumed to be its nature). All of them seem necessary for causal hypotheses to work, none being sufficient by itself. Probably further clauses would be found on closer inspection; but what is more important to our present concern is that none of these conditions can *exactly* be fulfilled in real cases.

In fact, external conditions are not efficient no matter what the internal ones are, but act always in combination with them and impinge upon the specific characteristics of the object concerned; consequently changes are never exclusively produced by extrinsic determiners, as required by (i). Besides, there are no perfect enclosures guaranteeing complete isolation even in a given respect, as required by (ii). Further, there is nowhere an inert stuff performing the role of perfect patient in the style of the peripatetic *materia prima*, as demanded in condition (iii). Finally, owing to the many-sidedness and changeability of interconnections, as well as to inner lawful spontaneity, real cause–effect connections are never exactly one-to-one, as demanded by (iv).

In short, the foregoing conditions can sometimes be fulfilled to a sufficient approximation—never exactly.

13.2.2. *Range of Validity of the Causal Principle*

We can now deal with the question concerning the range of validity of the causal principle. If the previous remarks are accepted, if at least one of the above-mentioned conditions is regarded as necessary, then the answer can be no other than this:

Strict and pure causation works nowhere and never. Causation works approximately in certain processes limited both in space and time—and, even so, only in particular respects. Causal hypotheses are no more (and no less) than rough, approximate, one-sided reconstructions of determination; they are often entirely dispensable, but they are sometimes adequate and indispensable.

To put it otherwise: in the external world there is always a wide class of processes the causal aspect of which is so important in certain respects and within limited contexts that they can be described as causal—although they are never *exactly* and *exclusively* causal.

A less vague delimitation of the domain of causal determinacy does not seem possible or even desirable. A few general conditions for the adequacy of causal hypotheses have been pointed out; but the decision about the range of every particular causal hypothesis should be the exclusive concern of science. To try to go much beyond the statement of the above-mentioned general conditions, in order to demarcate a priori and unambiguously the domain of causal determinacy in all fields of science, would be to come dangerously close to the traditional subjectivistic procedures. The adequacy of scientific and philosophic hypotheses has to be ascertained a posteriori, even though their probable adequacy can at times be assessed beforehand.

The notion of causal range of the laws of nature will next be clarified and illustrated, by analyzing in some detail a typical physical law with which every student of electricity is familiar. The few elementary mathematical symbols that will appear in the next section should not deter the nonmathematical reader, for they will be described in plain words, and are dispensable for a general grasp of the subject. (Mathematical signs do not replace physical concepts but ensure their clear, unambiguous, and accurate denotation.)

13.3. Delimitation of the Causal Range of a Particular Law

13.3.1. *Statement of the Problem*

Consider the elementary physical system constituted by an electrical circuit fed by a direct-current source, such as a battery (see Fig. 34). When the connection is established, an electric current of intensity i is set up, and this current in turn creates a magnetic field around the conductor. The relevant

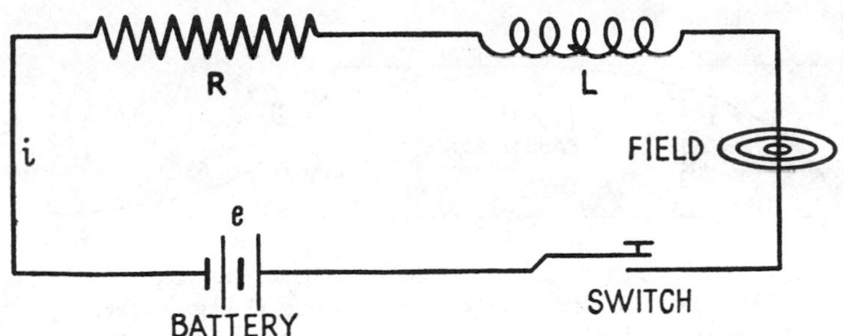

Fig. 34. The current i is set up under the action of a constant emf e impressed by a battery.

electrical properties of the wire are summarized in the constant R (resistance); the over-all magnetic properties of the circuit are condensed in the constant L (self-inductance). Notice that this is the simplest electrical circuit possible, for R and L never actually vanish, although in extreme cases either of them, or both, may become exceedingly small.

We make the usual assumption that our circuit is isolated enough from other physical systems that the above-mentioned parameters (e, i, R, and L) are the sole relevant ones and are moreover independent of external conditions. That is, we are assuming that condition (ii) above (isolation of process) holds. It is hardly necessary to point out that here, as everywhere else in scientific theory, we are concerned with an *ideal model* of an actual, concrete object; we are, in fact, neglecting a number of internal irregularities and external disturbances, such as small variations in the electromotive force, current losses, variations in the external magnetic field, random temperature fluctuations affecting e and R, and so on.

The law that sums up the dynamic interdependence among the essential features of our system (represented by a parameter each) is the following: the sum of the applied electromotive force e and the induced electromotive force e_i (produced by the magnetic field) is equal to the fall of potential Ri. In symbols,

$$e + e_i = R\,i.$$

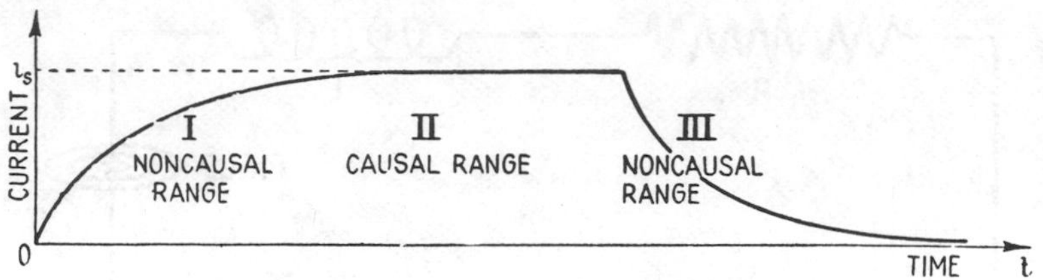

Fig. 35. Current as a function of time for the circuit of Fig. 34. The rise of the current (section I) begins when the switch is closed; the back emf induced by the growing magnetic field opposes the applied emf. The steady state (section II) is attained after the reaction of the field on the circuit is over. The dying away of the current (section III) corresponds to the self-movement of current after the applied emf has been switched off.

On the other hand, the law of induction reads,

$$e_i = -L\frac{di}{dt}.$$

Substitution of the second statement into the first yields

$$e = L\frac{di}{dt} + Ri. \qquad (1)$$

One is tempted at first sight to regard this as a typical causal law, stating that the *cause* e (applied electromotive force) produces the current i (effect). A closer examination will, however, show that this causal interpretation of law (I) is valid solely in connection with a limited domain, namely, when the current i has attained its final steady value i_s (given by Ohm's law, which is a particular case of law (I), applicable when the current does not vary). Before the steady state has been attained, and also after it has ceased, the actual connections among the relevant features of our system are *not* strictly causal, as will be seen presently.

13.3.2. *First Stage of the Process: Cycle of Determinants*

In fact, as soon as the battery is switched on, the current increases continuously from zero till its final steady value i_s (see

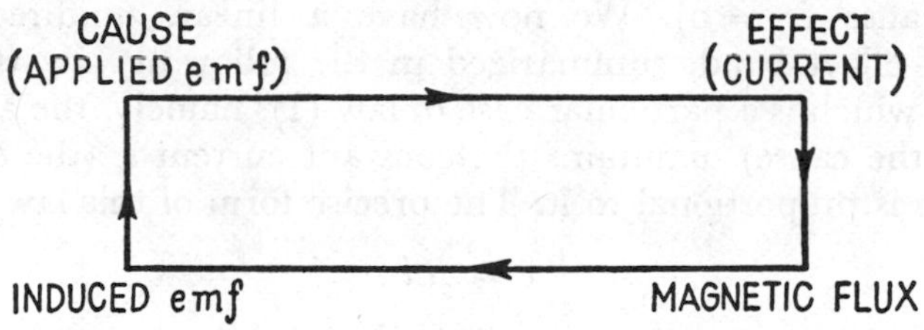

Fig. 36. Cycle of determinants in a simple electrical circuit while the current is increasing (stage I). The causal connection obtains when this cycle degenerates into a direct cause–effect bond (stage II).

Fig. 35); i does not attain this value i_s abruptly, the full effect does not follow immediately upon the cause, simply because in this case the effect i *opposes* the cause, so that condition (iii) above is violated. Indeed, while the current increases ($di/dt > 0$), it produces a magnetic field of increasing strength that surrounds the coil and reacts back on the current in it. The variable magnetic flux ϕ, due to the increasing current, produces in turn an induced emf ($e_i = -\, d\phi/dt = -\, L di/dt < 0$), which in this case opposes the applied emf. This back emf induced in the wire by the growing magnetic field is what prevents the current from attaining the final value i_s without any time lag.

This process is an instance of (natural) negative feedback with a variable total effect (see Fig. 36). From an ontological point of view, this stage of the process illustrates the *cycle of determinants*, in which the cause–effect nexus is but a side or a moment of a more complex loop of reciprocal actions. In this first stage of our process, the *interaction* category is the dominant determinant.

13.3.3. *Second Stage of the Process: Causal Nexus*

The first stage goes over continuously and asymptotically into the second stage, which is definitely causal. In this second case, the action of the cause is no longer counteracted by its effect; the magnetic flux produced by the current is now constant ($d\phi/dt = 0$), so that the effect does not react back on the cause ($e_i = 0$). We now have a linear, unidirectional cause–effect bond, summarized in the following law (Ohm's law), which is a particular case of law (I), namely, the applied emf (the cause) maintains the constant current i_s (the effect), which is proportional to it. The precise form of this law is

$$e = R\, i_s. \tag{II}$$

The steady state is, then, the *causal domain* of the process of electric conduction in metals; or, again, the special law (II) covers the *causal range* of the more general law (I). This does not mean, however, that the steady state of electric conduction is an *entirely* causal process. Quite the contrary, the theory of electronic conduction in solids—that is, the study of the lower-level motions giving rise to the macroscopic process covered by Ohm's macroscopic law (II)—shows that several categories of determination come into play; among them are self-determination (inertia of electrons and fields), interaction (mutual action among electrons, the crystal lattice, and the impurities), statistical determination (collective behavior of the electron gas), and so on. Still, these noncausal categories are finally masked in such a way that the average motion of the electrons "obeys" such a typically causal law as Ohm's—the causal nature of which is especially manifest in the form given to it by Drude, namely $E = (1/\sigma)\, j$, which is Aristotelianlike. The self-motion of the electrons again becomes apparent at very low temperatures, where instead of the law of Ohm and Drude we have London's law, $E = \Lambda \partial j / \partial t$, which accounts for superconductivity and is of the Newtonian type.

Only the *net effect* i_s is connected in the simple form (II) to the cause e. When we say that the law (II) that characterizes electric conduction in metals at ordinary temperatures is causal, we imply that it is causal *on its own level*, quite aside from its roots in the more complex laws characterizing electron conduction inside a crystal lattice subjected to an externally impressed electric field. Here we are, in short, analyzing the causal range of a definite law belonging to a definite level; we are concerned with the causal component of a given law, rather than with the causal aspects of a given event or string of events—as every single phenomenon is the locus of a whole *set* of laws belonging to various levels (see Sec. 10.4.3).

13.3.4. *Third Stage of the Process: Self-Determination*

Let us now turn to the third and last stage of our process. When the battery is switched off ($e = 0$), the current does not cease immediately but dies gradually, thus falsifying the well-known scholastic maxim *causa cessante cessat effectus*. The associated magnetic field decreases ($d\phi/dt < 0$), thereby producing an induced electromotive force acting now in the direction of the formerly applied emf (while the current decreases, that is, while $di/dt < 0$, $e_i = -L\,di/dt > 0$). This induced emf is what maintains the current i; it is a typical aftereffect, for the cause e has now vanished. The pattern of this stage of the process is the following particular case of law (I):

$$0 = L\frac{di}{dt} + R\,i. \qquad \text{(III)}$$

According to Maxwell's field theory (which explains the "phenomenological" circuit theory), in this last stage the causal component has entirely concentrated in the decreasing magnetic field—which is not, however, an *externally* acting cause but an *intrinsic* determinant of the whole system (of which the coil is only a part). But from the point of view of circuit theory, the process of dying out of an electric current has not even a causal component because, unlike stages I and II, in stage III the cause (e) has ceased to operate. From an ontological point of view this stage of the process is consequently dominated by the *self-movement* category, which characterizes predominantly or exclusively self-determined processes, whose ideal model is inertial motion.

To sum up, in the process of growth, maintenance of the steady state, and dying out of an electric current in the simplest possible metallic circuit, three stages must be distinguished, each characterized by a peculiar category of determination. These stages are:

(i) *Domain of reciprocal action* (*beginning of current*). The effect i has not yet fully developed, and reacts back on the cause e, in the manner of negative feedback. The differential law of this process of reciprocal action is (I); the corresponding integral law is

$$i = i_s(1 - e^{-(R/L)t}), \quad i_s = e/R.$$

As time passes, the effect reacts less and less on the cause (that is, di/dt decreases, and the corresponding back emf dies away) until finally the causal stage has fully developed.

(ii) *Domain of causation* (*steady state*). The (constant) cause e maintains the (constant) effect i_s in accordance with Ohm's law (II). This steady state lasts until, for one reason or another, the battery is switched off and the process goes over into its last stage.

(iii) *Domain of self-determination* (*dying away of the current*). The cause has disappeared ($e = 0$); but the effect does not vanish immediately; it dies away gradually, maintained by the decreasing magnetic field. The differential law of this process is (III); the corresponding integral law is

$$i = i_s\, e^{-(R/L)t}.$$

The foregoing example was *not* meant to illustrate the claim that the range of *every* law of nature splits unambiguously into nonoverlapping domains each dominated by a single category of determination; this contention is definitely untrue, as will be realized by recalling the various types of law discussed in the previous chapters. Some laws have no causal component at all (as is the case with the classificatory, kinematical, and conservation laws) and others have a causal component that is not additively combined with other categories of determination. What the consideration of the electrical circuit discussed is expected to support is the central thesis of this book, namely, that the causation category is indispensable in science but, like every other category of determination, has a limited range and, moreover, works in combination with other determination categories. The analysis is also expected to have shown that the causal range—if any—of the laws of nature cannot be determined a priori once and for all for all types of law, but must on the contrary be the outcome of detailed analyses of specific laws.

13.4. Any Causality Tomorrow?

13.4.1. *A Verbal Trap into Which Philosophers of Verb Have Fallen*

What will the future import of the causal principle be? According to a fashionable view, the domain of causal determinacy will continue to shrink, until nothing will remain of it: all laws of nature and society will ultimately be shown to be statistical, and the concept of causation will be seen to be a myth, a remnant of the prepositive stage of mankind. Russell, among others, prophesied that "in a sufficiently advanced science, the word 'cause' will not occur in any statement of invariable laws".[5] Now, this can easily be granted, and there hardly is any need to wait for the future. But it does not follow from it that the concept of cause will finally be extruded from *philosophy*, however scientific philosophy becomes. The *word* 'cause', which denotes a *generic* concept, need not occur explicitly in any *particular* scientific statement; the cause concept belongs to ontology, just as do the concepts of quality, change, connection, chance, and so on, which receive specific names in every chapter of science. Every science is concerned with *particular* qualities, changes, connections, types of chance, and so forth; and some particular changes fit causal patterns even if the *word* 'cause' does not appear explicitly in the conceptual reconstructions of such patterns. The fact that science employs less and less the word 'cause', which belongs to the philosophical vocabulary, cannot be regarded as a sign of decrepitude of the causal principle.[6] One should not allow oneself to become an easy prey of language.

Likewise, the fact that our philosophical ideas on determination all seem to bear the stamp of causality does not prove the latter's healthiness. Upon closer examination it is seen that in this case, as in so many others, we are not only deluded by the

[5] Russell (1914), *Our Knowledge of the External World*, p. 223.

[6] Lalande (1929), *Les théories de l'induction et de l'expérimentation*, chap. ix, makes the same mistake of believing that a *concept* is not employed when the *word* usually denoting it does not occur; he thinks that the concept of cause does not occur in theoretical physics, but that it still appears in experimental physics (as when the experimenter asks what is the cause preventing a current). Lalande, like Russell, believes that the concept of causation occurs only in underdeveloped disciplines.

survival of fossil theories into which we try to force every original finding; we are also deluded by the verbal form in which ideas on determination are usually expressed. We all tend to frame every sort of idea on determination, as well as every kind of explanation, in a causal *language* that often distorts the meaning we really wish to convey. Thus we usually cover the concepts of origin, ground, motive, and even reason with the single word 'cause'; and we readily confuse logical entailment with causation, and logical consequence with effect —a confusion that Western thought drags in from the Greek classical period.

Let us now turn away from problems of language, since neither the gradual disappearance of the *word* 'cause' from *science*, nor the survival of both the concept and the word in ordinary and in philosophical language, is a reliable sign of the evolution of the status of the causal principle. Since this is a factual question, let us take a look at the recent history of science.

13.4.2. *How Quantum Mechanics Finally Disappointed Acausalists*

Recent science seems to show neither an enlargement nor a progressive shrinking of the domain of validity of the causal principle. What is apparent is a complex, confused, and unpredictable intellectual process in which some phenomena are deprived of the causal character that had previously been attributed to them, whereas other phenomena are recognized as having a causal aspect. Besides, noncausal types of determination are shown to be somehow linked with causation. The over-all trend discernible in recent science in connection with the general problem of determinism is not so much an increasing departure from causality as a *progressive diversification of the types of determination*, with correlative changes in the meaning and scope of the causal principle.

The case of quantum mechanics is highly edifying. Until recently, most scientists and philosophers of science believed that quantum mechanics had given causality the death blow, by showing that quantum phenomena are inherently random,

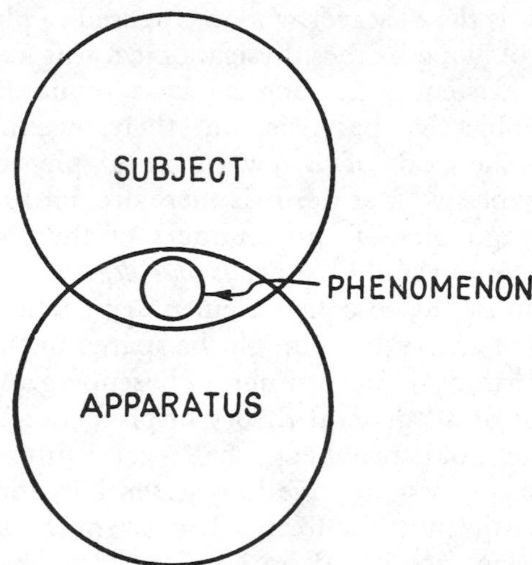

Fig. 37. The subject–object relation in the usual interpretation of the quantum theory: the two make up an unanalyzable unit, in the midst of which the phenomenon takes place (provided the experimenter deigns to conjure it up).

hence only statistically predictable. More exactly, "The partition of the world into observing and observed system prevents a sharp formulation of the law of cause and effect", as Heisenberg[7] explained. Such a partition is performed in every experiment, and the position of the cut (*Schnitt*) is arbitrary, being dependent upon the decision of the experimenter. Even if causality were assumed to hold on either side of the partition (as has sometimes been done), such a hypothesis could not be confirmed empirically, for what we observe is something lying precisely *at* the cut—and the behavior of this region is not "governed" by causal laws but by the laws of the quantum theory, which are (wrongly) supposed to be entirely noncausal. In other words, quantum phenomena are, according to the orthodox interpretation of the theory, located in the intersection of the observer and his observing setups (see Fig. 37);

[7] Heisenberg (1930), *The Physical Principles of the Quantum Theory*, p. 58.

in addition, it is the observer who is supposed to play the active part therein. As long as the physical object was thus denied an autonomous existence, as long as laws of nature were not regarded as objective patterns, but their meaning was confounded with the mode of their verification, physical causation could merrily be swept aside. It is therefore not surprising that a few years ago one of the founders of the theory[8] could prophesy that one should expect *further departures from causality* in the domain of the so-called elementary particles; not even the principle of antecedence might be spared in the next move toward indeterminism, according to Heisenberg. And the well-known author of a "general theory of prediction" went on to formulate the bold prophecy that every future quantum-mechanical theory would have to be essentially indeterministic and subjectivistic (and the former because of the latter).

But prophecies not based on laws are just prophecies—not scientific predictions (see Chapter 12). The above-mentioned prophecies on the progressive shrinkage of the domain of causality did not come true; one year after Heisenberg's article, Bohm[9] published his celebrated reinterpretation of elementary quantum mechanics, based on de Broglie's[10] old idea of the pilot wave. In this interpretation, which is so far empirically equivalent to the usual one, the object is conceived as existing autonomously, though as strongly interacting with its macroscopic surroundings (which may include measuring devices, but not the experimenter's mind). The usual quantum-mechanical variables (the "observables") are assigned to the zone of overlapping of the object and the apparatus; but the behavior of the object itself is described by means of new variables, the so-called hidden parameters, which are not subject to any uncertainty relation (see Fig. 38). Heisenberg's uncertainty principle is not regarded as an inherent limitation of precision, but rather as a technical limitation arising from

[8] Werner Heisenberg, "50 Jahre Quantentheorie", *Die Naturwissenschaften 38*, 49 (1951). See also the contribution of W. Pauli in A. George, ed., *Louis de Broglie, physicien et penseur* (1953), p. 42.

[9] David Bohm, "A Suggested Interpretation of the Quantum Theory in Terms of 'Hidden' Variables", *Physical Review 85*, 166, 180 (1952).

[10] De Broglie (1953), *La physique quantique, restera-t-elle indéterministe?*

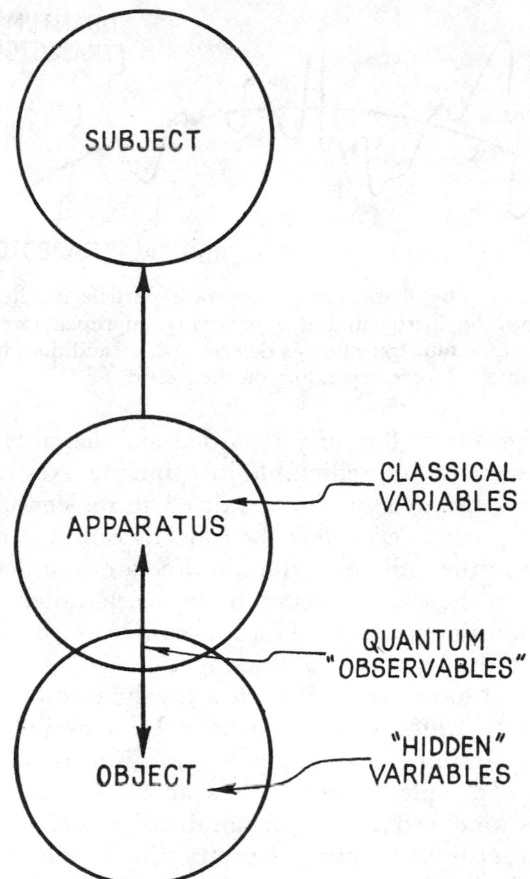

Fig. 38. The subject–object relation in the causal interpretation of quantum mechanics: the two physical systems (object and apparatus) that interact are external to the observer (subject), who does not conjure phenomena out of nothing but can control them statistically.

the objective object–apparatus interaction, the amount of which should in principle be computable with the help of a more detailed theory. Moreover, the de Broglie-Bohm interpretation affords a causal explanation of the quantum-mechanical fluctuations in the trajectories of atomic-scale "particles"—

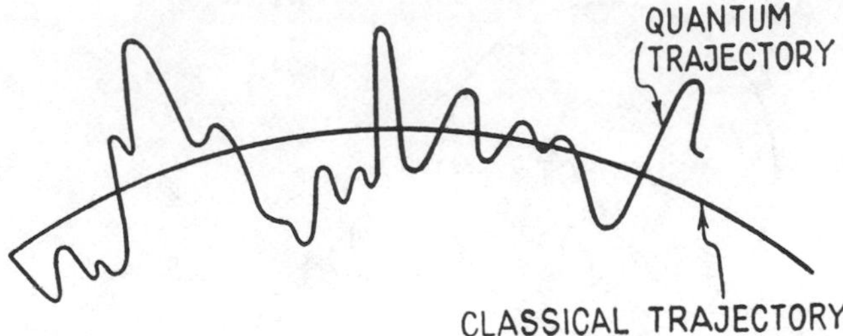

Fig. 39. The classical trajectory of a particle is determined by inertia and the externally impressed force. The quantum trajectory is determined, in addition, by an internal force depending on the ψ-field.

variations that were formerly regarded as inherently random, hence individually unpredictable in principle. Newton's force equation is reinstated in a generalized form, enabling us in principle to predict accurately the trajectory of the "particle"; in addition to the ordinary external force a new, internal force depending on the ψ-field occurs in the acceleration formula,[11] and this quantum-mechanical force accounts for the departures from the classical trajectories (see Fig. 39).

Who would have expected such a revaluation of the causation category, which seemed to have been excluded forever from the quantum level? Only a few disbelievers in linear progress and complete theories. The moral of this little story (which has produced a deep demoralization and confusion in the acausalist camp) seems to be pretty simple, and even known of old, to wit: Beware of prophesying in detail the future course of science along any other path than that of an increasing enrichment in both facts and concepts!

The future of the causal principle does not seem more uncertain than that of any other particular principle of determination subsumed under our general principle of determinacy (see Sec. 1.5.3). All of them seem *now* to be indispensable and are

[11] In the case of a Schroedinger "particle" acted on by the external force f_e, the new equation of motion reads $md^2x/dt^2 = f_e + f_i$, where x denotes the real ("hidden") position-co-ordinate, and f_i symbolizes the internal force, which depends on the location of the "particle" within its own ψ-field.

probably insufficient, but the best policy is not to bet on any of them.

13.5. General Conclusions

The net effect of the nihilistic criticism of causality was to encourage fortuitism and its epistemological partner, namely, irrationalism. On the other hand, the intention of the present book is to show that, just like every other category of determination, causation has a limited range of operation; that the causal principle holds a place in the broader context of general determinism; and that failures of the causal principle in certain domains by no means entail the failure of determinism *lato sensu*, or the breakdown of rational understanding.

The realization of the limited scope of the causal principle supports neither skepticism nor irrationalism; every failure of causality *stricto sensu* can be regarded as the victory of a different principle of determination, and it simply marks the breakdown of outdated ontologies that are too narrow to make room for the unlimited richness of reality, as progressively disclosed by the sciences. What in contemporary science has taken the commanding place once held by the causal principle is the broader *principle of determinacy*, or of lawful production. The two components of this principle, under which the general law of causation is subsumed, are the genetic principle (*Nothing springs out of nothing or goes into nothing*) and the principle of lawfulness (*Nothing unconditional, arbitrary, lawless occurs*). The principle of determinacy just states that reality is not a chaotic aggregate of isolated, unconditioned, arbitrary events that pop up here and there without connection with anything else; it states that events are produced and conditioned in definite ways, though not necessarily in a causal manner; and it asserts that things, their properties, and the changes of properties exhibit intrinsic patterns (objective laws) that are invariant in some respects.

The principle of determinacy, often mistaken for the law of causation, is the common ground of all forms of scientific determinism (from which fatalism is excluded, since it involves supernaturalistic elements violating the genetic principle). To reduce determinism to causal determinism is to have either a poor opinion of the resources of nature and culture, or too high an opinion of philosophical theories. Those who assign to causality the exclusive appurtenance of characteristics that are actually shared by all kinds of scientific determinism either fail to resist the attacks of indeterminism and irrationalism or—to the extent to which they succeed in the defense—inadvertently clothe noncausal types of determination in a causal language.

If there is some truth in what has been said in this book, the right policy with regard to the causal problem can be summarized in the following rules: (*a*) to employ the causation category whenever permissible, without fearing to be accused of fetishism, mechanism, or what not; (*b*) to recognize the limited character of causal hypotheses; (*c*) to make room for further categories of determination whenever they contribute to affording a truer account of being and becoming; and (*d*) to abstain from terming 'causal' all those categories which, like self-determination, reciprocal action, and so forth, clearly overflow causality and belong instead to general determinism.

According to Descartes,[12] the perfect science was the precise knowledge of effects by their causes, that is, the deduction or explanation of (observed) effects from their (assumed) causes. This Peripatetic norm may still be regarded as the paradigm of science—but, paradoxically enough, with the following essential qualifications: (*a*) the *link* between causes and effects need not always be causal (that is, unique, unsymmetrical, constant, external); (*b*) nothing warrants the presumption that we shall ever attain more than a *hypothetical* (but improvable) knowledge of causes, effects, and their links (whether causal or not). The causal principle reflects or reconstructs only a few aspects of determination. Reality is much too rich to be compressible once and for all into a framework of categories elaborated during an early stage of rational knowledge, which consequently cannot account for the whole variety of types of determination, the number of which is being increased by scientific research and by philosophical reflection upon it.

[12] Descartes (1644), *Principles of Philosophy*, part I, 24.

What has been rejected in this book is not the principle of causation, but its unlimited extrapolation, as asserted by the doctrine of causalism, or causal determinism—a primitive, rough, and one-sided version of what I took the liberty of calling *general determinism*. To use picture language, we may say that causal determinism is the ray-optics approximation of general determinism. Or, if a mathematical metaphor is preferred, determination is a vector in a space of a large, as yet unknown number of dimensions, causal determination being just one of its components or projections.

The causal principle is, in short, neither a panacea nor a myth; it is a general hypothesis subsumed under the universal principle of determinacy, and having an approximate validity in its proper domain.

Appendix: A Reply to Criticisms

Professors Schlegel's[1] and Morgenbesser's[2] thorough reviews of my book *Causality* offer me an opportunity to clarify several points. The first ten sections of this paper were prompted by Schlegel's remarks, the remainder by Morgenbesser's.

1. *Should determinism be equated with lawfulness?* Although such an equation is often suggested it is not commendable, because there is no equivalence between lawfulness and determination if the latter is defined as way of becoming. Something might conceivably emerge out of nothing in a lawful way or, conversely, it might be produced by something else but erratically and not in accordance with law. This is why in *Causality* the thesis of general determinism, or neo-determinism, is stated as the conjunction of the *genetic principle* ("Nothing can arise out of nothing or pass into nothing"), and the *principle of lawfulness* ("Nothing happens in a lawless, arbitrary manner"). The principle of lawfulness is presupposed by the search for laws of any type, and the genetic principle is both a generalization of available knowledge about modes of becoming, and a deterrent of magic. By the way, the cosmological hypothesis of the continuous creation of matter out of nothing should not be invoked against the genetic principle: if we wish to build a scientific ontology we should not employ *ad hoc* conjectures conflicting with well-established laws of conservation.

2. *Is any assumption of lawfulness extraneous to the quantum theory?* The core of the quantum theory is, like that of any other factual theory, a set of law statements, such as wave equations and commutation relations. Of course, these are not purely or even predominantly causal laws. But this, again, is not peculiar to the quantum

[1] Richard Schlegel, "Mario Bunge on Causalty," *Philosophy of Science*, 28 (1961), pp. 72–82.

[2] Sidney Morgenbesser, review of *Causality*, in *Scientific American*, 204, No. 2 (1961), pp. 175–8.

theory: there are few strictly causal laws in science[3] and, as *Causality* tries to show, rather than pure causal laws there are *causal ranges*, or domains in which *certain* laws can be said to be predominantly causal.

3. *What is the causal range of a scientific law?* Let us apply the analysis of causation proposed in *Causality* to the study of behavior. The program of the causal theory of behavior might be phrased thus: "Given a class of organisms (e.g., white rats) and a set of external stimuli or 'agents' (e.g., sounds), the responses of every member of the class of organisms, or 'patients,' will be uniquely determined by the external stimuli according to a set of laws. Consequently, the knowledge of such laws, conjoined with information about the relevant circumstances (kind and strength of stimuli, and initial state and previous history of the concerned organisms), will enable us to predict every single act of each member of the given class of organisms." Now, one such law expressing the causal stimulus-response nexus is the well-known psychophysical relation which, for the case of sound, is

$$R = k \cdot S^{0.3}. \tag{1}$$

We know, however, that this law is incomplete: it covers only overt responses, and it does not account for internal stimuli, which may be assumed to give rise to "spontaneous" responses, i.e., to responses that are random relative to the distribution of external stimuli. Suppose a neodeterminist pointed out this *noncausal* aspect of behavior and wished to correct Eq. (1) by including in it a spontaneous response term U, independent of the external stimulus S. This generalization can be done in either of the following ways: (a) *phenomenologically*: by regarding U as a random variable, without enquiring into the source of its variations, and (*b*) *representationally*: by regarding U as dependent on some lower-level, physiological variables. The former approach will be regarded as a temporary expedient by our determinist of the *nouvelle vague*, who will prefer the deeper theories to phenomenological summaries. Suppose he calls collectively 'x' these sub-behavioral, "hidden" parameters, which are not open to behavioral observation but are assumed to be scrutable by physiological (e.g., neurological) means. In the

[3] Mario Bunge, "Kinds and Criteria of Scientific Laws," *Philosophy of Science*, 28 (1961), p. 260.

spirit of determinism, a conceivable enlargement of Eq. (1) to cover internal stimuli (but not covert responses) is the following:[4]

$$(2) \qquad R = k \, . \, S^{0.3} + U(x)$$

Eq. (1), which is a special case of Eq. (2), may be called the *causal approximation* of Eq. (2); if preferred, the *causal range* of Eq. (2) is the region for which $U \ll R$. The qualification must be made that neither Eq. (1) nor Eq. (2) cover the *noncausal* range below the absolute threshold: below this threshold there is no response, hence no causal relation between stimulus and response; the deliberate neglect of this noncausal range is taken care of by the definition of '*S*,' which does not designate an arbitrary stimulus but the strength of stimuli above the absolute threshold. Let us finally notice that Eq. (2) is deterministic—in the sense explicated in sec. 1—but not strictly causal.

4. *Is the relation between the microlevel and the macrolevel causal?* The "hidden" variables lumped in '*x*' are occult from a macroscopic or behavioral point of view, but they may be investigated by the physiologist: he will probably imagine more or less fantastic models, will then express them mathematically—that is, he will hypothesize diverse functions $U(x)$—and will finally test such models. This he can do by investigating behavior in the absence of external stimuli (i.e., for $S = O$) or, better, in the presence of an approximately constant background of external stimuli (by means of Pavlov observation cages), and trying to control internal stimuli in an indirect way (e.g., by food deprivation). The possibility of setting up a reasonable function $U(x)$ will depend on the state of physiology, and the possibility of testing it will depend on the state of experimental techniques. In any case, the hypothesizing of new levels and of relations among different levels has usually paid off.

The relation between the microscopic or physiological variables x and the macroscopic or behavioral variable U would be described as a *reduction* by some, as *causal* by others. Yet both interpretations seem to be wrong: 'reduction' suggests explaining *away* by eliminating the macrolevel as ontologically and methodologically dispensable; and 'causation' does not apply to interlevel relations because it designates an unsymmetrical connection between an "agent" and

[4] Another conceivable generalization of Eq. (1) is, of course, $R = k.S^{0.3} [1 + U(x)]$.

a "patient." A relation such as '$U = U(x)$' connects two levels of one and the same system (an organism in the above example), each of them characterized by peculiar, irreducible (but explainable) properties. A distinguished representative of this kind of border-zone law[5] is Boltzmann-Planck's law, $S = k.\log W$, relating the number W of microscopic configurations to the entropy S characterizing a state of a closed system. Likewise, the imagined relation $U = U(x)$ is a functional interlevel law. Moreover, the responses U are spontaneous in the sense that they are not (immediately) elicited by external agents; but, since they are linked to definite inner (physiological) processes, they are perfectly *determined.*

5. *What is wrong with theories of hidden variables?* Bohm, de Broglie, Halbwachs, Schiller, Vigier, Weizel, Yevick, and others have been speculating during the last decade on the possibility of deriving the quantum-mechanical laws from assumptions regarding a substratum or sub-quantum-mechanical fluid or field, much in the same way as a psychologist would like to derive part of the overt behavior from physiological processes (see sec. 3). In trying to explain the seen by the unseen, and to split the latter into distinct but connected levels, these physicists are following an old tradition launched by the Greek atomists and embraced by modern science. Even though we may object to some of their speculations[6] we should stimulate their attempts. Science is the work of bold imagination no less than of patient work, and nobody can predict whence is the new light to come. As the mathematician Isaac Schoenberg used to say, scientists need be given moral credit for their fantasies as much as they need financial credits.

Let us only be clear that these new theories—contrary to their own author's claims—do not reinforce causality: they do support general determinism and realism, in the sense that they reject the anthropocentric view that whatever happens in the universe (e.g., the sun's flares) is the result of an act of observation, and assert on the other hand the objective existence of microsystems. The "hidden" variables occurring in these theories are supposed to describe objective properties of the microsystems or of the background fluids or fields that are supposed to explain the former's properties. On the other hand, the usual quantum-mechanical "observables" (which

[5] Mario Bunge, "On the Connections Among Levels," *Proc. XII Int. Congress Philosophy* (Florence: Sansoni, 1960), VI, 63.

[6] Mario Bunge, *Metascientific Queries* (Springfield, Ill.: Charles C. Thomas, 1959).

are observable in a Pickwickian way) may be interpreted as describing objective system-apparatus or, better, system-macroscopic surrounding interaction. (See Fig. 38, p. 349, of this book.) This basic realistic epistemological viewpoint is common to most attempts at deriving the quantum-mechanical laws from more basic equations; thus it is openly proclaimed by Landé,[7] who opposes both what he calls "the Copenhagen doublethink" and "Bohm's pseudo-causality."

6. *Has the hidden-variables revolt been sterile?* First of all, the various heterodox generalizations and reinterpretations of quantum mechanics have not been found *false*, even though some of them are *incomplete* (particularly those which do not account for the process of measurement). The only pertinent question is whether they add to our knowledge: do they provide a conceptual enrichment, and do they account for hitherto unexplained facts? The hidden-variables approach has, so far as I can see, two main aims: (*a*) the epistemological cleansing of the quantum theory: its purification from subjectivistic, anthropocentric assumptions; (*b*) the enlargement of the quantum theory with a view to encompassing experimental results such as those regarding the possible structure and mutual relations of the thirty-odd "fundamental" particles. Of course, the *practical* utility of the new approach will be measured by its ability to account, say, for the Nishijima-Gell-Mann scheme of "elementary" particles. Victories on this frontier are being reported.[8] But quite apart from such possible "positive" results, a new look at certain problems may be worth trying if only because it constitutes an exciting intellectual adventure, or at least because it enables us to evaluate the old view better. As Synge[9] says, "The lust for calculation must be tempered by periods of inaction, in which the mechanism is completely unscrewed and then put together again. It is the decarbonization of the mind." Of course, such critical work and tentative speculation may not find room in the *Physical Review* and may not count for academic promotion, but they are the salt of science and they should

[7] Alfred Landé, *From Dualism to Unity in Quantum Physics* (Cambridge: University Press, 1960).

[8] Incidentally, the whole new approach made quite a hit at the last International Congress for Logic, Methodology, and Philosophy of Science (Stanford, August-September, 1960), where Jean-Pierre Vigier's paper was enthusiastically received. See P. Hillion and J. P. Vigier, "New Isotopic Spin Space and Classification of Elementary Particles," *Nuovo Cimento*, 18, (1960), p. 209.

[9] J. L. Synge, *Relativity: The General Theory* (Amsterdam: North-Holland, 1960), p. vii.

command the respect of those who prefer the safe highway to the trail in the jungle.

But even if the new look were finally found insufficiently fruitful, its practical failure would not support the orthodox theory, just as eventual modifications of the relativity theory will not revive the hypothesis of absolute time. In formal science we are often faced with the decision between two contradictory propositions one of which is supposed to be altogether true. But in factual science, more often than not, competing hypotheses and theories are not contradictory but contrary or incompatible, and each of them contains an element of truth. In such cases, *tertium datur*. Furthermore, the criticism from apparent sterility presupposes that the subjectivist (or positivist, or operationalist) interpretation of the quantum theory *has* been fruitful—and this is at least controvertible, because the theory was first built and then made philosophically palatable. Heisenberg's operationalist requirement that only "observables" (measurable quantities) should enter physical theory was attached to it *a posteriori* (in his famous 1927 paper), as was recognized by Bridgman[10] with his characteristic frankness: "I have always wondered, however, whether perhaps this requirement of Heisenberg was not formulated after the event as a sort of philosophical justification for its success, rather than having played an indispensable part in the formulation of the theory." To sum up this section: the criticism of the orthodox view has been epistemologically fruitful, and it has stimulated creative imagination; progress along realistic and neodeterministic lines is being made, and some new look was needed anyhow to cope with the flood of new facts, as well as with old difficulties.

7. *Is the present quantum theory perfect?* Although computing physicists and dogmatizing philosophers are entirely satisfied with the present state of the quantum theory, the more open-minded and imaginative scientists believe that much could be improved both as regards the formalism and the interpretation. Do we know for certain what a coordinate and a velocity are at the quantum level? (Or, if preferred: Are we sure we have correctly identified the position and velocity operators in the theories of spinning particles?) Are we sure cut-offs and renormalizations are not *ad hoc* assumptions designed to save theories rather than phenomena? Are we to remain satisfied with the black-box approach (*S*-matrix, dispersion relations, and other phenomenological theories), instead of looking for deeper theories

[10] P. W. Bridgman, *The Nature of Physical Theory* (New York: Dover, 1936), p. 65.

that may tell us what goes on inside the box? Shall we continue thinking that the "elementary" particles are actually elementary (simple, undecomposable) no matter what the interaction energy? Shall we continue regarding them as point-masses and yet go on attributing to them more and more properties?[11] These and others are questions open-minded and seriously aimed scientists and meta-scientists ask despite the veto of the powerful yet declining conservative party.

8. *Does* Causality *advocate "a narrow kind of causal determinism at the level of elementary particle processes"?* What *Causality* does maintain is that the quantum theory is not indeterministic, in the sense of surrendering both the genetic principle and the principle of lawfulness (see sec. 1), even though it does set important restrictions on *causal* determinism. More specifically, the following points are made in this book:

(*a*) Quantum mechanics, whether in its orthodox interpretation or in any of the numerous tenable heterodox versions, is not only lawful but, moreover, respects the genetic principle. For instance, the theory does not maintain that light comes out of nothing in a lawless way, but assumes that light is emitted by microsystems (e.g., atoms) in accordance with definite nonmechanical "mechanisms" (e.g., quantum jumps) and definite laws (mostly of a statistical character).

(*b*) In all of its interpretations, quantum mechanics *restricts* causality without eliminating it entirely, in the sense that it "does not sweep out causes and effects, but rather the rigid causal nexus among them."[12]

(*c*) The subjectivist (positivist, operationalist) interpretation is neither deterministic nor indeterministic in an *ontological* sense, as was made clear by Philipp Frank long ago: how could it be, if positivism does not postulate the existence of an autonomous external world? What this school asserts, in connection with the quantum level, is what in *Causality* is called an *empirical indeterminacy*: it declares that there is no direct autonomous (objective) connection between any two successive states of a microsystem, and it asserts that only connections among *observations*—i.e., among results of acts

[11] This reminds me of the following delightful dialogue between Professor L. Garwin and a journalist at the New York meeting of the American Physical Society, February 4, 1961. PHYSICIST: We know now for certain that the mu-meson is a point-particle. JOURNALIST: How big a point? PHYSICIST: Well, surely less than 10^{-13} cm in diameter.

[12] *Causality*, pp. 14-15.

of observing subjects—are dealt with. A detailed discussion of the philosophy of quantum mechanics and electrodynamics will be found in my *Metascientific Queries.*[13]

(*d*) In heterodox interpretations of quantum mechanics, notably in the one advanced by Bohm in 1952,[14] "Newtonian determinism, with its causal ingredient, is partially restored."[15] Notice that it is not maintained that this is a *causal* interpretation, but an interpretation with a causal *ingredient;* the terminology of *Causality* differs considerably from that of Bohm and coworkers.

In short, *Causality* does not advocate causal determinism at any level; it does try to show that causal determinism is not as dead as had been supposed, and that it is sublated under a far richer determinism.

9. *Are the differences between the various meanings of 'law' secondary?* Consider (*a*) the objective pattern of free fall in a void, (*b*) the approximate reconstruction of the foregoing pattern by the law statement "If a body falls freely in a void, its acceleration is constant," and (*c*) one of the latter's possible translations for test purposes, namely, "If a body *is dropped* in a void, it *is found*—to within *experimental error*—that its acceleration is constant." In my terminology (*a*) is an instance of a law_1, (*b*) of a law_2, and (*c*) of a law_3. (There is even a fourth meaning of 'law,' that of metanomological statement, or lawlike statement about $laws_2$, which need not concern us here.[16] The acceptance of the *existence* of $laws_1$, or objective patterns, is peculiar to deterministic realists; but the need for a special *term*, such as 'law_1,' for the mere discussion of realist and nonrealist doctrines of scientific law, should be granted; how else can we explain the meaning of belief statements such as "There were laws before their discovery" and its contradictory? The further distinction, between 'law_2' or nomological statement, and 'law_3' or nomopragmatic statement, is needed as well: the latter contains pragmatic terms, which careful statements of $laws_2$ avoid. The terms 'is dropped,' 'is found,' and 'experimental error,' italicized in sentence (*c*) above, are relevant to the *study* of free fall, they arise in the *test* of its $laws_2$, but such terms are never regarded as a *part* of $laws_2$ themselves. For a further examination I must refer to my *Metascientific Queries*.

[13] Op. cit.

[14] David Bohm, "A Suggested Interpretation of Quantum Mechanics in Terms of 'Hidden' Variables," *Physical Review*, 85 (1952), pp. 166, 180.

[15] *Causality*, p. 16.

[16] See Mario Bunge, "Laws of Physical Laws," *American Journal of Physics*, 29, (1961), p. 518.

10. *Is causality in our minds alone?* It is a presupposition of *Causality* that a discussion of causal determination relevant to modern science should take into account the two basic traditions of Aristotle and Hume, but should depart from them in the sense that it should hinge on the concept of *law*, which is the heart of modern science. If there is causation in the world, it is because there are causal *laws* or, at least, laws with a causal range (see sec. 3). Hence the whole question of the nature and objectivity of causation cannot be discussed in connection with complex everyday phenomena (such as slipping on a banana peel), in which it is next to impossible to unravel the nomological net: it must be shifted to the question of the nature and objectivity of scientific laws.

If we maintain that causation is objective, it is because we think that causal laws are constructs ($laws_2$) that to some extent reconstruct, with conceptual units, corresponding objective patterns ($laws_1$). If the opposite belief is held, then the following puzzles must be solved: (*a*) why do we never seem to hit upon ultimately correct law statements, (*b*) why are law statements not only corrigible but eventually corrected, even though we normally presuppose the constancy of objective patterns ($laws_1$), and (*c*) why do we care for the test of those hypotheses we call '$laws_2$' instead of adopting them as convenient summaries of available information. On the other hand, to say that causation is objective at least to some extent, is like saying that electrons are real: in either case our mental picture is at best approximately true, and the *attribution* ('*x* is a causal law,' '*x* is an electron') is a mental act. So that, in a sense, causation *is* in our minds—but not in our minds *alone*. The same is true of our judgments about the capabilities of other people: or shall we deny them?

Unless they wish to look philosophically sophisticated, scientists adopt a realistic epistemology—including the hypothesis of the objective reality of $laws_1$—much for the same reason that we admit a certain humanoid pink patch as corresponding to a flesh-and-bone woman, and also because they have stated the great heuristic value of realism: whereas subjectivism leads to the sterile speculation of an Eddington, realism impels us to enrich experience and criticize theories. Scientists place such a high value on the assumption of the objectivity and immanence of $laws_1$, that they have made it part of physical theories. Two distinguished specimens of such metanomological statements recognizing the objectivity of laws vis-à-vis of the cognitive subject,[17] are "The laws of nature are invariant with respect

[17] Ibid.

to the conditions of observation" (or "with respect to the state of motion of the observer"), and "Transition probabilities are independent of the choice of representation" (or "do not depend on the mode of observation").

11. *Do all terms occurring in scientific laws name objective entities or properties?* While a naive realist would answer "Yes," a critical realist would reply "No," and would add that scientific theories introduce certain concepts which have no factual correlates, such as the components of vector quantities, lagrangians, and the average American. The fact that any physically significant function of the space and time coordinates can be Fourier-analyzed does not mean that the infinite ensuing components represent each a real wave; the fact that certain partial differential equations are separable does not mean that each of the resulting equations portrays a bit of reality. While a law statement *as a whole* is supposed to have a real referent, the way of building up such a whole depends not only on the world but also, perhaps mainly, on the available conceptual tools. (It might even be argued that only whole bundles of law statements have a real referent.) This is all a realist would claim, and all *Causality* (Ch. 11) and *Metascientific Queries* (Chs. 4, 5, and 10) maintain: to claim more would be—unrealistic.

12. *Is Ohm's law causal?* The law '$e = Ri$' (the electromotive force is proportional to the electric current) is causal at its own, macroscopic level, on condition that the customary interpretation of the electromotive force as the cause, and of the current as its effect, be retained—and on condition that the magnetic field does not react on the current, which is true in the case of the stationary state. Certainly, Ohm's law asserts an interdependence among certain variables; but so does every other functional law statement. When studying scientific laws from an ontological point of view we are little interested in their mathematical form, if any: thus, one and the same integral equation may be used to describe a physical phenomenon of hysteresis and a biological or psychological heredity or memory effect. The fact that not only the electromotive force e and the intensity i, but also the resistance R appear in Ohm's law, does not show that two agents are at work instead of one, so that no simple causation operates. Resistance, like permeability or solubility, is a disposition, so that there is no ambiguity in saying "e causes i" when a given circuit (characterized by a given value of R) is referred to. The causal nature of Ohm's law is better seen upon recalling the meaning of 'e' in field theory; translated into the language of field theory our law takes the form $E = (1/\sigma)j$, where 'E' denotes the electric field strength, which is a

specific force, and 'j' the electric current density and 'σ' the conductivity of the material. Since σ is constant for every conductor in a given state, we may introduce the concept of specific current density, designating it by 'J,' whence the last law statement would read '$E = J$,' a form which facilitates the causal interpretation. But it is no less interesting to discover the limitations of this causal interpretation of Ohm's law; thus, e.g., both E and J are statistical *averages* rather than irreducible quantities. Such limitations suggest the inadequacy of the theory of efficient causation and the need to conceive a neodeterministic theory containing a broad spectrum of determination categories.

13. *Is every ordered time sequence causal?* Laws which state changes of state without pointing to causes *producing* unambiguously and *ab extrinseco* such changes are noncausal according to the terminology of *Causality*. Of course, if 'state' is misconstrued as *event*, then every regular nonbranching sequence of states will be called causal. The trouble is that a state is not an event but a *cluster of properties*—what is elsewhere[18] called a poistem—of a system. Thus, e.g., the thermodynamic state of a gas contained in an insulating enclosure with a given volume will be "defined" (as physicists usually say) by the internal pressure and the temperature of the gas. An event is a *change of state*, such as a gas expansion, or a particle collision. An *affaire* can be a cause, a *state* of affairs cannot. And if scientists occasionally, when philosophizing, construe states as events, the metascientist has to notice the semantic shift.

Now, if only events—e.g., changes of properties—can be physically efficient, whether in causal or in noncausal ways, then it is clear that relations among *states*—such as those specified by state equations, like the law of ideal gases or the wave equations of the quantum theory—cannot be causal. A necessary condition for a causal nexus to obtain is that some change be involved. Yet this condition is not sufficient. Consider the elementary equation of state of the perfect gas at a constant temperature, namely, $pV = \text{const}$. We may rephrase it in search for a causal interpretation. Thus, we may differentiate, obtaining $p.\, dV + V.\, dp = 0$, whence $-dV/V = dp/p$. This equation may be read thus: The relative decrease in volume, $-dV/V$, equals the relative increase in pressure, dp/p, *accompanying* the compression. This is not a causal statement: it expresses a symmetrical (reversible) relation between compression and increase of internal pressure. A

[18] Mario Bunge, "Levels: A Semantical Preliminary," *Review of Metaphysics*, 13 (1960), p. 396.

purely formal transformation did not—and could not—have the virtue of infusing causation into our formula. But causal statements will be obtained if we enlarge the context, by including in it references to the surroundings of the gas. For example, we may establish an interaction between the gas and its environment and, in the treatment of this reciprocal action, we may neglect the reaction of the gas on the environment, since we are still interested in the gas rather than in the world as a whole. We now may add, for instance, that the previously considered compression was *caused* by an external force and that, in turn, the decrease in volume *caused* an increase of the internal pressure.

In short, changes of state, but not states themselves, can be produced, and sometimes linked, in a causal way; consequently, not every pattern of succession in time is causal: it may simply be a self-unfolding sequence of states.

14. *Are contiguity and antecedence essential to causality?* That action at a distance violates causality has been maintained by many. That this contention is mistaken is shown historically by the fact that causality was accepted during the reign of Newtonian physics (roughly 1700-1850), which was field-free and admitted actions at a distance. Contiguity, or nearby action, is—like antecedence—consistent with causality but logically external to it. In fact, the principle of nearby action and the principle of antecedence can be formulated independently from one another and from the principle of causality. Quite general possible formulations of the former are as follows:

$$(3) \qquad E(x) = \int_{-\infty}^{\infty} K_1(x - x') \, . \, C(x') \, . \, dx' \quad (\textit{Nearby action})$$

$$(4) \qquad E(t) = \int_{-\infty}^{t} K_2(t - t') \, . \, C(t') \, . \, dt' \quad (\textit{Antecedence}),$$

where 'C' designates the cause, input, or stimulus, impinging on the concerned system at the place x' different from, and at a time t' earlier than, the place x and the time t at which the effect, output, or response E occurs (or is calculated).[19]

It is *logically* possible to build theories respecting nearby action but violating antecedence; thus, by superposing retarded and advanced electromagnetic waves, we obtain (fictitious!) fields which propagate instantaneously. (Incidentally, it is a mistake to suppose

[19] For a more detailed discussion of these principles, and a generalization of them, see Mario Bunge, *The Myth of Simplicity* (Englewood Cliffs, N. J.: Prentice-Hall, Inc., 1963), Ch. 11.

that some theories admit an influence of the future upon the present: all scientific theories, even the overtly noncausal quantum theories, admit or presuppose the principle of antecedence, which is often called "causality condition.") Likewise, it is *logically* possible to build theories respecting antecedence but violating nearby action; however, this would be pointless, because retardation is precisely explained in terms of traveling along a distance.

In short, contrary to what Hume and his followers have supposed, nearby action and antecedence are logically independent from one another and from causality. The occurrence of the terms '*C*' and '*E*' in equations (3) and (4) is not enough to ensure that they designate *causally* related events. Those formulations of nearby action and antecedence do not involve specific mechanisms: they are consistent with both phenomenological and causal theories. On the other hand, there is no general mathematical formulation of the causal principle, of the kind of equations (3) and (4).

15. *If strict and pure causation works never and nowhere, how is it possible to confirm causal hypotheses?* Is this a contradiction, or does it point to an ambiguous use of 'cause'? Neither the one nor the other. If we say that the Galileo-Newton equations of ballistics have been verified many times, we do not thereby mean that the *phenomena* they partially and approximately describe are entirely mechanical: we know that a part of the bullet's kinetic energy is transferred to the air in the form of heat, that electric charges are generated by friction, that the earth's magnetic field has an influence, and so on. Every law, whether of physics or of sociology, is valid exactly only with respect to the corresponding theoretical model, which disregards complications. The confirmation of an empirical hypothesis by means of a set of observations does not enable us to equate our model with its real referent: there is and there will always be a residue. Thus, e.g., if we state the causal hypothesis "Flames burn the skin," we make a statement which, on a first level of analysis, can be termed causal. But a deeper analysis will show that the *phenomenon* itself is not strictly causal: further categories will be needed to account for it, such as interaction (e.g., the deformation of the flame by the very process of tissue burning), teleological determination (e.g., the action of cicatrizants in the burned zone), chance (at the molecular level), and so on. Strict and pure causation is as ideal—yet as useful—as any model can be.

16. *Should a definition of 'cause' be vindicated by usage?* It is possible to justify definitions of some terms, like 'state,' when usage is precise or unambiguous. This is not the case of 'cause' and its cognates, which

have almost as many meanings as there are philosophical schools. In such cases it is better to attempt to elucidate the term by clarifying, expanding, or building a *theory*. In *Causality* the theory of efficient causation is chosen for this purpose, because of its richness and long standing, even though it is concluded once and again that this theory is, strictly speaking, false. Much in the same way, a physicist who wishes to elucidate the term 'force' will refer, e.g., to Newton's mechanics even knowing that the theory contains blatantly false assumptions. Divergences with usage will of course originate in this way; but the aim of a monograph on causality could hardly be the recording of the ways people use the term 'cause'; this is a task for dictionaries. One such remarkable deviation of the philosophic from the scientific usage of 'causality' occurs in recent physics, where the phrase "causality condition" occurs frequently to designate the hypothesis that the cause is never posterior to the effect; this same condition is called hypothesis of antecedence in *Causality*, where it is shown that antecedence and causality are logically independent principles (see sec. 14). After all, usage can be wrong, and the job of philosophers is not to perpetuate but to straighten wrongs, even if by so doing they may look Quixotic.

From Aristotle we have inherited not only the most influential theory of causation but also the definitionist mania, consisting in the belief that all key terms should be defined. Since 'production' is such a term in *Causality*, the definitionist expects its definition much as the beginning student of mathematics asks for an explicit definition of 'number.' Yet logicians have taught us long ago that primitives must be accepted in every context unless one is prepared to run in circles. The meaning of such primitives can, of course, be explained extra-systematically, e.g., by way of examples. Thus we will say "Ionizing radiation *produces* tissue damage", and not simply "Tissue damage is *followed by* or *conjoined with* ionizing radiation." With the exception of Hume and his followers, everyone knows the difference between a relation of production, on the one hand, and the various nongenetic relations, such as functional interdependence, statistical correlation, and accidental coincidence, on the other. The interesting problem is to distinguish the causal from the noncausal patterns, and this is attempted in *Causality*, where a variety of determinants and kinds of law are distinguished.

In conclusion, I am glad that Professors Schlegel and Morgenbesser have perceived the need for a revaluation of causality in the present age, and I am grateful to them for having elicited the above complements and illustrations.

Bibliography

This bibliography is not a guide to the enormously wealthy literature on the problem of causality; it is merely a list of the books quoted in the text. Books regarded as especially relevant to the problems dealt with in this work are marked with an asterisk.

Abro, A. d', **The Decline of Mechanism in Modern Physics* (New York: Van Nostrand, 1939).

Alembert, Jean [Le Rond] d', *Traité de dynamique* (1743) (Paris: Gauthier-Villars, 1921), 2 vols.

Amadou, Robert, *L'occultisme: Esquisse d'un monde vivant* (Paris: Julliard, 1950).

Ampère, André Marie, *Essai sur la philosophie des sciences*, 2 vols. (Paris: Bachelier, 1834 and 1843).

Aquinas, Thomas, **Summa theologiae* (c. 1272), trans. by Fathers of the English Dominican Province (London: Burns Oates & Washbourne, s.d.), 24 vols.

—— *Introduction to*, selections ed. by A. C. Pegis (New York: Modern Library, 1948).

Aristotle, **Physics*, in *Works*, ed. by W. D. Ross (Oxford: Clarendon Press, 1930), vol. II.

—— **Metaphysics*, trans. by H. Tredennick (London: Heinemann; Cambridge, Mass.: Harvard University Press, 1947), Loeb Classical Library, 2 vols.

—— *Organon*, in *Works*, ed. by W. D. Ross (Oxford: Clarendon Press, 1921), vol. X.

Ashby, William Ross, *An Introduction to Cybernetics* (New York: Wiley, 1956).

Ayer, Alfred Jules, **Language, Truth and Logic* (1936; 2nd ed., London: Victor Gollancz, 1946).

—— *The Foundations of Empirical Knowledge* (1940; London: Macmillan, 1951).

—— *Philosophical Essays* (London: Macmillan, 1954).

Bacon, Francis, *Works*, ed. by J. Spedding, R. L. Ellis and D. N. Heath (London: Longmans, 1857–1874), 14 vols.

Bailey, Cyril, *The Greek Atomists and Epicurus* (Oxford: Clarendon Press, 1928).

Bayer, Raymond, ed., **Nature des problèmes en philosophie* (Entretiens d'été, Lund, 1947) (Paris: Hermann, 1949), Actualités Scientifiques et Industrielles, No. 1077.

Bergson, Henri, *Essai sur les données immédiates de la conscience* (1888; Paris: Presses Universitaires de France, 1948).

—— **L'évolution créatrice* (1907; Paris: Presses Universitaires de France, 1948).

Berkeley, George, *Works*, ed. by A. C. Fraser (Oxford: Clarendon Press, 1901), 4 vols.

Bernal, J. D., **The Freedom of Necessity* (London: Routledge & Kegan Paul, 1949).

Bernard, Claude, **Introduction à l'étude de la médecine expérimentale* (1865; 2nd ed., Paris: Charles Delagrave, 1903).

Boole, George, *Studies in Logic and Probability* (London: Watts, 1952).

Borel, Émile, *Le hasard* (1914; Paris: Presses Universitaires de France, 1948).

Born, Max, **Natural Philosophy of Cause and Chance* (1949; 2nd ed., Oxford: Clarendon Press, 1951).

Boutroux, Émile, **De la contingence des lois de la nature* (1874; Paris: Alcan, 1898).

—— **De l'idée de loi naturelle* (1895; Paris: Vrin, 1950).

Braithwaite, Richard Bevan, **Scientific Explanation* (Cambridge: Cambridge University Press, 1953).

Bridgman, P. W., **The Logic of Modern Physics* (1927; New York: Macmillan, 1948).

Broglie, Louis de, **La physique quantique, restera-t-elle indéterministe?* (Paris: Gauthier-Villars, 1953).

Bruno, Giordano, *Opere italiane*, ed. by G. Gentile (Bari: Laterza, 1925, 1927), 2 vols.

Brunschvicg, Léon, **L'expérience humaine et la causalité physique* (1922; Paris: Presses Universitaires de France, 1949).

Bunge, Mario, **Metascientific Queries* (Springfield, Ill.: Charles C. Thomas, 1959).

Burtt, E. A., *The Metaphysical Foundations of Modern Physical Science* (1932; rev. ed., London: Routledge & Kegan Paul, 1950).

Campbell, Norman, **What is Science?* (1921; New York: Dover, 1952).

Carnap, Rudolf, **Physikalische Begriffsbildung* (Karlsruhe: Braun, 1926).

—— *Foundations of Logic and Mathematics*, in *International Encyclopedia of Unified Science* (Chicago: University of Chicago Press, 1939), vol. I, no. 3.

—— *Logical Foundations of Probability* (1950; London: Routledge & Kegan Paul, 1951).

[Carnot, Lazare N. M.], *Oeuvres mathématiques du Citoyen Carnot* (Basle: Decker, 1797).

Cassirer, Ernst, **Determinismus und Indeterminismus in der modernen Physik* (Göteberg: Elanders, 1937), being no. 3, vol. XLII, of *Götebergs Högskolas Årsskrift* (1936).

Childe, V. Gordon, *History* (London: Cobbet Press, 1947).

Cicero, **De fato* [On destiny], trans. by H. Rackham (Cambridge: Harvard University Press, 1942; Loeb Classical Library).

Clymer, R. Swinburne, *A Compendium of Occult Law* (Quakertown, Pa.: Philosophical Publishing Co., 1938).

Cohen, Morris R., and Nagel, Ernest, **An Introduction to Logic and Scientific Method* (New York: Harcourt, Brace, 1934).

Comte, Auguste, **Cours de philosophie positive* (1830; Paris: Costes, 1934), 6 vols.

Conrad, J., *et al.*, ed., *Handwörterbuch der Staatswissenschaften* (3rd ed.; Jena: Fischer, 1909–1911), 8 vols.

Cournot, Antoine-Augustin, **Exposition de la théorie des chances et des probabilités* (Paris, 1843).

—— *Traité de l'enchaînement des idées fondamentales dans les sciences et dans l'histoire* (Paris: Hachette, 1861), 2 vols.

Darwin, Charles, **The Origin of Species* (1859, 1872; London: Oxford University Press, 1951).

Descartes, René, **Oeuvres*, ed. by V. Cousin (Paris: Levrault, 1824–1826), 10 vols. In the national ed. by C. Adam and P. Tannery (Paris, 1897–1913), 14 vols., the Latin works of Descartes are not translated.

Destouches, Jean-Louis, *La mécanique ondulatoire* (Paris: Presses Universitaires de France, 1948).

Diderot, Denis, *Oeuvres*, ed. by A. Billy (Paris: N.R.F. Bibliothèque de la Pléiade, 1951).

Dirac, P. A. M., *The Principles of Quantum Mechanics* (3rd ed.; Oxford: Clarendon Press, 1947).

Duhem, Pierre, **La théorie physique: son objet et sa structure* (1906; 2nd ed., Paris: Rivière, 1914).

Duhem, Pierre, *Essai sur la notion de théorie physique* (Paris: Hermann, 1908), repr. from *Annales de philosophie chrétienne.*

—— *Le système du monde* (Paris: Hermann, 1913–1917), 5 vols.

Eddington, Arthur S., **The Philosophy of Physical Science* (Cambridge: Cambridge University Press; New York: Macmillan, 1939).

Emerson, Ralph Waldo, *Works* (London: Routledge, 1903).

Engels, Friedrich, **Anti-Dühring* (1878; London: Lawrence & Wishart, 1955).

—— **Dialectics of Nature* (1872–1882), trans. by C. Dutt (London: Lawrence & Wishart, 1940).

Enriques, F[ederigo], **Causalité et déterminisme dans la philosophie et l'histoire des sciences* (Paris: Hermann, 1941).

Feigl, Herbert, and Brodbeck, May, ed., **Readings in the Philosophy of Science* (New York: Appleton-Century-Crofts, 1953).

—— and Wilfrid Sellars, ed., **Readings in Philosophical Analysis* (New York: Appleton-Century-Crofts, 1949).

Feuerbach, Ludwig, *Zur Kritik der Hegelschen Philosophie* (1839; Berlin: Aufbau Verlag, 1955).

Fourier, [J. B. J.], *Oeuvres*, ed. by G. Darboux (Paris: Gauthier-Villars, 1888).

Frank, Philipp, **Modern Science and its Philosophy* (Cambridge, Mass.: Harvard University Press, 1949).

—— **Le principe de causalité et ses limites*, trans. by J. du Plessis de Grénedan (Paris: Flammarion, 1937), from *Das Kausalgesetz und seine Grenzen* (Vienna: Springer, 1932).

—— **Foundations of Physics*, in *International Encyclopedia of Unified Science* (Chicago: University of Chicago Press, 1946), vol. I, no. 7.

Frankfort, H., and H. A., Wilson, J. A., and Jacobsen, T., *Before Philosophy: The Intellectual Adventure of Ancient Man* (1946; London: Penguin, 1949).

Galilei, Galileo, **Opere*, Edizione Nazionale (Florence, 1890–1909), 20 vols. See also the selection by F. Flora entitled *Opere* (Milan and Naples: Ricciardi, s.d.).

Gardiner, Patrick, **The Nature of Historical Explanation* (London: Oxford University Press, 1952).

Geiger, H., and Scheel, K., ed., *Handbuch der Physik* (Berlin: Springer, 1926 ff.), 23 vols, esp. vol. IV, containing Hans Reichenbach's **Ziele und Wege der physikalischen Erkenntnis* (1929).

George, André, ed., **Louis de Broglie, physicien et penseur* (Paris: Albin Michel, 1953).

Gilson, Étienne, *La philosophie au moyen âge* (2nd ed.; Paris: Payot, 1952).

—— *The Spirit of Mediaeval Philosophy* (New York: Scribner, 1940).

Gorce, M., and Bergounioux, F., *Science moderne et philosophie médiévale* (Paris: Alcan, 1938).

Hartmann, Nicolai, **Neue Wege der Ontologie* (3rd ed.; Stuttgart: Kohlhammer Verlag, 1949).

—— *Einführung in die Philosophie* (1949; 3rd ed., Osnabrück: Luise Hanckel, 1954).

Hegel, [Georg Wilhelm Friedrich], **Science of Logic* (1812, 1816), trans. by W. H. Johnston and L. G. Struthers (London: Allen & Unwin; New York: Macmillan, 1929) 2 vols.

—— **Encyklopädie der philosophischen Wissenschaften im Grundrisse*, ed. by Karl Rosenkranz (Berlin: Heimann, 1870).

—— **Selections*, ed. by J. Loewenberg (New York: Scribner, 1929).

Heidegger, Martin, *Was ist Metaphysik?* (1929; 2nd ed., Frankfurt: Vittorio Klostermann, s.d.).

Heisenberg, Werner, **The Physical Principles of the Quantum Theory*, trans. by C. Eckart and F. C. Hoyt (Chicago: University of Chicago Press, 1930).

Heitler, W[alter], *The Quantum Theory of Radiation* (3rd ed.; Oxford: Clarendon Press, 1954).

Helmholtz, Hermann, *Über die Erhaltung der Kraft* (1847; 2nd ed., Leipzig: Engelmann, 1889), Ostwald's Klassiker der Exakten Naturwissenschaften.

Helvetius, *Oeuvres complètes* (Paris: Durand, 1776), 4 vols.; contains also d'Holbach's *Système de la nature*.

Hertz, Heinrich, *The Principles of Mechanics*, trans. by P. E. Jones and J. T. Walley (New York: Dover, 1956).

Hilbert, David, and Ackermann, W., *Principles of Mathematical Logic*, trans. by L. M. Hammond, G. G. Leckie, and F. Steinhardt, ed. by R. E. Luce (1938; New York: Chelsea, 1950).

Hinneberg, Paul, ed., *Die Kultur der Gegenwart, Physik* (Leipzig and Berlin: Teubner, 1915).

Hiriyanna, M., *The Essentials of Indian Philosophy* (London: Allen & Unwin, 1949).

Hobbes, Thomas, *Selections*, ed. by F. J. E. Woodbridge (New York: Scribner, Sons, 1930).

Hoyle, Fred, *The Nature of the Universe* (London: Blackwell, 1952).

Hume, David, **A Treatise of Human Nature* (1739–1740; London: Dent; New York: Dutton, 1911), Everyman's Library, 2 vols.

—— **An Enquiry Concerning Human Understanding* (1748), in *Theory of Knowledge*, ed. by D. C. Yalden-Thomson (Edinburgh: Nelson, 1951).

Hutten, Ernest H., **The Language of Modern Physics* (London: Allen & Unwin; New York: Macmillan, 1956).

Inge, William Ralph, *The Philosophy of Plotinus* (3rd ed., 1928); London: Longmans, Green, 1949), 2 vols.

International Union for the Philosophy of Science, *Proceedings of the Second International Congress* (Neuchatel: Editions du Griffon, 1955), 5 vols.

Jacobs, Melville, and Stern, Bernhard J., *General Anthropology* (1947; 2nd ed., New York: Barnes & Noble, 1952).

James, William, *Pragmatism* (1907; New York: Meridian, 1955).

—— **Some Problems of Philosophy* (1911; London: Longmans, Green, 1940).

Jeffreys, Harold, *Theory of Probability* (2nd ed.; Oxford: Clarendon Press, 1948).

Jordan, Pascual, *Physics of the 20th Century*, trans. by E. Oshry (New York: Philosophical Library, 1944).

Kant, Immanuel, **Kritik der reinen Vernunft* (1781: A–1787: B; Hamburg: Meiner, 1952).

Kroeber, A[lfred], ed., *Anthropology Today* (Chicago: University of Chicago Press, 1953).

Lalande, André, ed., *Vocabulaire technique et critique de la philosophie* (3rd ed.; Paris: Alcan, 1938), 3 vols.

—— *Les théories de l'induction et de l'expérimentation* (Paris, Boivin, 1929).

Langer, Susanne K., *An Introduction to Symbolic Logic* (2nd ed.; New York: Dover, 1953).

Laplace, Pierre Simon, **Essai philosophique sur les probabilités* (1814; Paris: Gauthier-Villars, 1921), 2 vols.

Le Chatelier, Henri, *De la méthode dans les sciences expérimentales* (1936; 2nd ed., Paris: Dunod, 1947).

Le Dantec, Felix, **Les lois naturelles* (Paris: Alcan, 1904).

Leibniz, [Gottfried Wilhelm von], **Philosophical Papers and Letters*, trans. and ed. by L. E. Loemker (Chicago: University of Chicago Press, 1956), 2 vols.

Leibniz, [Gottfried Wilhelm von], *Nouveaux essais sur l'entendement humain* (1703; Paris: Flammarion, s.d.).

—— *Principes de la nature et de la grace fondés en raison* (1714), ed. by A. Robinet (Paris: Presses Universitaires de France, 1954).

Lenzen, Victor F., *Procedures of Empirical Science*, in *International Encyclopedia of Unified Science* (Chicago: University of Chicago Press, 1938), vol. I, no. 5.

—— **Causality in Natural Science* (Springfield, Ill.: Thomas, 1954).

Leonardo, *Notebooks*, ed. by E. MacCurdy (New York: Braziller, 1955).

Lévy-Bruhl, Lucien, *Les fonctions mentales dans les sociétés inférieures* (Paris: Alcan, 1910).

—— *La mentalité primitive* (Paris: Alcan, 1922).

Lindsay, Robert B., *Introduction to Physical Statistics* (New York: Wiley, 1941).

Locke, John, **An Essay Concerning Human Understanding* (1690; London: Routledge; New York: Dutton, s.d.).

Lovejoy, Arthur O., The *Great Chain of Being* (1936; Cambridge, Mass.: Harvard University Press, 1953).

Lucretius, *On the Nature of Things*, trans. by W. E. Leonard (London: Dent; New York: Dutton, 1921), Everyman's Library.

Mach, Ernst, *Die Geschichte und die Wurzel des Satzes von der Erhaltung der Arbeit* (1872; 2nd ed., Leipzig: Barth, 1909).

—— **Die Mechanik in ihrer Entwicklung historisch-kritisch dargestellt* (1883; 4th ed., Leipzig: Brockhaus, 1901); *The Science of Mechanics*, trans. by T. J. McCormack (La Salle, Ill., and London: Open Court, 1942).

—— **Die Principien der Wärmelehre* (1896; 2nd ed., Leipzig: Barth, 1900).

—— **Erkenntnis und Irrtum* (Leipzig: Barth, 1905).

Machiavelli, Niccolò, *The Prince and the Discourses*, trans. by L. Ricci and C. E. Detmold (New York: Modern Library, 1940).

MacIver, R. M., **Social Causation* (Boston: Ginn, 1942).

McKeon, Richard, ed., *Selections from Mediaeval Philosophers* (New York: Scribner, 1929), 2 vols.

McWilliams, James A., S. J., **Physics and Philosophy: A Study of Saint Thomas' Commentary on the Eight Books of Aristotle's Physics* (Washington: American Catholic Philosophical Association, 1945).

Magie, William Francis, ed., *A Source Book in Physics* (New York and London: McGraw-Hill, 1935).

Malebranche, [Nicolas], **Entretiens sur la métaphysique* (1688), ed. by A. Cuvillier (Paris: Vrin, 1948), 2 vols.

Margenau, Henry, **The Nature of Physical Reality* (New York, Toronto and London: McGraw-Hill, 1950).

Marx, Karl, and Engels, Friedrich, **Selected Works* (London: Lawrence & Wishart, 1950), 2 vols.

—— *Correspondence* (1846–1895), trans. by D. Torr (London: Lawrence and Wishart, 1936).

Maxwell, James Clerk, *A Treatise on Electricity and Magnetism* (1873; 3rd ed., Oxford: University Press, 1937), 2 vols.

—— **Matter and Motion* (1877; New York: Dover, s.d.).

Meyerson, Émile, **Identité et réalité* (Paris: Alcan, 1908).

—— **De l'explication dans les sciences* (Paris: Payot, 1921), 2 vols.

—— **Du cheminement de la pensée* (Paris: Alcan, 1931), 3 vols.

—— *Essais* (Paris: Vrin, 1936).

Mill, John Stuart, **A System of Logic* (1843, 1875; London: Longmans, Green, 1952).

—— **An Examination of Sir William Hamilton's Philosophy* (1865; New York: Holt, 1873), 2 vols.

Mises, Richard von, **Positivism: A Study in Human Understanding*, trans. by J. Bernstein and R. G. Newton (Cambridge, Mass.: Harvard University Press, 1951), from *Kleines Lehrbuch des Positivismus* (1939).

Nagel, Ernest, *Logic Without Metaphysics* (Glencoe, Ill.; The Free Press, 1956).

Neumann, Johann von, *Mathematische Grundlagen der Quantenmechanik* (Berlin: Springer, 1932; New York: Dover, 1943).

Newton, Isaac, **Mathematical Principles of Natural Philosophy* (1687), trans. by A. Motte, ed. by F. Cajori (Berkeley: University of California Press, 1947).

Nigris, Leone G. B., *Crisi nella scienza* (Milan: Soc. Ed. Vita e Pensiero, 1939).

Northrop, F. S. C., **The Logic of the Sciences and the Humanities* (New York: Macmillan, 1947).

Ostwald, Wilhelm, **Grundriss der Naturphilosophie* (Leipzig: Reclam, 1908).

Pauli, W., Rosenfeld, L., and Weisskopf, V., *Niels Bohr and the Development of Physics* (London: Pergamon, 1955).

Pearson, Karl, **The Grammar of Science* (1892; 3rd ed., London: Adam and Charles Black, 1911).

Peirce, Charles S., **Philosophical Writings*, ed. by J. Buchler (New York: Dover, 1955).

Petzold, Joseph, **Das Weltproblem von positivistischen Standpunkte aus* (Leipzig: Teubner, 1906).

Planck, Max, *Introduction to Theoretical Physics*, trans. by H. L. Brose (London: Macmillan, 1932), 5 vols. esp. vol. V.

—— **Where is Science Going?* trans. by J. Murphy (London: Allen & Unwin, 1933).

—— **The Philosophy of Physics*, trans. by W. H. Johnston (London: Allen & Unwin, 1936).

—— **Vorträge und Erinnerungen* (5th ed.; Stuttgart: Hirzel, 1949).

Plato, **Dialogues*, trans. by B. Jowett (New York: Random House, 1937), 2 vols.

Pohlenz, Max, *Die Stoa* (Göttingen: Vandenhoeck & Ruprecht, 1948), 2 vols.

Poincaré, Henri, *Thermodynamique* (1908; 2nd ed., Paris: Gauthier-Villars, 1923).

—— **Calcul des probabilités* (2nd ed.; Paris, Gauthier-Villars, 1912).

Popper, Karl R., **The Open Society and its Enemies* (rev. ed.; Princeton: Princeton University Press, 1950).

[Pseudo-] Dionysius the Areopagite, *The Mystical Theology and The Celestial Hierarchies* (6th century; London: Shrine of Wisdom, 1949).

Radhakrishnan, S., *Indian Philosophy* (2nd ed.; London: Allen & Unwin, 1934), 2 vols.

Ranzoli, C., *Dizionario di scienze filosofiche* (3rd ed.; Milan: Hoepli, 1926).

Rapoport, Anatol, **Operational Philosophy* (New York: Harper, 1954).

Reichenbach, Hans, **Philosophic Foundations of Quantum Mechanics* (1944; Berkeley and Los Angeles: University of California Press, 1946).

—— *Elements of Symbolic Logic* (New York: Macmillan, 1947).

—— **The Rise of Scientific Philosophy* (Berkeley and Los Angeles: University of California Press, 1951).

—— **The Direction of Time* (Berkeley and Los Angeles: University of California Press, 1956). See also Geiger and Scheel.

Renouvier, Charles, *Les dilemmes de la métaphysique pure* (1901; Paris: Alcan, 1927).

Rey, Abel, *La théorie de la physique chez les physiciens contemporains* (1907; 2nd ed., Paris: Alcan, 1923).

Rosenbloom, Paul C., *The Elements of Mathematical Logic* (New York: Dover, 1950).

Ruckhaber, Erich, *Des Daseins und Denken: Mechanik und Metamechanik* (Hirschberg in Schlesien, 1910).

Russell, Bertrand, **Our Knowledge of the External World* (1914; London: Allen & Unwin, 1952).

—— **Mysticism and Logic* (1918; London: Penguin Books, 1953).

—— *An Outline of Philosophy* (London: Allen & Unwin, 1927).

—— **Human Knowledge: Its Scope and Limits* (London: Allen & Unwin; New York: Simon and Schuster, 1948).

Russell, E. S., **The Directiveness of Organic Activities* (Cambridge: Cambridge University Press, 1945).

Sánchez, Francisco, *Que nada se sabe* [*De multum nobili et prima universali scientia quod nihil scitur*] (1581; Buenos Aires: Emecé, s.d.).

Schlick, Moritz, **Philosophy of Nature* (1936), trans. by A. v. Zeppelin (New York: Philosophical Library, 1949).

Schilpp, Paul Arthur, ed., **Albert Einstein: Philosopher-Scientist* (Evanston, Ill.: Library of Living Philosophers, 1949).

Schopenhauer, Arthur, *Sämtliche Werke* (2nd ed.; Wiesbaden: Eberhard-Brockhaus, 1947–1950), esp. vol. I, containing **Über die vierfache Wurzel des Satzes vom zureichenden Grunde* (1813).

Schrödinger, Erwin, *Science: Theory and Man* [formerly titled *Science and the Human Temperament*, 1935] (New York: Dover, 1957).

Sextus Empiricus, **Works*, trans. by R. G. Bury (London: Heinemann; Cambridge, Mass.: Harvard University Press, 1939–49) Loeb Classical Library, 4 vols.

Spencer, Herbert, *First Principles* (1862, 1900; London: Watts, 1937).

Spinoza, *Traité de la réforme de l'entendement* (1661), bilingual text, trans. by A. Koyré (Paris: Vrin, 1951).

—— **Éthique* (c. 1666), bilingual text, trans. and ed. by C. Appuhn (Paris: Garnier, 1909).

Thompson, D'Arcy Wentworth, *On Growth and Form* (2nd ed.; Cambridge: Cambridge University Press, 1942).

Törnebohm, Håkan, *A Logical Analysis of the Theory of Relativity* (Stockholm: Almqvist & Wiksell, 1952).

Toulmin, Stephen, *The Philosophy of Science* (London: Hutchinson House, 1953).

Vaihinger, Hans, *Die Philosophie des Als Ob* (4th ed.; Leipzig: Meiner, 1920).

Verein Ernst Mach, *Wissenschaftliche Weltauffassung der Wiener Kreis* (Vienna: Artur Wolf Verlag, 1929).

Voltaire, *Oeuvres complètes* (Paris: Hachette, 1859–1862), 35 vols.

Weizsäcker, Carl Friedrich von, **Zum Weltbild der Physik* (1943; 5th ed., Stuttgart: Hirzel, 1951).

Weyl, Hermann, *Philosophy of Mathematics and Natural Science* (1927), trans. by O. Helmer (Princeton: Princeton University Press, 1949).

Wiener, Philip P., ed., **Readings in Philosophy of Science* (New York: Scribner, 1953).

Wisdom, John Oulton, **Foundations of Inference in Natural Science* (London: Methuen, 1952).

Wittgenstein, Ludwig, *Tractatus Logico-Philosophicus* (1922; rev. ed., London: Routledge & Kegan Paul, 1951).

—— *Philosophical Investigations*, trans. by G. E. M. Anscombe (New York: Macmillan, 1953).

Index

Dr. Mario Bunge is professor of philosophy and head of the Foundations and Philosophy of Science Unit at McGill University, Montreal. He has been a full professor of theoretical physics, as well as of philosophy, in his native Argentina and in the U.S.A. and has held visiting positions at the Universities of Pennsylvania, Texas, Temple, Delaware, Freiburg, Aarhus, and Mexico, as well as at the E.T.H. in Zürich. Dr. Bunge has been awarded Ernesto Santamarina, Alexander von Humboldt, and Guggenheim fellowships, as well as Killam grants. He has published more than 250 articles or books, including *Metascientific Queries* (1959), *Intuition and Science* (1962), *The Myth of Simplicity* (1963), *Foundations of Physics* (1967), *Scientific Research* (in two volumes, 1967), *Method, Model, and Matter* (1973), *Philosophy of Physics* (1973), *Sense and Reference* (1974), *Interpretation and Truth* (1974), and *The Furniture of the World* (1977). In addition, Professor Bunge founded the Universidad Obrera Argentina (1938), and cofounded the Asociación Física Argentina (1944), the Asociación Ríoplatense de Lógica y Filosofía Científica, the Society for Exact Philosophy, and the Asociación Mexicana de Epistemología. He is currently working on a seven-volume treatise on philosophy.

A CATALOGUE OF SELECTED DOVER BOOKS IN ALL FIELDS OF INTEREST

A CATALOGUE OF SELECTED DOVER BOOKS IN ALL FIELDS OF INTEREST

CELESTIAL OBJECTS FOR COMMON TELESCOPES, T. W. Webb. The most used book in amateur astronomy: inestimable aid for locating and identifying nearly 4,000 celestial objects. Edited, updated by Margaret W. Mayall. 77 illustrations. Total of 645pp. 5⅜ x 8½.
20917-2, 20918-0 Pa., Two-vol. set $9.00

HISTORICAL STUDIES IN THE LANGUAGE OF CHEMISTRY, M. P. Crosland. The important part language has played in the development of chemistry from the symbolism of alchemy to the adoption of systematic nomenclature in 1892. ". . . wholeheartedly recommended,"—Science. 15 illustrations. 416pp. of text. 5⅝ x 8¼. 63702-6 Pa. $6.00

BURNHAM'S CELESTIAL HANDBOOK, Robert Burnham, Jr. Thorough, readable guide to the stars beyond our solar system. Exhaustive treatment, fully illustrated. Breakdown is alphabetical by constellation: Andromeda to Cetus in Vol. 1; Chamaeleon to Orion in Vol. 2; and Pavo to Vulpecula in Vol. 3. Hundreds of illustrations. Total of about 2000pp. 6⅛ x 9¼.
23567-X, 23568-8, 23673-0 Pa., Three-vol. set $27.85

THEORY OF WING SECTIONS: INCLUDING A SUMMARY OF AIRFOIL DATA, Ira H. Abbott and A. E. von Doenhoff. Concise compilation of subatomic aerodynamic characteristics of modern NASA wing sections, plus description of theory. 350pp. of tables. 693pp. 5⅜ x 8½.
60586-8 Pa. $8.50

DE RE METALLICA, Georgius Agricola. Translated by Herbert C. Hoover and Lou H. Hoover. The famous Hoover translation of greatest treatise on technological chemistry, engineering, geology, mining of early modern times (1556). All 289 original woodcuts. 638pp. 6¾ x 11.
60006-8 Clothbd. $17.95

THE ORIGIN OF CONTINENTS AND OCEANS, Alfred Wegener. One of the most influential, most controversial books in science, the classic statement for continental drift. Full 1966 translation of Wegener's final (1929) version. 64 illustrations. 246pp. 5⅜ x 8½. 61708-4 Pa. $4.50

THE PRINCIPLES OF PSYCHOLOGY, William James. Famous long course complete, unabridged. Stream of thought, time perception, memory, experimental methods; great work decades ahead of its time. Still valid, useful; read in many classes. 94 figures. Total of 1391pp. 5⅜ x 8½.
20381-6, 20382-4 Pa., Two-vol. set $13.00

YUCATAN BEFORE AND AFTER THE CONQUEST, Diego de Landa. First English translation of basic book in Maya studies, the only significant account of Yucatan written in the early post-Conquest era. Translated by distinguished Maya scholar William Gates. Appendices, introduction, 4 maps and over 120 illustrations added by translator. 162pp. 5⅜ x 8½. 23622-6 Pa. $3.00

THE MALAY ARCHIPELAGO, Alfred R. Wallace. Spirited travel account by one of founders of modern biology. Touches on zoology, botany, ethnography, geography, and geology. 62 illustrations, maps. 515pp. 5⅜ x 8½. 20187-2 Pa. $6.95

THE DISCOVERY OF THE TOMB OF TUTANKHAMEN, Howard Carter, A. C. Mace. Accompany Carter in the thrill of discovery, as ruined passage suddenly reveals unique, untouched, fabulously rich tomb. Fascinating account, with 106 illustrations. New introduction by J. M. White. Total of 382pp. 5⅜ x 8½. (Available in U.S. only) 23500-9 Pa. $4.00

THE WORLD'S GREATEST SPEECHES, edited by Lewis Copeland and Lawrence W. Lamm. Vast collection of 278 speeches from Greeks up to present. Powerful and effective models; unique look at history. Revised to 1970. Indices. 842pp. 5⅜ x 8½. 20468-5 Pa. $8.95

THE 100 GREATEST ADVERTISEMENTS, Julian Watkins. The priceless ingredient; His master's voice; 99 44/100% pure; over 100 others. How they were written, their impact, etc. Remarkable record. 130 illustrations. 233pp. 7⅞ x 10 3/5. 20540-1 Pa. $5.95

CRUICKSHANK PRINTS FOR HAND COLORING, George Cruickshank. 18 illustrations, one side of a page, on fine-quality paper suitable for watercolors. Caricatures of people in society (c. 1820) full of trenchant wit. Very large format. 32pp. 11 x 16. 23684-6 Pa. $5.00

THIRTY-TWO COLOR POSTCARDS OF TWENTIETH-CENTURY AMERICAN ART, Whitney Museum of American Art. Reproduced in full color in postcard form are 31 art works and one shot of the museum. Calder, Hopper, Rauschenberg, others. Detachable. 16pp. 8¼ x 11. 23629-3 Pa. $3.00

MUSIC OF THE SPHERES: THE MATERIAL UNIVERSE FROM ATOM TO QUASAR SIMPLY EXPLAINED, Guy Murchie. Planets, stars, geology, atoms, radiation, relativity, quantum theory, light, antimatter, similar topics. 319 figures. 664pp. 5⅜ x 8½. 21809-0, 21810-4 Pa., Two-vol. set $11.00

EINSTEIN'S THEORY OF RELATIVITY, Max Born. Finest semi-technical account; covers Einstein, Lorentz, Minkowski, and others, with much detail, much explanation of ideas and math not readily available elsewhere on this level. For student, non-specialist. 376pp. 5⅜ x 8½. 60769-0 Pa. $4.50

CATALOGUE OF DOVER BOOKS

AMERICAN BIRD ENGRAVINGS, Alexander Wilson et al. All 76 plates. from Wilson's *American Ornithology* (1808-14), most important ornithological work before Audubon, plus 27 plates from the supplement (1825-33) by Charles Bonaparte. Over 250 birds portrayed. 8 plates also reproduced in full color. 111pp. 9⅜ x 12½. 23195-X Pa. $6.00

CRUICKSHANK'S PHOTOGRAPHS OF BIRDS OF AMERICA, Allan D. Cruickshank. Great ornithologist, photographer presents 177 closeups, groupings, panoramas, flightings, etc., of about 150 different birds. Expanded *Wings in the Wilderness.* Introduction by Helen G. Cruickshank. 191pp. 8¼ x 11. 23497-5 Pa. $6.00

AMERICAN WILDLIFE AND PLANTS, A. C. Martin, et al. Describes food habits of more than 1000 species of mammals, birds, fish. Special treatment of important food plants. Over 300 illustrations. 500pp. 5⅜ x 8½. 20793-5 Pa. $4.95

THE PEOPLE CALLED SHAKERS, Edward D. Andrews. Lifetime of research, definitive study of Shakers: origins, beliefs, practices, dances, social organization, furniture and crafts, impact on 19th-century USA, present heritage. Indispensable to student of American history, collector. 33 illustrations. 351pp. 5⅜ x 8½. 21081-2 Pa. $4.50

OLD NEW YORK IN EARLY PHOTOGRAPHS, Mary Black. New York City as it was in 1853-1901, through 196 wonderful photographs from N.-Y. Historical Society. Great Blizzard, Lincoln's funeral procession, great buildings. 228pp. 9 x 12. 22907-6 Pa. $8.95

MR. LINCOLN'S CAMERA MAN: MATHEW BRADY, Roy Meredith. Over 300 Brady photos reproduced directly from original negatives, photos. Jackson, Webster, Grant, Lee, Carnegie, Barnum; Lincoln; Battle Smoke, Death of Rebel Sniper, Atlanta Just After Capture. Lively commentary. 368pp. 8⅜ x 11¼. 23021-X Pa. $8.95

TRAVELS OF WILLIAM BARTRAM, William Bartram. From 1773-8, Bartram explored Northern Florida, Georgia, Carolinas, and reported on wild life, plants, Indians, early settlers. Basic account for period, entertaining reading. Edited by Mark Van Doren. 13 illustrations. 141pp. 5⅜ x 8½. 20013-2 Pa. $5.00

THE GENTLEMAN AND CABINET MAKER'S DIRECTOR, Thomas Chippendale. Full reprint, 1762 style book, most influential of all time; chairs, tables, sofas, mirrors, cabinets, etc. 200 plates, plus 24 photographs of surviving pieces. 249pp. 9⅞ x 12¾. 21601-2 Pa. $7.95

AMERICAN CARRIAGES, SLEIGHS, SULKIES AND CARTS, edited by Don H. Berkebile. 168 Victorian illustrations from catalogues, trade journals, fully captioned. Useful for artists. Author is Assoc. Curator, Div. of Transportation of Smithsonian Institution. 168pp. 8½ x 9½. 23328-6 Pa. $5.00

THE CURVES OF LIFE, Theodore A. Cook. Examination of shells, leaves, horns, human body, art, etc., in "*the* classic reference on how the golden ratio applies to spirals and helices in nature . . . . "—Martin Gardner. 426 illustrations. Total of 512pp. 5⅜ x 8½. 23701-X Pa. $5.95

AN ILLUSTRATED FLORA OF THE NORTHERN UNITED STATES AND CANADA, Nathaniel L. Britton, Addison Brown. Encyclopedic work covers 4666 species, ferns on up. Everything. Full botanical information, illustration for each. This earlier edition is preferred by many to more recent revisions. 1913 edition. Over 4000 illustrations, total of 2087pp. 6⅛ x 9¼. 22642-5, 22643-3, 22644-1 Pa., Three-vol. set $25.50

MANUAL OF THE GRASSES OF THE UNITED STATES, A. S. Hitchcock, U.S. Dept. of Agriculture. The basic study of American grasses, both indigenous and escapes, cultivated and wild. Over 1400 species. Full descriptions, information. Over 1100 maps, illustrations. Total of 1051pp. 5⅜ x 8½. 22717-0, 22718-9 Pa., Two-vol. set $15.00

THE CACTACEAE,, Nathaniel L. Britton, John N. Rose. Exhaustive, definitive. Every cactus in the world. Full botanical descriptions. Thorough statement of nomenclatures, habitat, detailed finding keys. The one book needed by every cactus enthusiast. Over 1275 illustrations. Total of 1080pp. 8 x 10¼. 21191-6, 21192-4 Clothbd., Two-vol. set $35.00

AMERICAN MEDICINAL PLANTS, Charles F. Millspaugh. Full descriptions, 180 plants covered: history; physical description; methods of preparation with all chemical constituents extracted; all claimed curative or adverse effects. 180 full-page plates. Classification table. 804pp. 6½ x 9¼. 23034-1 Pa. $12.95

A MODERN HERBAL, Margaret Grieve. Much the fullest, most exact, most useful compilation of herbal material. Gigantic alphabetical encyclopedia, from aconite to zedoary, gives botanical information, medical properties, folklore, economic uses, and much else. Indispensable to serious reader. 161 illustrations. 888pp. 6½ x 9¼. (Available in U.S. only) 22798-7, 22799-5 Pa., Two-vol. set $13.00

THE HERBAL or GENERAL HISTORY OF PLANTS, John Gerard. The 1633 edition revised and enlarged by Thomas Johnson. Containing almost 2850 plant descriptions and 2705 superb illustrations, Gerard's *Herbal* is a monumental work, the book all modern English herbals are derived from, the one herbal every serious enthusiast should have in its entirety. Original editions are worth perhaps $750. 1678pp. 8½ x 12¼. 23147-X Clothbd. $50.00

MANUAL OF THE TREES OF NORTH AMERICA, Charles S. Sargent. The basic survey of every native tree and tree-like shrub, 717 species in all. Extremely full descriptions, information on habitat, growth, locales, economics, etc. Necessary to every serious tree lover. Over 100 finding keys. 783 illustrations. Total of 986pp. 5⅜ x 8½. 20277-1, 20278-X Pa., Two-vol. set $11.00

"OSCAR" OF THE WALDORF'S COOKBOOK, Oscar Tschirky. Famous American chef reveals 3455 recipes that made Waldorf great; cream of French, German, American cooking, in all categories. Full instructions, easy home use. 1896 edition. 907pp. 6⅝ x 9⅜. 20790-0 Clothbd. $15.00

COOKING WITH BEER, Carole Fahy. Beer has as superb an effect on food as wine, and at fraction of cost. Over 250 recipes for appetizers, soups, main dishes, desserts, breads, etc. Index. 144pp. 5⅜ x 8½. (Available in U.S. only) 23661-7 Pa. $2.50

STEWS AND RAGOUTS, Kay Shaw Nelson. This international cookbook offers wide range of 108 recipes perfect for everyday, special occasions, meals-in-themselves, main dishes. Economical, nutritious, easy-to-prepare: goulash, Irish stew, boeuf bourguignon, etc. Index. 134pp. 5⅜ x 8½. 23662-5 Pa. $2.50

DELICIOUS MAIN COURSE DISHES, Marian Tracy. Main courses are the most important part of any meal. These 200 nutritious, economical recipes from around the world make every meal a delight. "I . . . have found it so useful in my own household,"—*N.Y. Times*. Index. 219pp. 5⅜ x 8½. 23664-1 Pa. $3.00

FIVE ACRES AND INDEPENDENCE, Maurice G. Kains. Great back-to-the-land classic explains basics of self-sufficient farming: economics, plants, crops, animals, orchards, soils, land selection, host of other necessary things. Do not confuse with skimpy faddist literature; Kains was one of America's greatest agriculturalists. 95 illustrations. 397pp. 5⅜ x 8½. 20974-1 Pa.$3.95

A PRACTICAL GUIDE FOR THE BEGINNING FARMER, Herbert Jacobs. Basic, extremely useful first book for anyone thinking about moving to the country and starting a farm. Simpler than Kains, with greater emphasis on country living in general. 246pp. 5⅜ x 8½. 23675-7 Pa. $3.50

PAPERMAKING, Dard Hunter. Definitive book on the subject by the foremost authority in the field. Chapters dealing with every aspect of history of craft in every part of the world. Over 320 illustrations. 2nd, revised and enlarged (1947) edition. 672pp. 5⅜ x 8½. 23619-6 Pa. $7.95

THE ART DECO STYLE, edited by Theodore Menten. Furniture, jewelry, metalwork, ceramics, fabrics, lighting fixtures, interior decors, exteriors, graphics from pure French sources. Best sampling around. Over 400 photographs. 183pp. 8⅜ x 11¼. 22824-X Pa. $6.00

ACKERMANN'S COSTUME PLATES, Rudolph Ackermann. Selection of 96 plates from the *Repository of Arts,* best published source of costume for English fashion during the early 19th century. 12 plates also in color. Captions, glossary and introduction by editor Stella Blum. Total of 120pp. 8⅜ x 11¼. 23690-0 Pa. $4.50

SECOND PIATIGORSKY CUP, edited by Isaac Kashdan. One of the greatest tournament books ever produced in the English language. All 90 games of the 1966 tournament, annotated by players, most annotated by both players. Features Petrosian, Spassky, Fischer, Larsen, six others. 228pp. 5⅜ x 8½. 23572-6 Pa. $3.50

ENCYCLOPEDIA OF CARD TRICKS, revised and edited by Jean Hugard. How to perform over 600 card tricks, devised by the world's greatest magicians: impromptus, spelling tricks, key cards, using special packs, much, much more. Additional chapter on card technique. 66 illustrations. 402pp. 5⅜ x 8½. (Available in U.S. only) 21252-1 Pa. $4.95

MAGIC: STAGE ILLUSIONS, SPECIAL EFFECTS AND TRICK PHOTOGRAPHY, Albert A. Hopkins, Henry R. Evans. One of the great classics; fullest, most authorative explanation of vanishing lady, levitations, scores of other great stage effects. Also small magic, automata, stunts. 446 illustrations. 556pp. 5⅜ x 8½. 23344-8 Pa. $6.95

THE SECRETS OF HOUDINI, J. C. Cannell. Classic study of Houdini's incredible magic, exposing closely-kept professional secrets and revealing, in general terms, the whole art of stage magic. 67 illustrations. 279pp. 5⅜ x 8½. 22913-0 Pa. $4.00

HOFFMANN'S MODERN MAGIC, Professor Hoffmann. One of the best, and best-known, magicians' manuals of the past century. Hundreds of tricks from card tricks and simple sleight of hand to elaborate illusions involving construction of complicated machinery. 332 illustrations. 563pp. 5⅜ x 8½. 23623-4 Pa. $6.00

MADAME PRUNIER'S FISH COOKERY BOOK, Mme. S. B. Prunier. More than 1000 recipes from world famous Prunier's of Paris and London, specially adapted here for American kitchen. Grilled tournedos with anchovy butter, Lobster a la Bordelaise, Prunier's prized desserts, more. Glossary. 340pp. 5⅜ x 8½. (Available in U.S. only) 22679-4 Pa. $3.00

FRENCH COUNTRY COOKING FOR AMERICANS, Louis Diat. 500 easy-to-make, authentic provincial recipes compiled by former head chef at New York's Fitz-Carlton Hotel: onion soup, lamb stew, potato pie, more. 309pp. 5⅜ x 8½. 23665-X Pa. $3.95

SAUCES, FRENCH AND FAMOUS, Louis Diat. Complete book gives over 200 specific recipes: bechamel, Bordelaise, hollandaise, Cumberland, apricot, etc. Author was one of this century's finest chefs, originator of vichyssoise and many other dishes. Index. 156pp. 5⅜ x 8. 23663-3 Pa. $2.75

TOLL HOUSE TRIED AND TRUE RECIPES, Ruth Graves Wakefield. Authentic recipes from the famous Mass. restaurant: popovers, veal and ham loaf, Toll House baked beans, chocolate cake crumb pudding, much more. Many helpful hints. Nearly 700 recipes. Index. 376pp. 5⅜ x 8½. 23560-2 Pa. $4.50

HISTORY OF BACTERIOLOGY, William Bulloch. The only comprehensive history of bacteriology from the beginnings through the 19th century. Special emphasis is given to biography-Leeuwenhoek, etc. Brief accounts of 350 bacteriologists form a separate section. No clearer, fuller study, suitable to scientists and general readers, has yet been written. 52 illustrations. 448pp. 5⅝ x 8¼. 23761-3 Pa. $6.50

THE COMPLETE NONSENSE OF EDWARD LEAR, Edward Lear. All nonsense limericks, zany alphabets, Owl and Pussycat, songs, nonsense botany, etc., illustrated by Lear. Total of 321pp. 5⅜ x 8½. (Available in U.S. only) 20167-8 Pa. $3.95

INGENIOUS MATHEMATICAL PROBLEMS AND METHODS, Louis A. Graham. Sophisticated material from Graham *Dial,* applied and pure; stresses solution methods. Logic, number theory, networks, inversions, etc. 237pp. 5⅜ x 8½. 20545-2 Pa. $4.50

BEST MATHEMATICAL PUZZLES OF SAM LOYD, edited by Martin Gardner. Bizarre, original, whimsical puzzles by America's greatest puzzler. From fabulously rare *Cyclopedia,* including famous 14-15 puzzles, the Horse of a Different Color, 115 more. Elementary math. 150 illustrations. 167pp. 5⅜ x 8½. 20498-7 Pa. $2.75

THE BASIS OF COMBINATION IN CHESS, J. du Mont. Easy-to-follow, instructive book on elements of combination play, with chapters on each piece and every powerful combination team—two knights, bishop and knight, rook and bishop, etc. 250 diagrams. 218pp. 5⅜ x 8½. (Available in U.S. only) 23644-7 Pa. $3.50

MODERN CHESS STRATEGY, Ludek Pachman. The use of the queen, the active king, exchanges, pawn play, the center, weak squares, etc. Section on rook alone worth price of the book. Stress on the moderns. Often considered the most important book on strategy. 314pp. 5⅜ x 8½. 20290-9 Pa. $4.50

LASKER'S MANUAL OF CHESS, Dr. Emanuel Lasker. Great world champion offers very thorough coverage of all aspects of chess. Combinations, position play, openings, end game, aesthetics of chess, philosophy of struggle, much more. Filled with analyzed games. 390pp. 5⅜ x 8½. 20640-8 Pa. $5.00

500 MASTER GAMES OF CHESS, S. Tartakower, J. du Mont. Vast collection of great chess games from 1798-1938, with much material nowhere else readily available. Fully annoted, arranged by opening for easier study. 664pp. 5⅜ x 8½. 23208-5 Pa. $7.50

A GUIDE TO CHESS ENDINGS, Dr. Max Euwe, David Hooper. One of the finest modern works on chess endings. Thorough analysis of the most frequently encountered endings by former world champion. 331 examples, each with diagram. 248pp. 5⅜ x 8½. 23332-4 Pa. $3.75

THE AMERICAN SENATOR, Anthony Trollope. Little known, long unavailable Trollope novel on a grand scale. Here are humorous comment on American vs. English culture, and stunning portrayal of a heroine/villainess. Superb evocation of Victorian village life. 561pp. 5⅜ x 8½.
23801-6 Pa. $6.00

WAS IT MURDER? James Hilton. The author of *Lost Horizon* and *Goodbye, Mr. Chips* wrote one detective novel (under a pen-name) which was quickly forgotten and virtually lost, even at the height of Hilton's fame. This edition brings it back—a finely crafted public school puzzle resplendent with Hilton's stylish atmosphere. A thoroughly English thriller by the creator of Shangri-la. 252pp. 5⅜ x 8. (Available in U.S. only)
23774-5 Pa. $3.00

CENTRAL PARK: A PHOTOGRAPHIC GUIDE, Victor Laredo and Henry Hope Reed. 121 superb photographs show dramatic views of Central Park: Bethesda Fountain, Cleopatra's Needle, Sheep Meadow, the Blockhouse, plus people engaged in many park activities: ice skating, bike riding, etc. Captions by former Curator of Central Park, Henry Hope Reed, provide historical view, changes, etc. Also photos of N.Y. landmarks on park's periphery. 96pp. 8½ x 11. 23750-8 Pa. $4.50

NANTUCKET IN THE NINETEENTH CENTURY, Clay Lancaster. 180 rare photographs, stereographs, maps, drawings and floor plans recreate unique American island society. Authentic scenes of shipwreck, lighthouses, streets, homes are arranged in geographic sequence to provide walking-tour guide to old Nantucket existing today. Introduction, captions. 160pp. 8⅞ x 11¾. 23747-8 Pa. $6.95

STONE AND MAN: A PHOTOGRAPHIC EXPLORATION, Andreas Feininger. 106 photographs by *Life* photographer Feininger portray man's deep passion for stone through the ages. Stonehenge-like megaliths, fortified towns, sculpted marble and crumbling tenements show textures, beauties, fascination. 128pp. 9¼ x 10¾. 23756-7 Pa. $5.95

CIRCLES, A MATHEMATICAL VIEW, D. Pedoe. Fundamental aspects of college geometry, non-Euclidean geometry, and other branches of mathematics: representing circle by point. Poincare model, isoperimetric property, etc. Stimulating recreational reading. 66 figures. 96pp. 5⅝ x 8¼.
63698-4 Pa. $2.75

THE DISCOVERY OF NEPTUNE, Morton Grosser. Dramatic scientific history of the investigations leading up to the actual discovery of the eighth planet of our solar system. Lucid, well-researched book by well-known historian of science. 172pp. 5⅜ x 8½. 23726-5 Pa. $3.50

THE DEVIL'S DICTIONARY. Ambrose Bierce. Barbed, bitter, brilliant witticisms in the form of a dictionary. Best, most ferocious satire America has produced. 145pp. 5⅜ x 8½. 20487-1 Pa. $2.25

TONE POEMS, SERIES II: TILL EULENSPIEGELS LUSTIGE STREICHE, ALSO SPRACH ZARATHUSTRA, AND EIN HELDENLEBEN, Richard Strauss. Three important orchestral works, including very popular *Till Eulenspiegel's Marry Pranks,* reproduced in full score from original editions. Study score. 315pp. 9⅜ x 12¼. (Available in U.S. only)
23755-9 Pa. $8.95

TONE POEMS, SERIES I: DON JUAN, TOD UND VERKLARUNG AND DON QUIXOTE, Richard Strauss. Three of the most often performed and recorded works in entire orchestral repertoire, reproduced in full score from original editions. Study score. 286pp. 9⅜ x 12¼. (Available in U.S. only)
23754-0 Pa. $7.50

11 LATE STRING QUARTETS, Franz Joseph Haydn. The form which Haydn defined and "brought to perfection." (*Grove's*). 11 string quartets in complete score, his last and his best. The first in a projected series of the complete Haydn string quartets. Reliable modern Eulenberg edition, otherwise difficult to obtain. 320pp. 8⅜ x 11¼. (Available in U.S. only)
23753-2 Pa. $7.50

FOURTH, FIFTH AND SIXTH SYMPHONIES IN FULL SCORE, Peter Ilyitch Tchaikovsky. Complete orchestral scores of Symphony No. 4 in F Minor, Op. 36; Symphony No. 5 in E Minor, Op. 64; Symphony No. 6 in B Minor, "Pathetique," Op. 74. Bretikopf & Hartel eds. Study score. 480pp. 9⅜ x 12¼.
23861-X Pa. $10.95

THE MARRIAGE OF FIGARO: COMPLETE SCORE, Wolfgang A. Mozart. Finest comic opera ever written. Full score, not to be confused with piano renderings. Peters edition. Study score. 448pp. 9⅜ x 12¼. (Available in U.S. only)
23751-6 Pa. $11.95

"IMAGE" ON THE ART AND EVOLUTION OF THE FILM, edited by Marshall Deutelbaum. Pioneering book brings together for first time 38 groundbreaking articles on early silent films from *Image* and 263 illustrations newly shot from rare prints in the collection of the International Museum of Photography. A landmark work. Index. 256pp. 8¼ x 11.
23777-X Pa. $8.95

AROUND-THE-WORLD COOKY BOOK, Lois Lintner Sumption and Marguerite Lintner Ashbrook. 373 cooky and frosting recipes from 28 countries (America, Austria, China, Russia, Italy, etc.) include Viennese kisses, rice wafers, London strips, lady fingers, hony, sugar spice, maple cookies, etc. Clear instructions. All tested. 38 drawings. 182pp. 5⅜ x 8.
23802-4 Pa. $2.50

THE ART NOUVEAU STYLE, edited by Roberta Waddell. 579 rare photographs, not available elsewhere, of works in jewelry, metalwork, glass, ceramics, textiles, architecture and furniture by 175 artists—Mucha, Seguy, Lalique, Tiffany, Gaudin, Hohlwein, Saarinen, and many others. 288pp. 8⅜ x 11¼.
23515-7 Pa. $6.95

THE COMPLETE BOOK OF DOLL MAKING AND COLLECTING, Catherine Christopher. Instructions, patterns for dozens of dolls, from rag doll on up to elaborate, historically accurate figures. Mould faces, sew clothing, make doll houses, etc. Also collecting information. Many illustrations. 288pp. 6 x 9. 22066-4 Pa. $4.50

THE DAGUERREOTYPE IN AMERICA, Beaumont Newhall. Wonderful portraits, 1850's townscapes landscapes; full text plus 104 photographs. The basic book. Enlarged 1976 edition. 272pp. 8¼ x 11¼. 23322-7 Pa. $7.95

CRAFTSMAN HOMES, Gustav Stickley. 296 architectural drawings, floor plans, and photographs illustrate 40 different kinds of "Mission-style" homes from *The Craftsman* (1901-16), voice of American style of simplicity and organic harmony. Thorough coverage of Craftsman idea in text and picture, now collector's item. 224pp. 8⅛ x 11. 23791-5 Pa. $6.00

PEWTER-WORKING: INSTRUCTIONS AND PROJECTS, Burl N. Osborn. & Gordon O. Wilber. Introduction to pewter-working for amateur craftsman. History and characteristics of pewter; tools, materials, step-by-step instructions. Photos, line drawings, diagrams. Total of 160pp. 7⅞ x 10¾. 23786-9 Pa. $3.50

THE GREAT CHICAGO FIRE, edited by David Lowe. 10 dramatic, eyewitness accounts of the 1871 disaster, including one of the aftermath and rebuilding, plus 70 contemporary photographs and illustrations of the ruins—courthouse, Palmer House, Great Central Depot, etc. Introduction by David Lowe. 87pp. 8¼ x 11. 23771-0 Pa. $4.00

SILHOUETTES: A PICTORIAL ARCHIVE OF VARIED ILLUSTRATIONS, edited by Carol Belanger Grafton. Over 600 silhouettes from the 18th to 20th centuries include profiles and full figures of men and women, children, birds and animals, groups and scenes, nature, ships, an alphabet. Dozens of uses for commercial artists and craftspeople. 144pp. 8⅜ x 11¼. 23781-8 Pa. $4.50

ANIMALS: 1,419 COPYRIGHT-FREE ILLUSTRATIONS OF MAMMALS, BIRDS, FISH, INSECTS, ETC., edited by Jim Harter. Clear wood engravings present, in extremely lifelike poses, over 1,000 species of animals. One of the most extensive copyright-free pictorial sourcebooks of its kind. Captions. Index. 284pp. 9 x 12. 23766-4 Pa. $8.95

INDIAN DESIGNS FROM ANCIENT ECUADOR, Frederick W. Shaffer. 282 original designs by pre-Columbian Indians of Ecuador (500-1500 A.D.). Designs include people, mammals, birds, reptiles, fish, plants, heads, geometric designs. Use as is or alter for advertising, textiles, leathercraft, etc. Introduction. 95pp. 8¾ x 11¼. 23764-8 Pa. $3.50

SZIGETI ON THE VIOLIN, Joseph Szigeti. Genial, loosely structured tour by premier violinist, featuring a pleasant mixture of reminiscenes, insights into great music and musicians, innumerable tips for practicing violinists. 385 musical passages. 256pp. 5⅝ x 8¼. 23763-X Pa. $4.00

THE PHILOSOPHY OF HISTORY, Georg W. Hegel. Great classic of Western thought develops concept that history is not chance but a rational process, the evolution of freedom. 457pp. 5⅜ x 8½. 20112-0 Pa. $4.50

LANGUAGE, TRUTH AND LOGIC, Alfred J. Ayer. Famous, clear introduction to Vienna, Cambridge schools of Logical Positivism. Role of philosophy, elimination of metaphysics, nature of analysis, etc. 160pp. 5⅜ x 8½. (Available in U.S. only) 20010-8 Pa. $2.00

A PREFACE TO LOGIC, Morris R. Cohen. Great City College teacher in renowned, easily followed exposition of formal logic, probability, values, logic and world order and similar topics; no previous background needed. 209pp. 5⅜ x 8½. 23517-3 Pa. $3.50

REASON AND NATURE, Morris R. Cohen. Brilliant analysis of reason and its multitudinous ramifications by charismatic teacher. Interdisciplinary, synthesizing work widely praised when it first appeared in 1931. Second (1953) edition. Indexes. 496pp. 5⅜ x 8½. 23633-1 Pa. $6.50

AN ESSAY CONCERNING HUMAN UNDERSTANDING, John Locke. The only complete edition of enormously important classic, with authoritative editorial material by A. C. Fraser. Total of 1176pp. 5⅜ x 8½. 20530-4, 20531-2 Pa., Two-vol. set $16.00

HANDBOOK OF MATHEMATICAL FUNCTIONS WITH FORMULAS, GRAPHS, AND MATHEMATICAL TABLES, edited by Milton Abramowitz and Irene A. Stegun. Vast compendium: 29 sets of tables, some to as high as 20 places. 1,046pp. 8 x 10½. 61272-4 Pa. $14.95

MATHEMATICS FOR THE PHYSICAL SCIENCES, Herbert S. Wilf. Highly acclaimed work offers clear presentations of vector spaces and matrices, orthogonal functions, roots of polynomial equations, conformal mapping, calculus of variations, etc. Knowledge of theory of functions of real and complex variables is assumed. Exercises and solutions. Index. 284pp. 5⅝ x 8¼. 63635-6 Pa. $5.00

THE PRINCIPLE OF RELATIVITY, Albert Einstein et al. Eleven most important original papers on special and general theories. Seven by Einstein, two by Lorentz, one each by Minkowski and Weyl. All translated, unabridged. 216pp. 5⅜ x 8½. 60081-5 Pa. $3.50

THERMODYNAMICS, Enrico Fermi. A classic of modern science. Clear, organized treatment of systems, first and second laws, entropy, thermodynamic potentials, gaseous reactions, dilute solutions, entropy constant. No math beyond calculus required. Problems. 160pp. 5⅜ x 8½. 60361-X Pa. $3.00

ELEMENTARY MECHANICS OF FLUIDS, Hunter Rouse. Classic undergraduate text widely considered to be far better than many later books. Ranges from fluid velocity and acceleration to role of compressibility in fluid motion. Numerous examples, questions, problems. 224 illustrations. 376pp. 5⅝ x 8¼. 63699-2 Pa. $5.00

THE SENSE OF BEAUTY, George Santayana. Masterfully written discussion of nature of beauty, materials of beauty, form, expression; art, literature, social sciences all involved. 168pp. 5⅜ x 8½. 20238-0 Pa. $3.00

ON THE IMPROVEMENT OF THE UNDERSTANDING, Benedict Spinoza. Also contains *Ethics, Correspondence,* all in excellent R. Elwes translation. Basic works on entry to philosophy, pantheism, exchange of ideas with great contemporaries. 402pp. 5⅜ x 8½. 20250-X Pa. $4.50

THE TRAGIC SENSE OF LIFE, Miguel de Unamuno. Acknowledged masterpiece of existential literature, one of most important books of 20th century. Introduction by Madariaga. 367pp. 5⅜ x 8½. 20257-7 Pa. $4.50

THE GUIDE FOR THE PERPLEXED, Moses Maimonides. Great classic of medieval Judaism attempts to reconcile revealed religion (Pentateuch, commentaries) with Aristotelian philosophy. Important historically, still relevant in problems. Unabridged Friedlander translation. Total of 473pp. 5⅜ x 8½. 20351-4 Pa. $6.00

THE I CHING (THE BOOK OF CHANGES), translated by James Legge. Complete translation of basic text plus appendices by Confucius, and Chinese commentary of most penetrating divination manual ever prepared. Indispensable to study of early Oriental civilizations, to modern inquiring reader. 448pp. 5⅜ x 8½. 21062-6 Pa. $5.00

THE EGYPTIAN BOOK OF THE DEAD, E. A. Wallis Budge. Complete reproduction of Ani's papyrus, finest ever found. Full hieroglyphic text, interlinear transliteration, word for word translation, smooth translation. Basic work, for Egyptology, for modern study of psychic matters. Total of 533pp. 6½ x 9¼. (Available in U.S. only) 21866-X Pa. $5.95

THE GODS OF THE EGYPTIANS, E. A. Wallis Budge. Never excelled for richness, fullness: all gods, goddesses, demons, mythical figures of Ancient Egypt; their legends, rites, incarnations, variations, powers, etc. Many hieroglyphic texts cited. Over 225 illustrations, plus 6 color plates. Total of 988pp. 6⅛ x 9¼. (Available in U.S. only)
22055-9, 22056-7 Pa., Two-vol. set $16.00

THE STANDARD BOOK OF QUILT MAKING AND COLLECTING, Marguerite Ickis. Full information, full-sized patterns for making 46 traditional quilts, also 150 other patterns. Quilted cloths, lame, satin quilts, etc. 483 illustrations. 273pp. 6⅞ x 9⅝. 20582-7 Pa. $4.95

CORAL GARDENS AND THEIR MAGIC, Bronsilaw Malinowski. Classic study of the methods of tilling the soil and of agricultural rites in the Trobriand Islands of Melanesia. Author is one of the most important figures in the field of modern social anthropology. 143 illustrations. Indexes. Total of 911pp. of text. 5⅝ x 8¼. (Available in U.S. only)
23597-1 Pa. $12.95

HOUSEHOLD STORIES BY THE BROTHERS GRIMM. All the great Grimm stories: "Rumpelstiltskin," "Snow White," "Hansel and Gretel," etc., with 114 illustrations by Walter Crane. 269pp. 5⅜ x 8½. 21080-4 Pa. $3.50

SLEEPING BEAUTY, illustrated by Arthur Rackham. Perhaps the fullest, most delightful version ever, told by C. S. Evans. Rackham's best work. 49 illustrations. 110pp. 7⅞ x 10¾. 22756-1 Pa. $2.50

AMERICAN FAIRY TALES, L. Frank Baum. Young cowboy lassoes Father Time; dummy in Mr. Floman's department store window comes to life; and 10 other fairy tales. 41 illustrations by N. P. Hall, Harry Kennedy, Ike Morgan, and Ralph Gardner. 209pp. 5⅜ x 8½. 23643-9 Pa. $3.00

THE WONDERFUL WIZARD OF OZ, L. Frank Baum. Facsimile in full color of America's finest children's classic. Introduction by Martin Gardner. 143 illustrations by W. W. Denslow. 267pp. 5⅜ x 8½. 20691-2 Pa. $3.50

THE TALE OF PETER RABBIT, Beatrix Potter. The inimitable Peter's terrifying adventure in Mr. McGregor's garden, with all 27 wonderful, full-color Potter illustrations. 55pp. 4¼ x 5½. (Available in U.S. only) 22827-4 Pa. $1.25

THE STORY OF KING ARTHUR AND HIS KNIGHTS, Howard Pyle. Finest children's version of life of King Arthur. 48 illustrations by Pyle. 131pp. 6⅛ x 9¼. 21445-1 Pa. $4.95

CARUSO'S CARICATURES, Enrico Caruso. Great tenor's remarkable caricatures of self, fellow musicians, composers, others. Toscanini, Puccini, Farrar, etc. Impish, cutting, insightful. 473 illustrations. Preface by M. Sisca. 217pp. 8⅜ x 11¼. 23528-9 Pa. $6.95

PERSONAL NARRATIVE OF A PILGRIMAGE TO ALMADINAH AND MECCAH, Richard Burton. Great travel classic by remarkably colorful personality. Burton, disguised as a Moroccan, visited sacred shrines of Islam, narrowly escaping death. Wonderful observations of Islamic life, customs, personalities. 47 illustrations. Total of 959pp. 5⅜ x 8½. 21217-3, 21218-1 Pa., Two-vol. set $12.00

INCIDENTS OF TRAVEL IN YUCATAN, John L. Stephens. Classic (1843) exploration of jungles of Yucatan, looking for evidences of Maya civilization. Travel adventures, Mexican and Indian culture, etc. Total of 669pp. 5⅜ x 8½. 20926-1, 20927-X Pa., Two-vol. set $7.90

AMERICAN LITERARY AUTOGRAPHS FROM WASHINGTON IRVING TO HENRY JAMES, Herbert Cahoon, et al. Letters, poems, manuscripts of Hawthorne, Thoreau, Twain, Alcott, Whitman, 67 other prominent American authors. Reproductions, full transcripts and commentary. Plus checklist of all American Literary Autographs in The Pierpont Morgan Library. Printed on exceptionally high-quality paper. 136 illustrations. 212pp. 9⅛ x 12¼. 23548-3 Pa. $12.50

AN AUTOBIOGRAPHY, Margaret Sanger. Exciting personal account of hard-fought battle for woman's right to birth control, against prejudice, church, law. Foremost feminist document. 504pp. 5⅜ x 8½.
20470-7 Pa. $5.50

MY BONDAGE AND MY FREEDOM, Frederick Douglass. Born as a slave, Douglass became outspoken force in antislavery movement. The best of Douglass's autobiographies. Graphic description of slave life. Introduction by P. Foner. 464pp. 5⅜ x 8½. 22457-0 Pa. $5.50

LIVING MY LIFE, Emma Goldman. Candid, no holds barred account by foremost American anarchist: her own life, anarchist movement, famous contemporaries, ideas and their impact. Struggles and confrontations in America, plus deportation to U.S.S.R. Shocking inside account of persecution of anarchists under Lenin. 13 plates. Total of 944pp. 5⅜ x 8½.
22543-7, 22544-5 Pa., Two-vol. set $12.00

LETTERS AND NOTES ON THE MANNERS, CUSTOMS AND CONDITIONS OF THE NORTH AMERICAN INDIANS, George Catlin. Classic account of life among Plains Indians: ceremonies, hunt, warfare, etc. Dover edition reproduces for first time all original paintings. 312 plates. 572pp. of text. 6⅛ x 9¼. 22118-0, 22119-9 Pa.. Two-vol. set $12.00

THE MAYA AND THEIR NEIGHBORS, edited by Clarence L. Hay, others. Synoptic view of Maya civilization in broadest sense, together with Northern, Southern neighbors. Integrates much background, valuable detail not elsewhere. Prepared by greatest scholars: Kroeber, Morley, Thompson, Spinden, Vaillant, many others. Sometimes called Tozzer Memorial Volume. 60 illustrations, linguistic map. 634pp. 5⅜ x 8½.
23510-6 Pa. $10.00

HANDBOOK OF THE INDIANS OF CALIFORNIA, A. L. Kroeber. Foremost American anthropologist offers complete ethnographic study of each group. Monumental classic. 459 illustrations, maps. 995pp. 5⅜ x 8½.
23368-5 Pa. $13.00

SHAKTI AND SHAKTA, Arthur Avalon. First book to give clear, cohesive analysis of Shakta doctrine, Shakta ritual and Kundalini Shakti (yoga). Important work by one of world's foremost students of Shaktic and Tantric thought. 732pp. 5⅜ x 8½. (Available in U.S. only)
23645-5 Pa. $7.95

AN INTRODUCTION TO THE STUDY OF THE MAYA HIEROGLYPHS, Syvanus Griswold Morley. Classic study by one of the truly great figures in hieroglyph research. Still the best introduction for the student for reading Maya hieroglyphs. New introduction by J. Eric S. Thompson. 117 illustrations. 284pp. 5⅜ x 8½. 23108-9 Pa. $4.00

A STUDY OF MAYA ART, Herbert J. Spinden. Landmark classic interprets Maya symbolism, estimates styles, covers ceramics, architecture, murals, stone carvings as artforms. Still a basic book in area. New introduction by J. Eric Thompson. Over 750 illustrations. 341pp. 8⅜ x 11¼.
21235-1 Pa. $6.95